BIOLOGY AND CHEMISTRY
OF BASEMENT MEMBRANES

Proceedings of the First International Symposium
on the Biology and Chemistry
of Basement Membranes
November 29– December 1, 1976
Philadelphia, Pennsylvania

International Symposium on the Biology and Chemistry of Basement Membranes, 1st, Philadelphia, 1976

BIOLOGY AND CHEMISTRY OF BASEMENT MEMBRANES

EDITED BY

Nicholas A. Kefalides

Connective Tissue Research Institute
University of Pennsylvania
University City Science Center
Philadelphia, Pennsylvania

Academic Press   New York   San Francisco   London   1978
A Subsidiary of Harcourt Brace Jovanovich, Publishers

ACADEMIC PRESS, INC.
111 Fifth Avenue, New York, New York 10003

United Kingdom Edition published by
ACADEMIC PRESS, INC. (LONDON) LTD.
24/28 Oval Road, London NW1 7DX

Library of Congress Cataloging in Publication Data

International Symposium on the Biology and
    Chemistry of Basement Membranes, 1st, Philadelphia,
    Pa., 1976.
    Biology and chemistry of basement membranes.

    1. Kidneys—Diseases—Congresses.   2. Membrane,
Basement—Congresses.   3. Diabetes—Congresses.
I. Kefalides, Nicholas A.   II. Title.
RC904.I57 1976   599'.01'49   78-607
ISBN 0-12-403150-1

CONTENTS

III. PATHOLOGY OF BASEMENT MEMBRANES

LIST OF SENIOR AUTHORS

Numbers in parentheses indicate the pages on which the authors' contributions begin.

ROBERT ALPER (239), Connective Tissue Research Section, University of Pennsylvania, University City Science Center, Philadelphia, Pennsylvania

B. ALBINI (577), Department of Pathology, State University of New York, Buffalo, New York

GIUSEPPE A. ANDRES (577), Department of Pathology, State University of New York, Buffalo, New York

ROBERT ASKENASI (443), Hôpital Brugmann, Université Libre de Bruxelles, Bruxelles, Belgium

MERTON R. BERNFIELD (137), Department of Pediatrics, Stanford University School of Medicine, Stanford, California

KLAUS BRENDEL (177), Department of Pharmacology, University of Arizona School of Medicine, Tucson, Arizona

R. A. CAMERINI-DAVALOS (511), Department of Medicine, Diabetes and Metabolism Division, New York Medical College, New York

EDWARD C. CARLSON (31), Department of Anatomy, University of Arizona Medical Center, Tucson, Arizona

JOHN P. CAULFIELD (81),* Department of Cell Biology and Pathology, Yale University School of Medicine, New Haven, Connecticut

CHARLES C. CLARK (287), Connective Tissue Research Section, University of Pennsylvania, University City Science Center, Philadelphia, Pennsylvania

MARGO P. COHEN (523), Wayne State University School of Medicine, Detroit, Michigan

GERSON COTTA-PEREIRA (111), Department of Histology and Embryology, State University of Rio de Janeiro, Rio de Janeiro, Brazil

PETER DEHM (235), Connective Tissue Research Section, University of Pennsylvania, University City Science Center, Philadelphia, Pennsylvania

*Present address: Robert B. Brigham Hospital, Harvard Medical School, Boston, Massachusetts.

MARILYN G. FARQUHAR (43), Department of Cell Biology and Pathology, Yale University School of Medicine, New Haven, Connecticut

WIJNHOLT FERWERDA (403), Department of Medical Chemistry, Faculty of Medicine, Vrije Universiteit, Amsterdam, The Netherlands

JOHN H. FESSLER (373), Molecular Biology Institute, University of California, Los Angeles, California

RICHARD J. GLASSOCK (421), Department of Medicine, UCLA School of Medicine, Harbor General Hospital, Torrance, California

MICHAEL E. GRANT (335), Department of Medical Biochemistry, University of Manchester Medical School, Manchester, England

DIANE E. GUNSON (397), New Bolton Center, University of Pennsylvania, Kennett Square, Pennsylvania

ELIZABETH D. HAY (119), Department of Anatomy, Harvard Medical School, Boston, Massachusetts

J. G. HEATHCOTE (335), Department of Medical Biochemistry, University of Manchester Medical School, Manchester, England

BILLY G. HUDSON (253), Department of Biochemistry, University of Kansas Medical Center, Kansas City, Kansas

ERIC A. JAFFE (355), Department of Medicine, Division of Hematology-Oncology, The New York Hospital-Cornell Medical Center, New York

LEWIS D. JOHNSON (299), Department of Pathology, Emory University School of Medicine, Atlanta, Georgia

MARLENE W. KARAKASHIAN (383), Connective Tissue Research Section, University of Pennsylvania, University City Science Center, Philadelphia, Pennsylvania

NICHOLAS A. KEFALIDES (215), Connective Tissue Research Section, University of Pennsylvania, University City Science Center, Philadelphia, Pennsylvania

CECIL A. KRAKOWER (1), The Abraham Lincoln School of Medicine of the University of Illinois, Chicago, Illinois

EDWARD J. MACARAK (343), Connective Tissue Research Section, University of Pennsylvania, University City Science Center, Philadelphia, Pennsylvania

PH. MAHIEU (569), Department of Medicine, University Hospital, Liège, Belgium

M. E. MARAGOUDAKIS (309), Research Department, Pharmaceuticals Division, CIBA-GEIGY Corporation, Ardsley, New York

ANTONIO MARTINEZ-HERNANDEZ (99), Department of Pathology, University of Colorado Medical Center, Denver, Colorado

ELIAS MEEZAN (17), Department of Pharmacology, University of Arizona School of Medicine, Tucson, Arizona

ALFRED F. MICHAEL (463), Department of Pediatrics, University of Minnesota Medical Center, Minneapolis, Minnesota

EDWARD J. MILLER (265), Department of Biochemistry and Institute of Dental Research, University of Alabama Medical Center, Birmingham, Alabama

RUTH ØSTERBY (495), University Institute of Pathology, Kommunehospitalet, Aarhus, Denmark

A. M. ROBERT (195), Laboratoire de Biochimie du Tissu Cônjonctif, Faculté de Medecine, Université de Paris, Créteil, France

L. ROBERT (503), Laboratoire de Biochimie du Tissu Cônjonctif, Faculté de Medecine, Université de Paris, Créteil, France

EVELINE E. SCHNEEBERGER (561), Department of Pathology, Children's Hospital Medical Center and Harvard Medical School, Boston, Massachusetts

SEIICHI SHIBATA (535), 3rd Department of Internal Medicine, Faculty of Medicine, University of Tokyo, Tokyo, Japan

G. E. STRIKER (319), School of Medicine, University of Washington, Seattle, Washinton

MARVIN L. TANZER (367), Department of Biochemistry, University of Connecticut Health Center, Farmington, Connecticut

RUPERT TIMPL (413), Max-Planck-Institut für Biochemie, Munich, West Germany

ROBERT L. TRELSTAD (229), Experimental Pathology Laboratory, Shriners Burns Institute, Massachusetts General Hospital and Harvard Medical School, Boston, Massachusetts

IVO VAN DE RIJN (589), The Rockefeller University, New York

ARTHUR VEIS (279), Department of Biochemistry, Northwestern University Medical School, Chicago, Illinois

MANJERI A. VENKATACHALAM (149), Department of Pathology, Peter Bent Brigham Hospital, Boston, Massachusetts

RUDOLF VRACKO (165, 483), Department of Pathology, University of Washington School of Medicine, Seattle, Washington

N. GUNNAR WESTBERG (205), Section of Nephrology, Medical Clinic I, Sahlgren's Hospital, Göteborg, Sweden

JOHN B. ZABRISKIE (453), The Rockefeller University, New York

PREFACE

Knowledge about the nature of basement membranes remained in a state of dormant tranquility for almost 100 years after they were first described by Todd and Bowman until Krakower and Greenspon isolated this extracellular matrix from renal glomeruli and showed it to contain collagenous and noncollagenous glycoprotein components. Since 1951 a series of studies in various laboratories resulted in important contributions in the study of ultrastructure, chemistry, metabolism, immunology, function, and pathology of these structures, to the degree that we can safely state that basement membranes are functional tissue units having a complex organization of several protein and carbohydrate moieties. These protein and carbohydrate components are synthesized by the cells with which the basement membrane is associated. The relative proportions of the various glycoprotein components, as well as their mode of interaction, determine not only the ultrastructural morphology but the functional behavior of these structures as well.

Considerable information has been obtained about the chemistry and synthesis of one of the protein components of basement membranes, namely, collagen and procollagen, but the exact nature and number of the noncollagen glycoprotein components remain unknown.

The functional properties of basement membranes, their role in development and differentiation, their rate of synthesis and turnover, and their immunologic behavior are still incompletely understood.

Although the morphological and functional changes of basement membranes in disease have been well documented, the presence of any associated chemical changes in their glycoprotein components remains unclear.

The purpose of the First International Symposium on the Biology and Chemistry of Basement Membranes was to bring together the leading investigators in the field so that new data could be discussed and old and new concepts on the nature of basement membranes could be critically evaluated.

This book is a multifaceted approach to the question of structure and function of basement membranes in health and disease.

ACKNOWLEDGMENTS

The organizing committee of the First International Symposium on the Biology and Chemistry of Basement Membranes acknowledges with gratitude the generous financial support received from the following, whose kind cooperation made the arrangements possible for this symposium:

The Philadelphia General Hospital Research Fund, *Roland A. Bosee, Director*

The National Institute of Arthritis, Metabolism, and Digestive Diseases

The National Heart, Lung, and Blood Institute

Burroughs Wellcome Company

Ciba-Geigy Corporation

Eli Lilly and Company

Hoffmann-La Roche, Inc.

Sandoz Pharmaceuticals

The Upjohn Company

I. Biology of Basement Membranes
A. Preparation

THE ISOLATION OF BASEMENT MEMBRANES

Cecil A. Krakower

and

Seymour A. Greenspon

*The Abraham Lincoln School of Medicine of the
University of Illinois
Chicago, Illinois*

*SUMMARY: Methods for the isolation of basement membranes
are described.*

I.   INTRODUCTION

     This review deals with the isolation of basement mem-
branes (BM) in their natural state. It is exceedingly difficult to obtain pure BM without adherent tissues. The ideal is
to use mild mechanical means to achieve this end. However,
few BM, in fact perhaps only one, namely, lens capsule, can be
obtained by these means. The use of chemical agents, enzymes,
or harsher mechanical methods to rid the BM of its contaminants are all fraught with the danger of altering these membranes so that subsequent determinations, chemical, physical,
or immunologic, may not be entirely valid. This should be
borne in mind in the consideration of the methods used to
isolate the different BM discussed below. It is to be emphasized that there is a whole range of BM in different locales
of mesenchymal (about fat cells, smooth and cardiac muscle
fibers), epithelial, endothelial, mesothelial, schwannian,
and glial orgin which have not been isolated. It is to be

hoped that means will be found to isolate these if we are to
determine to what extent BM have chemical and immunologic
properties in common and what are their differences. In that
context, some of the answers may come from the culture of
cells from different regions and of different types which are
capable of synthesizing BM.

Non-Glomerular Blood Vessels

BM of Choroid Plexuses

     The choroid plexuses of fresh frozen and thawed brains
can be removed from the lateral and fourth ventricles. They
are placed in a petri dish with 0.15 M NaCl. The capillary-
bearing portions of the plexuses are then excised under a dissecting microscope with the use of iris scissors and dissecting needles. After all large vessels and extraneous tissues
have been dissected away, the excised vessels are repeatedly
washed in 0.15 M NaCl, shaved on a freezing microtome to yield
10-15 micra thick sections. The latter are then sonically
vibrated in 1.4 M NaCl. The vibrated material is centrifuged
at 1000 X g for 20 min., resuspended and washed repeatedly in
0.15 M NaCl by centrifugation. All the above procedures are
performed in a cold room at 5°C. Kefalides and Denduchis (1)
wash the choroid plexuses in 0.15 M NaCl. They then treat
them with 0.3 M acetic acid with gentle stirring at 4°C for
25 hours followed by centrifugation at 1400 X g for 20 min.
The extraction of the sediment is repeated three more times.
The sediment is then sonicated for 15 min. in 0.15 M NaCl,
centrifuged at 1400 X g for 20 min. at 4°C, and the final
sediment is washed by centrifugation with distilled water and
lyophilized. The latter is suspended in 0.1 M sodium acetate
(pH 5) and incubated with DNase at room temperature for 16
hours with slow stirring. The suspension is centrifuged at
34,800 X g for 30 min. and the residue is now digested with
RNase under the same conditions.

BM of Ciliary Processes

     The highly vascular ciliary processes of the eye can be
used as a source of vascular basement membrane, albeit with a
mixture of interstitial collagen and melanin pigment in animals which are not albinos. Fresh frozen and partially thawed eyes are trimmed of extraocular tissues. The corneae are
excised. Lens and vitreous are gently extruded. The eyes
are inverted and gently washed in 0.15 M NaCl. With a stout
forceps inserted into the inverted cavity as a holder, the

eye is immersed in a shallow basin with saline. The ciliary
processes processes float and can be plucked with iris forceps,
using a spotlight focused on the ciliary area and a jeweler's
loupe to help visualize them. The isolated processes are
sonically vibrated in 1.4 M saline and further treated as with
the choroid plexuses.

BM of Cerebral Cortical and Retinal Vessels

The vessels of the cortex of the brain have been isolated
by Brendel and co-workers (2). Fresh cortical tissue is homogenized in Earle's balanced salt solution buffered with Hepes
(1:1 by vol.) in a loosely fitting smooth glass tube with a
teflon pestle. The homogenate is poured over a 153 micron
nylon sieve and washed with buffer. The vessels remain on the
sieve. They can be rehomogenized, resieved, and washed. They
can also be suspended in a petri dish, swept up with a piece
of 210 micron nylon mesh, washed off in another dish with a
stream of buffer to rid the vessels of granular contaminants.
The purity of the vessels can be monitored with a dissecting
microscope. Retinal vessels can be prepared in the same way
(3) using an 86 micron nylon sieve for small vessels. If
larger vessels are wanted, the sieving can be done first
through a 210 micron sieve followed by an 86 micron one.
These isolated vessels can be treated with detergents triton
X-100 and sodium deoxycholate to obtain their BM (4).

Eye

Capsule of the Lens

Freshly removed eyes are washed in 0.15 M NaCl. They are
stored at -15°C. Although the capsule surrounds the whole
lens, it is thicker and is separated from the lenticular fibers
by an epithelium only on its anterior aspect. Posteriorly it
merges with the lenticular fibers. Hence, only the anterior
portion of the capsule should be removed. There is a ready
cleavage plane through the epithelial layer. Guided by a
jeweler's loupe, the capsule can be stripped quite readily
with forceps. It is a highly refractile tough membrane. Adherent epithelial cells can be removed by thoroughly scraping
its inner surface with a scalpel blade. It can then be shaved
into 10-15 micra thick sections or even thinner on a frozen
section microtome. These shavings can be sonically vibrated
so as to render the capsule more particulate. The stripped
capsules are washed in 0.15 M NaCl and the shaved vibrated

products are centrifuged at 1000 X g for 15 min. and likewise washed in sale or distilled water. All procedures are best carried out in a cold room at 5°C. It is to be noted that the suspensory ligaments of the lens insert into the capsule and are an integral part of it.

Bowman's and Descemet's Membranes of the Cornea

Since attempts have been made to isolate both these membranes at the same time, they will be considered together. In the method of Dohlman and Balazs (5) the endothelium of the cornea is wiped off in the fresh state. The cornea is frozen and thawed and Descemet's membrane is scraped off with a scoop and suspended in distilled water. The membranes can be identified and distinguished from the fragments containing corneal stroma i.e., the substantia propria since the former remain glassy whilst the fragments with corneal stroma become swollen and opaque. The glassy fragments can be picked out. Attempts to remove Bowman's membrane in this way have not been described. Dardenne and co-workers (6) have modified Dohlman and Balasz's method as follows. Both epithelium and endothelium are mechanically separated from the freshly removed and chilled eyes. The cornea is then immersed in distilled water at 2°C for 6 to 12 hours. Descemet's membrane can now be stripped from the swollen substantia propria of the cornea. It is less easy to obtain a clear separation of Bowman's membrane. The authors suggest that more prolonged immersion in water may be necessary. Paulini and Beneke (7) remove epithelium and endothelium mechanically after immersion of the cornea in distilled water at the temperature and for the time interval indicated above. Thiele and co-workers (8) have suggested the following approach to isolate these membranes. The corneae are immersed in 0.8 N lactic acid at 20°C for 5 hours on a mechanically operated mixer. Epithelium and endothelium are loosened and come away in shreds. These surface membranes are therefore separated. The corneae are then further treated with 0.8 N lactic acid for an additional 24 hours at 20°C. At this stage remaining epithelial remnants can be removed. Descemet's membrane rolls up but is not completely loosened from the markedly swollen substantia propria. After appropriate aqueous rinsing, additional treatment of the corneae with 0.5 N sodium hydroxide for 48 hours at 20°C completes the separation of the rolled-up membrane. Bowman's membrane does not come away in the same manner. It would appear, therefore, that one can isolate Descemet's membrane with a fair degree of purity. This has not been accomplished, however, for Bowman's membrane.

Kidney

Glomerular BM (GBM)

To obtain GBM, it is first essential to isolate the glomeruli in as pure a form as possible. Freshly removed kidneys are stacked in test tubes or for larger kidneys in larger containers immersed in 0.15 M NaCl. They can be perfused with cold 0.15 M NaCl prior to removal. The kidneys are stored at -15°C. For larger kidneys such as those from dogs, the containers at -15°C are placed in a refrigerator at 4°C overnight prior to processing. Small kidneys such as those from rats can be processed directly. The processing is best done in a cold room held at 4°C or 5°C. For small kidneys, however, the initial steps can be carried out at room temperature. The tubes with the kidneys, however, are kept packed in ice, and once the sievings have been obtained all further steps are performed in the cold. No matter what sized kidneys are used, adjustments should be made so that the kidneys at the time they are being processed are not too frozen, yet not soft or mushy. The softer they are, the more tubular contaminants will come through the sieve. The mesh or pore size of the sieve should be adjusted in accordance with the glomerular size of the particular species of animal whose kidney is being processed. The capsules of the kidneys are stripped. The kidneys are bisected from superior to inferior pole along their convex border. The medullae are dissected away with forceps and curved ophthalmic scissors. The cortex is buttered through a 150 mesh monel metal screen held in a special holder. The design of the latter is illustrated in Figure 1. A rigid angled paraffin embedding spatula is used

Fig. 1. The dismantled components of the screen-holder are seen with the threaded spikes exposed in the male component. To the right the screen-holder is assembled with the screen held taut by wing nuts.

to do the buttering. The sievings are gently removed from the opposite surface of the screen with a separate spatula and mixed with 0.15 M NaCl in a bowl held under the screen. The suspension thus obtained is poured into a test tube. The latter is corked and shaken briskly. Each kidney is processed through a fresh screen or portion of screen and numbered. The screens are prepared by scouring them with a brush, placing them in cleaning solution followed by a rinse in water. They can then be sterilized in the oven at 120°C. Screens are not reused. Each numbered tube after shaking is centrifuged at 800 X g for 5 min. The supernatant is discarded. The sediment is resuspended with 0.15 M NaCl. The suspension is allowed to stand and the supernatant withdrawn by suction. This process is repeated as many times as necessary until the supernatant is clear. With a jeweler's loupe and bright light one can make out the characteristic shape of glomeruli. Particles which do not conform to that shape can be removed with a Pasteur pipet. The purity of the preparation is checked by mounting a sample on a slide with a coverslip and examining it microscopically. Preparations with tubular contaminants can be further cleaned by removing them with a pipet. As each numbered tube is processed and examined separately, the clean ones are retained and combined. Those tubes which cannot be cleaned adequately are discarded. The combined cleaned preparations are centrifuged. The firmly packed glomeruli are drained and weighed, resuspended in a given volume of 0.15 M NaCl and counts can be made using a hemacytometer and methylene blue as a diluent of the number of glomeruli with or without parietal capsules, free parietal capsules, cortical and tubular contaminants. The glomeruli are repacked and resuspended in 0.15 M, 1.0 M, or 1.4 M NaCl and sonically vibrated. The voltage and time will depend on the type of sonicator that is used. Small amounts of glomeruli should be vibrated at any one time and precautions should be taken that the material being vibrated is not overheated. The BM obtained in the sediment of the sonicate by centrifuging at 800 X g for 10 min. should be made up of microscopic refractile plates often with spiral or looped configuration. No cells should be identified. The sediment can be washed repeatedly by centrifugation in 1.4 M, 1.0 M, or 0.15 M NaCl or in distilled water, particularly if the material is to be lyophilized.

A well-prepared product of GBM at the very best has ultrastructurally some cytoplasmic membranes attached to it (Figure 2). This can be ignored depending upon the use the GBM is to be put. At the expense of the solution of portions of the membrane a product can be obtained which is as pure a GBM as

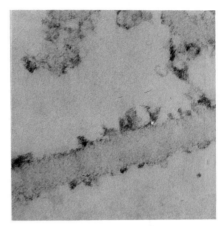

*Fig. 2. Untreated soni-
cally vibrated GBM with adherent bits of cytoplasm.*

one can wish for. Such GBM is made up of its lamina densa.
An example of treatment of isolated glomeruli with 60% aqueous
trichloroacetic acid solution (w/w) at 25°C for 4 to 5 hours
(for details see Reference 9) is represented in Figure 3.
Similarly, treatment of GBM with 5% deoxycholate in Tris-HCl
buffer, pH 7.4, for 20 hours at 5°C in a wheel mixer, results
in a loss of about 50% of the original wet weight of the sonicated GBM but with the ultrastructural appearance presented in
Figure 4.

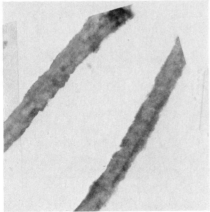

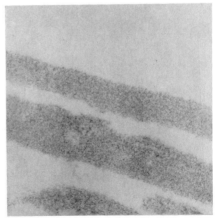

*Fig. 3. GBM treated with
60% Trichloracetic acid.
Lamina densa only.*

*Fig. 4. GBM treated with
5% Deoxycholate. Lamina densa
only.*

There have been modifications of the above method, particularly in the use of a series of sieves of different mesh to isolate the glomeruli. Many investigators have used Spiro's (10) modification. He forces cortical slices through a 115 mesh sieve using the bottom surface of a beaker. The material emerging through the sieve is collected in 0.15 M saline, gently shaken, and poured onto an 80 mesh sieve which is positioned in turn over one of 150 mesh. This is followed by a thorough wash with abundant ice-cold 0.15 M saline. The glomeruli collect on top of the 150 mesh screen. Since a variable percentage of the glomeruli obtained by sieving retain their parietal or Bowman's capsule, Gang (11) rebutters the isolated glomeruli through a 200 mesh sieve in order to obtain decapsulated glomeruli. Westberg and Michael (12) filter the sonicate through a 250 mesh stainless steel sieve to remove insufficiently disrupted glomeruli and tubular segments. Sato and co-workers (13) use a sieve of 325 mesh for the same purpose. Glomeruli can be isolated from a renal cortical homogenate or pressed pulp by using an osmotic gradient. Sucrose has been used by Richterich and Franz (14), Nagano and co-workers (15), and Helwig and co-workers (16). Nørgaard uses a Ficoll gradient to isolate glomeruli from cortical tissue which has been pressed through a teflon net with 250 micra openings (17).

In place of sonic vibration of isolated glomeruli, GBM can also be obtained by using osmotic gradients. Lidsky and co-workers (18) suspend the isolated glomeruli in 35% potassium tartrate, centrifuge and remove the supernatant. The latter is then spun in a potassium tartrate gradient ranging from 35-45%. Nagano and co-workers (19) homogenize the glomeruli in sucrose with polyoxyethylene sorbital monooleate. The glomeruli are then placed on a sucrose gradient with specific gravity from 1021-1081 and allowed to settle in the cold for 30-40 min. Three layers are formed. GBM is found in the lowest layer. Kibel and co-workers (20) homogenize the glomeruli in 0.25 M sucrose. The homogenate is layered on top of a sucrose gradient ranging from 1.8 to 1.4 M and centrifuged. The GBM obtained is treated with DNase. Meezan and co-workers (21) treat isolated glomeruli obtained by the iron-oxide method (vide infra) with distilled water containing 0.1% sodium azide. A pellet is obtained on centrifugation which is treated with DNase and subsequently with 4% sodium deoxycholate containing 0.1% sodium azide. The GBM so obtained is either washed with water by centrifugation and resuspension or by irrigation on a 44 micron nylon sieve.

A number of investigators have used Cook and Pickering's method (22) to help to isolate glomeruli. The kidneys are

perfused with a freshly prepared warm (37°C) suspension of
magnetic oxide of iron in 0.9% NaCl. The cortices are pressed
through a 150 mesh sieve. The sievings are received in physiological saline containing 2 g dextrose and 4.0 g sucrose/liter
or in Krebs' solution. They are centrifuged and rewashed and
poured through a system of tubes which at one point lies close
to the pole of an electromagnet. By a series of runs and
washes, glomeruli are obtained with or without attached fragments of tubules and arterioles. These glomeruli are then
pressed through a 150 mesh sieve which allows the non-attached
glomeruli to go through, retaining the glomeruli with attached
fragments.

Renal Tubular BM (TBM)

In order to make sure that glomeruli are not included,
one can prepare TBM either from the cortex corticis or from
the medulla of the kidney. In the dog, the cortex corticis
measures about 0.5 mm in thickness. There are no glomeruli
in this layer. Fresh dog kidneys are removed, their capsules
are stripped, and they are wrapped in aluminum foil with an
outer wrap of heavy waxed paper. The kidneys are stored at
-15°C. In preparing the cortex corticis, the kidneys are
directly moved one at a time from storage to a room kept at
0°C. The wraps are removed and the outer 0.5 mm of cortex is
shaved by single sweeps with a Bard-Parker blade, the bevel of
which is 0.5 mm wide. The kidneys must not be allowed to
thaw to the slightest degree and the depth of shaving must not
exceed the width of the bevel of the blade. The shavings are
collected in 0.15 M saline, washed, and then sonically vibrated as in obtaining GBM. If precautions are taken, no glomeruli are included in the preparation but there are collagenous
fibers from the true capsule as well as from the interstitium
of the renal tubules as well as capillary BM.

In preparation of TBM from the renal medulla, only the
more pelvic pyramidal portion is used to avoid including the
juxtamedullary glomeruli. The tissue is cut into thin shavings
on a freezing microtome and these shavings can be recut one or
more times. The shaved fragments are washed in 0.15 M saline
and sonically vibrated as in the case of GBM. Here too,
collagen and capillary BM contaminate the preparation. Sato
and co-workers (13) mince and homogenize the medulla in a
Waring blendor and filter the homogenate through sieves with
80, 115, 200, 250, 325, and 400 mesh. The tubules are found
on the last two sieves. The tubules are then sonicated in
1M saline.

Most investigators obtain TBM in the course of preparing
GBM. Mahieu and Winand (23) use the material remaining on the

sieve after pressing renal cortex through it. The material on the undersurface of the sieve is used to prepare GBM. The crude tubular material is homogenized in a Potter-Elvehjem homogenizer and filtered through a 150 mesh sieve. The sieved material is sonically vibrated. Steblay and Rudofsky (24) select the supernatants with approximately 90% tubules. These supernatants are from the individually processed kidneys after their cortices have been buttered through a sieve, shaken and centrifuged, and the glomeruli are now being allowed to settle. The collected supernatants are spun. The sediments are vibrated in phosphate buffered saline, pH 7.1. Tubules have been isolated in the course of isolating glomeruli using multiple sieves of different pore size. Thus, Ferwerda and co-workers (25) obtained uncontaminated tubules on the 166 mesh sieve in a series of sieves one over the other ranging from 50, 70, 120, 140 to 166 mesh. The tubules so obtained were sonically vibrated in 1M saline. Helwig and co-workers (16) obtain tubules in a three-layered sucrose gradient at the interface between 58.6% and 85.6%. The glomeruli are found at the interface of the 75% and 58.6% layers. The tubes are pierced from the bottom. The tubules are collected with the 58.6% layer and recentrifuged. The tubular fraction is collected with a pipet while the residual glomeruli coat the bottom of the tube. Meezan and co-workers (21) treat the renal tubules with DNase and 4% deoxycholate containing 0.1% sodium azide to obtain their basement membranes just as they do for isolated glomeruli. The tubules are obtained by processing the cortices of kidneys perfused with magnetic iron oxide through nylon sieves of varying mesh sizes. The glomeruli are separated from the tubules by use of a magnet (26).

Bowman's Capsule of the Renal Glomeruli

Bowman's capsule of the renal glomeruli are prepared by Kefalides and Denduchis (1) by crushing isolated glomeruli between microscope slides. The capsules are separated from the capillary tuft with the aid of tungsten wire probes under a dissecting microscope.

Lung

Alveolar Epithelial and Vascular Membranes

Fresh frozen lung is used. The pleura is stripped or shaved off while the lung is still in the frozen state. Thin subpleural blocks are removed from the denuded surface. By

hugging the denuded surface larger bronchial and vascular
branches with their associated connective tissue sheaths can
be avoided. The excised fragments are thoroughly washed with
0.15 M saline, shaved on a freezing microtome into 10-15 micra
sections, washed in 0.15 M saline by centrifugation at 1000 X
g, and sonically vibrated in 0.15 M saline. The sonicate is
centrifuged at 1000 X g for 15 min. and washed by recentrifugation with saline or distilled water. All operations are
carried out at 5°C. Both vascular and alveolar epithelial BM
are obtained including collagen and elastic tissue intervening
between them. Kefalides and Denduchis (1) perfuse the lung
with 0.15 M saline and inflate it with saline through the
trachea. The inflated lung is quickly frozen and treated as
above to obtain subpleural fragments. The latter are minced,
extracted with 0.3 M acetic acid at 4°C for 24 hours. The
extraction is performed four times. The insoluble residue is
sonicated in 0.85 M saline. The sonicate is centrifuged at
1400 X g for 10 minutes at 4°C. The supernatant is allowed
to stand for three hours at 4°C. The sediment is washed
three times with 0.15 M saline followed by three washes with
distilled water. All washings are performed by centrifugation
at 1400 X g for 20 min. at 4°C. The sediment is again extracted with 0.3 M acetic acid at 4°C for 24 hours and the remaining
sediment is treated with DNase in 0.1 M sodium acetate buffer
(pH 5) at room temperature for 16 hours with slow stirring.
The residue is recovered by centrifugation at 34,800 X g for
30 min. and treated with RNase under the same conditions.

Skeletal Muscle

     The sarcolemma is made up of an external layer of fibril-
lar and banded collagen, a middle homogenous BM and an internal
plasma membrane. In the methods to be described the muscle
fibers free of myofibrillar and sarcoplasmic contents retain
the above three layers in the isolated sarcolemmae.

The Method of Kono and Colowick (27)

     Fourty g of skeletal muscle in 200 ml 0.01 Tris buffer,
pH 8.2-8.4, are minced and filtered through cheesecloth. The
filtrate is spun at 12,000 X g for 10 min. The sediment is
suspended in 250 ml 0.4 M LiBr in a 0.01 M Tris buffer, stirred
for 4 hours and spun at 20,000 X g for 25 min. This procedure
is repeated with stirring for 3 hours. The sediment is now
suspended in 200 ml of 1.0 M KCl in 0.01 M Tris buffer stirred
and left standing overnight. The mixture is spun at 8000 x g

for 15 min. The sediment is suspended in 150 ml 25% KBr (w/w/) (d = 1.210) and spun at 25,000 X g for 30 min. The bottom layer is suspended in 150 ml 31% KB4 (w/w) (d = 1.275) and again spun at 25,000 X g for 30 min. The top and superior layers are mixed and centrifuged again. The top and superior layers are diluted to a density of 1.238 or approximately 27% KBr spun at 25,000 X g for 30 min. The bottom layer yields the cleanest product. All processes are performed at 0°-6°C.

The Method of McCollester (28)

Twenty-five to 35 g of excised muscle are placed in a beaker containing 200 ml of cold 150 mM NaCl solution. The muscle is transferred to a 100 ml of previously chilled 50 mM $CaCl_2$ solution and homogenized in a Waring blender at high speed for 10 sec. The homogenate is poured over a terylene net stretched across the top of a 600 ml polythene beaker. Flow through the mesh is facilitated by stirring and scraping with a piece of perspex. Material failing to pass through the net is re-homogenized. It is important to keep the temperature of the blendor as close to 0°C as possible. Four or five such treatments are sufficient to permit most of the muscle but not the gross connective tissue to pass through the net. The homogenates are pooled and are refiltered through clean terylene nets. All subsequent operations are performed at 0°C or room temperature. The homogenates are concentrated by centrifugation. They are resuspended by gentle agitation and inversions in 25 mM NaCl. The washing is repeated three times. The suspension is then incubated at 37°C for 0.5 hour and again washed five times. The sediment is now resuspended in doubly distilled water, adjusted to pH 7.4 to 7.8 with Tris buffer, and centrifuged for 10-20 seconds. The supernatants are decanted and more buffered water is added. The swollen sediments are divided into several bottles, resuspended by vigorous shaking, and spun at 2000 X g for 7 min. A white opaque bottom layer is formed. The supernatant over this is aspirated, fresh buffered water is added, and the process of suspension and centrifugation repeated once or twice. The bottom layer is withdrawn and expelled three or four times through a 20 ml nylon syringe fitted with a 10 cm length of polythene tubing. The material is centrifuged and the cloudy supernatant decanted. The white, fluffy pellets forming the sediment are resuspended in buffered water. Resuspension and recentrifugation are continued until the supernatant is clear.

The Method of Rosenthal, Edelman, and Schwartz (29)

All procedures are performed at 0°-4°C unless otherwise indicated. Thirty-five g of muscle are placed in a 100 ml of 50 mM $CaCl_2$ solution in a Waring blendor with reversed blades. The blending is performed at high speed for 10 seconds, filtered through a 16-18 mesh terylene net. The residue on the filter is homogenized repeatedly until it is devoid of red color. The pooled homogenates are refiltered through terylene netting and collected. They are spun in 50 ml cellulose nitrate tubes at 3000 X g for 5 seconds. The sediments are combined, resuspended to a volume of 80 ml in fresh KCl buffer (45 mM KCl, 30 mM $KHCO_3$, 2.5 mM DL-Histidine monohydrochloride, adjusted to pH 7.8 with 1 M Tris) and centrifuged. The resulting sediment is washed two times in this manner. The sediment is brought to 80 ml with KCl buffer, divided among four centrifuge tubes, and incubated at 37°C for 0.5 hour in a water-bath shaker at 100 oscillations/min. The test tubes are then placed in ice and allowed to settle for 5 minutes. The supernatant is aspirated and the sediment resuspended in the KCl buffer to a volume of 320 ml. This suspension is allowed to settle in 50 ml centrifuge tubes in ice for 5 min. The settling procedure is repeated four times. The sediments are now combined and resuspended in 320 ml of glass-distilled water and allowed to settle. The sediment is resuspended in 160 ml of 2.5×10^{-7} N NaOH. Settling and resuspension of the sediment in 2.5×10^{-7} N NaOH are continued until there is no further swelling of the sediment. The sediment is now resuspended in 150 ml of the NaOH, centrifuged at 3000 X g for 15 sec. After discarding the supernatant, the process is repeated prolonging the centrifugation for 5 min. The upper portion of the supernatant is clear and is discarded. The lower cloudy portion is saved. The process is repeated two more times and the cloudy supernatants are saved. Two ml of 2.0 mM ATP are added to 20 ml aliquots of the cloudy supernatant fractions. The suspensions are gently stirred, diluted with distilled water to a volume of 500 ml each, and centrifuged at 1500 X g for 1 min. The supernatants are discarded and the pellets of relatively pure sarcolemmal tubules are combined.

Placenta

Trophoblastic BM

Gang and co-workers (30) have isolated trophoblastic BM
from fresh, chilled, full-term, human placentae. Decidual and
chorionic plates are removed. The middle zone of the central
region of the placenta is cut into small fragments with a
knife. The small villi are scraped off the larger trunks with
a razor blade. The preparation of small villi is gently forced
through an 80 mesh stainless steel wire cloth. The sieved
material is collected with an angular spatula and transferred
to ice-cold 0.15 M saline. The suspension is washed through
the same sieve. The material collected is poured onto a 325
mesh sieve and washed with an excess of cold 0.15 M saline.
The decellulated villi retained on the surface of the screen
are collected with a Pasteur pipet, transferred to a centrifuge tube with cold saline, and centrifuged at 1060 X g for
15 min. The supernatant is discarded. The residual pellet
is gently buttered through a 200 mesh sieve resting on a
beaker placed in an ice bath. The sieve is washed with cold
saline. The collected sievings are washed by centrifugation
at 1060 X g for 15 min. three times with ice-cold saline and
five times with ice-cold distilled water. The final sediment
is lyophilized. The latter is suspended in cold 1 M saline
and homogenized in a loosely fitting Potter-Elvehjem hand
homogenizer. The homogenized suspension is then sonically
vibrated. The authors use a Bronwell Biosonic III Sonifier
with a 1.0 inch probe terminating about 0.5 cm below the
liquid surface. Vibration is performed at 60% of maximum
power output for a total of 7 min. with 1 min. vibration and
3 min. cooling. The sonicate is centrifuged at 250 X g for
2 min. The supernatant is centrifuged at 1500 X g for 10 min.
The residue is washed 5 times in distilled water and lyophilized. The product at this stage is extracted with 0.3 M acetic
acid at 4°C for 24 hours. The sediment obtained after centrifugation at 1500 X g for 20 minutes is re-extracted. The
process is repeated three times. The insoluble residue is
then washed in distilled water and lyophilized. The latter is
incubated in 0.1 M sodium acetate buffer, pH 5.0, containing
0.002% DNase and 0.002% RNase for 16 hours at room temperature
with slow stirring. The suspension is then centrifuged at
5000 X g for 30 min. and the residue washed with distilled
water. The residue examined electronmicroscopically is described as made up of amorphous BM free from collagen fibers
and plasma membranes.

Yolk Sac BM

It is extremely difficult to separate the parietal yolk
sac membrane in mice from the rest of the placenta. In its
stead Pierce and co-workers (31) have used the ascitic form
of a teratocarcinoma carried in mice of the 129 J strain. The
tumor produces small spherical aggregates of neoplastic BM
surrounded by carcinomatous cells. The ascitic fluid with
its aggregates are vibrated in 8% NaCl in a sonicator for
15-30 min. The insoluble fraction is collected by centrifugation in the cold at 1400 X g for 30 minutes, washed five
times in 8% NaCl and subsequently in distilled water. The
ascitic aggregates can also be separated by centrifugation
from the ascitic fluid and disrupted by sonication in 5
volumes of 8 M NaCl. The BM secured by centrifugation are
washed repeatedly in decreasing concentrations of NaCl,
finally in distilled water, and then lyophilized (32).

REFERENCES

1. Kefalides, N.A. and Denduchis, B., *Biochemistry* 8:4613,
    1969.
2. Brendel, K., Meezan, E., and Carlson, E.C., *Science* 185:
    953, 1974.
3. Meezan, E., Brendel, K., and Carlson, E.C., *Nature* 251:
    65, 1974.
4. Meezan, E., Brendel, K., Hjelle, J.T., and Carlson, E.C.
    Biochem. Biophys. Res. Comm. (This volume), 1977.
5. Dohlman, C.H. and Balazs, E.A., *Arch. Biochem. Biophys.*
    57:445, 1955.
6. Dardenne, M.U., Iwangoff, P., and Diotallevi, M.,
    Ophthalmologica 156:385, 1968.
7. Paulini, K. and Beneke, G., *Virchows Archiv. Abt. B.* 4:
    208, 1970.
8. Thiele, H., Flasch, R., and Joraschky, W., *Albrecht v.
    Graefes Arch. Klin. Exp. Ophthal.* 179:157, 1969.
9. Nicholes, B.K., Krakower, C.A., and Greenspon, S. A.,
    Proc. Soc. Exper. Biol. Med. 142:1316, 1973.
10. Spiro, R.G., *J. Biol. Chem.* 242:1915, 1967.
11. Gang, N.F., *Am. J. Clin. Path.* 53:267, 1970.
12. Westberg, N.G. and Michael, A. F., *Biochemistry* 9:3837,
    1970.
13. Sato, T., Munakata, H., Yoshinaga, K., and Yosizawa, Z.,
    Clin. Chim. Acta 61:145, 1975.

14. Richterich, R. and Franz, H.E., *Nature* 188:498, 1960.
15. Nagano, M., Kogure, T., Kawamura, M., and Kawanishi, M., *Jap. Circulation J.* 32:1579, 1968.
16. Helwig, J.J., Zachary, D., and Bollack, C., *Urol. Res.* 2:55, 1974.
17. Nørgaard, J.O. Rytter, *Kidney Internat.* 9:278, 1976.
18. Lidsky, M.D., Sharp, J.T., and Rudee, M.L., *Arch. Biochem. Biophys.* 121:491, 1967.
19. Nagano, M., Kawamura, M., Kawanishi, M., and Suzuki, A., *Res. Exp. Med.* 165:191, 1975.
20. Kibel, G., Heilhecker, A., and von Bruchhausen, F., *Biochem. J.* 155:535, 1976.
21. Meezan, E., Hjelle, J.T., and Brendel, K., *Life Sci.* 17:1721, 1976.
22. Cook, W.F. and Pickering, G.W., *Nature* 182:1103, 1958.
23. Mahieu, P. and Winand, R.J., *Eur. J. Biochem.* 12:410, 1970.
24. Steblay, R.W. and Rudofsky, U., *J. Immunol.* 107:589, 1971.
25. Ferwerda, W., Meijer, J.F.M., van den Eijnden, D.H., and van Dijk, W., *Hoppe-Seyler's Z. Physiol. Chem.* 355:976, 1974.
26. Brendel, K. and Meezan, E., *Fed. Proc.* 34:803, 1975.
27. Kono, T. and Colowick, S.P., *Arch. Biochem. Biophys.* 93:520, 1961.
28. McCollester, D.L., *Biochim. Biophys. Acta* 57:427, 1962.
29. Rosenthal, S.L., Edelman, P.M. and Schwartz, I.L., *Biochim. Biophys. Acta* 109:512, 1965.
30. Gang, N.F. and Gelfand, M.M., *Proc. Soc. Exp. Biol. and Med.* 140:188, 1972.
31. Pierce, G.B. Jr., Midgley, A.R. Jr., and Ram, J. Sri., *J. Exper. Med.* 117:339, 1963.
32. Pierce, G.B. Jr. and Nakane, P.K., *Lab. Invest.* 17:499, 1967.

A VERSATILE METHOD FOR THE
ISOLATION OF ULTRASTRUCTURALLY AND CHEMICALLY
PURE BASEMENT MEMBRANES WITHOUT SONICATION

Elias Meezan, Klaus Brendel,
J. Thomas Hjelle* and Edward C. Carlson

Arizona Health Sciences Center
Tucson, Arizona

SUMMARY: A method has been developed for the isolation of basement membranes from kidney glomeruli and tubules and brain and retinal microvessels which relies on the selective solubilization of cell membranes, intracellular protein and plasma proteins with the detergents Triton X-100 and sodium deoxycholate, leaving behind the basement membranes which are insoluble in these reagents. This procedure yields preparations of basement membranes which are ultrastructurally intact, and in contrast to those obtained by sonication, which are non-fragmented and retain their boundary architecture. The chemical composition of the isolated basement membranes from lens capsule and renal glomeruli which have been well characterized previously compared well with reported values, indicating that the method of detergent isolation gives membrane preparations which are both ultrastructurally and chemically intact. The versatility of the isolation method has been demonstrated by its use to obtain the basement membranes of retinal and brain cortical microvessels, which had not been previously isolated or characterized.

*Present Address: Department of Physiological Chemistry, Roche Institute of Molecular Biology, Nutley, New Jersey.

17

INTRODUCTION

        Basement membranes have been studied extensively
both morphologically and biochemically since they were
first isolated from the renal glomerulus 25 years ago by
Krakower and Greenspon (1). Since then, these extracellular
boundary structures have been isolated and studied from
several tissues with most investigations concentrating on the
basement membranes of the renal glomerulus and the lens
capsule of the eye (2-5). The general approach to basement
membrane isolation has been one of obtaining a pure organ
subfraction, e.g., renal glomeruli, disrupting the cells
lining the basement membranes by extensive sonication, and
separating the basement membranes from cell membranes and
intracellular material by low speed centrifugation. There
are several problems associated with these procedures.
Obtaining a pure organ subfraction as a starting material for
basement membrane isolation while simple for a macroscopic
tissue component such as the lens capsule of the eye (4), is
more difficult for tissue fractions such as renal glomeruli
which must be separated with variable success from numerous
and closely associated tubules (2,3), and until recently has
not been possible for the microvessels of the retina and
cerebral cortex (6,7). The elimination of contamination by
cell membrane fragments which adhere to or co-sediment with
the basement membranes following sonication has also been
difficult to achieve (8), as well as the removal of plasma
proteins which have penetrated into the basement membrane
matrix (9). Finally, and most importantly, sonication
severely fragments the basement membranes destroying the normal spatial arrangement which they occupy in intact tissue
and limiting the usefulness of isolated preparations for
ultrastructural investigation (9). We have attempted to
minimize these problems and to extend basement membrane
isolation to previously unexamined tissues such as the retinal and cerebral cortical microvasculature by the development
of a method for basement membrane isolation (10) which relies
on the selective solubilization of cell membranes. Intracellular protein and plasma proteins with the detergents
Triton X-100 and sodium deoxycholate. The resulting basement
membrane preparations are non-fragmented, ultrastructurally
intact and retain their boundary architecture as well as
being chemically compatible with the composition of isolated
basement membranes.

METHODS

        Retinal and brain microvessels were isolated from
bovine tissue as described previously (7,7). Renal glomeruli
and tubules were obtained from rats and rabbits, respectively,

by a procedure based on that of Cook and Pickering (11-13)
which is outlined in more detail in the following paper (14).
Lens capsules were obtained by dissection from bovine eyes.
The purified organ subfractions were used as the starting
material for basement membrane isolation.

The organ subfractions were first osmotically lysed by
suspension in a large volume (100:1) of distilled water for
1-2 hrs followed by separation of the water insoluble cellular residue containing the basement membrane fraction from
the soluble cell contents by centrifugation. All steps were
carried out at room temperature with solutions containing
0.05% sodium azide to prevent bacterial growth. The pelleted
material was suspended in 40 ml of 3% Triton X-100 and
stirred for 2-4 hrs, to solubilize cell and intracellular
membranes with the exception of those of the nuclei. The
crude basement membrane fraction plus adhering nuclei was
separated from the Triton X-100 soluble material by centrifugation or sieving on a 44 μm nylon sieve. This insoluble
fraction was suspended in 40 ml of 1 M NaCl and treated with
2000 Kunitz units of deoxyribonuclease I (Sigma) for 1-2 hrs
in order to disrupt intact nuclei and digest released DNA.
The now relatively pure basement membrane fraction obtained
by centrifugation or sieving was suspended in 40 ml of 4%
sodium deoxycholate and stirred for 2-4 hrs to solubilize any
remaining cellular contaminants. The resulting pure preparation of basement membrane was washed extensively with distilled water and processed for electron microscopic examination
or dried over $CaCl_2$ for chemical analysis.

Light and electron microscopic studies of the isolated
organ subfractions and their basement membranes were carried
out as described previously (10) and detailed in the accompanying paper (14).

Carbohydrate analyses of the isolated basement membrane
samples were performed by gas chromatography of the sugar
alcohol acetates (10). Amino acid analyses of the N-tri-
fluoroacetyl-butylester derivatives were obtained by the gas
chromatographic procedure of Gehrke et al. (15).

RESULTS

Examination by light and electron microscopy of the
microvessel preparations obtained from bovine retinas and
cerebral cortex indicated that they were remarkably clean and,
with the exception of erythrocytes trapped in the vessel
lumens, free of non-vascular tissue contamination. They were
thus suitable organ subfractions for the isolation of basement membrane. Cross sections through retinal vessel walls
clearly revealed the basement membrane surrounding endothelial cells and sometimes splitting to enclose intramural

pericytes (Fig. la). Treatment of the vessels by the detergent procedure for basement membrane isolation resulted in

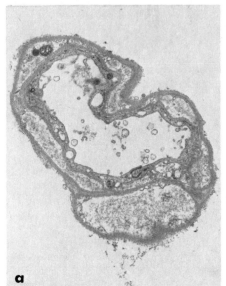

Fig. la. Electron micro-
graph of cross section through
isolated bovine retinal vessel.
The structural integrity of
endothelium lining the vessel
lumen is maintained. Subendothelial basement membrane
completely surrounds the
vessel and frequently splits
to enclose intramural pericytes. Almost no non-vascular
tissues are seen in these preparations. X 9,800.

solubilization of all cellular contents leaving ghost-like
profiles of the original vessel walls remaining, due to the
insolubility in the detergents of the basement membrane forming these structures (Fig. lb). These basement membrane

Fig. lb. Electron micro-
graph of cross section through
a "tube" of a bovine retinal
vessel treated with Triton
X-100 and sodium deoxycholate.
Ghost-like profiles of vessel
wall are evident. Sub-endo-
thelial basement membrane lines
the vessel lumen and is clearly
distinguishable from underlying
basement membranes of pericytes.
X 11,400.

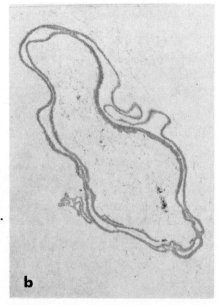

preparations were free of cell membrane and other cellular debris contamination, and were ultrastructurally indistinguishable from their intact tissue counterparts. The histoarchitecture of the basement membrane of the vessel wall was maintained after isolation and at higher magnifications the luminal side and the adventitial side of the membrane could be clearly distinguished (Fig. 1c). The sharply demarcated surface of the luminal side of the basement membrane contrasts with the less distinct fuzzy surface of the abluminal side which was frequently associated with masses of amorphous material. Very high magnifications of the isolated retinal vessel basement membrane revealed a granular matrix intermingled with very fine fibrils 30-50 Å in diameter (Fig. 1d).

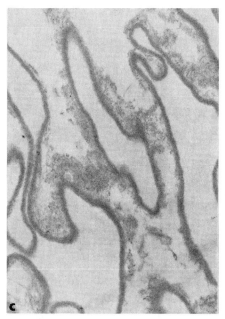 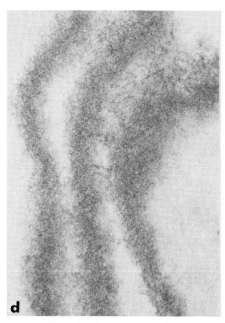

Fig. 1c. Electron micrograph of isolated bovine retinal basement membranes. Clear areas represent vessel lumina and electron dense material connecting adjacent vessels demonstrates obliquely sectioned basement membrane material. Cellular debris is not present in these preparations. X 16,400.

Fig. 1d. High resolution electron micrograph of adjacent layers of isolated bovine retinal vessel basement membrane. Fine fibrils are poorly resolved within the granular matrix. Surfaces of these basement membranes are not sharply demarcated but become progressively less dense as the distance from their centers is increased. X 69,500.

The greatest electron density occurred in the central portion
of the basement membrane and was decreased at both surfaces
corresponding to the *in vivo* lamina densa and lamina rarae.

Basement membrane isolated from bovine cerebral cortical
microvessels morphologically resembled that obtained from
retinal vessels (10), but was more difficult to obtain in
ultrastructurally pure form.

Pure preparations of renal glomeruli and tubules were
obtained by perfusion of kidneys with a suspension of magnetic
iron oxide (11-14) and made possible the isolation of the
basement membranes of these organ subfractions with almost no
cross-contamination. The isolated glomeruli exhibited an
intact ultrastructure by electron microscopy with a continuous
basement membrane separating fenestrated capillary endothelium
from foot processes of podocytes (Fig. 2a). After detergent
treatment all cellular material was solubilized leaving ultrastructurally intact glomerular basement membrane whose spatial
arrangement within the glomerular histoarchitecture was preserved (Fig. 2b). Mesangial areas could be readily distinguished central to capillary lumina which radiated from these
sites. In the mesangial region, the capillary basement membrane was observed to be incomplete with patches of unstruc-
tured basement membrane material located most often on the
tissue space side. At high magnifications, it was strikingly
apparent that the ultrastructure of the isolated glomerular

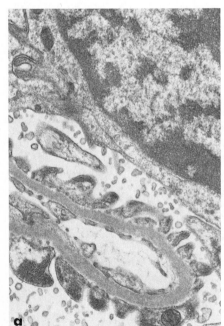

*Fig. 2a. Electron micro-
graph of portion of isolated
rat renal glomerulus. Upper
right shows adjacent parietal
epithelial cells of Bowman's
capsule. Glomerular basement
membrane (lower left) inter-
venes between foot processes of
podocytes and inner fenestrated
capillary endothelium.
X 28,300.*

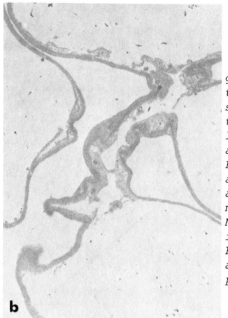

b

Fig. 2b. Electron micrograph of rat renal glomerulus treated with Triton X-100 and sodium deoxycholate. The histoarchitecture of the glomerulus is maintained. Clear areas represent the in vivo Bowman's space and in most areas are located directly across glomerular basement membrane from capillary lumina. Mesangial matrix is retained in the central mesangial area but unit collagenous fibrils are virtually excluded by this procedure. X 10,500.

basement membrane (Fig. 2d) was identical to that seen in intact tissue sections (Fig. 2c). The isolated basement membrane was a composite of fibrillar and amorphous materials; the fibrillar component being an aggregate of 40-50 Å fibrils and the interfibrillar matrix consisting of a flocculent material of medium electron density.

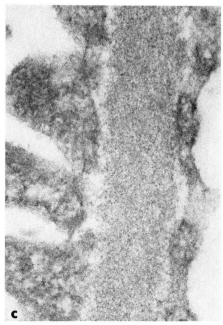

c

Fig. 2c. High resolution electron micrograph of in vivo rat renal glomerular basement membrane. On the left side of the membrane, slit membranes connect foot processes of podocytes which are separated from lamina densa by lamina rara externa. The lamina rara interna is shown on the right separating fenestrated capillary endothelium from the lamina densa. X 139,000.

Fig. 2d. High resolution
electron micrograph of isolated rat renal glomerular basement membrane. The basement
membrane is shown at a similar
magnification, section angle
and thickness as that seen in
Fig. 2c. and is ultrastructurally indistinguishable from
its in vivo counterpart.
X 135,000.

Kidney tubular basement membrane isolated by the detergent procedure was remarkably free of contamination and uniform in appearance, and was characterized by long continuous
folded sheets which formed large loops. The ultrastructure
of this basement membrane is described in detail in the
accompanying paper (14).

Amino acid and carbohydrate analyses of the isolated
basement membrane preparations described in this study are
shown in Table I along with those of lens capsule basement
membrane isolated by the identical procedure of detergent
extraction. All of the basement membranes are notable for
their high content of glycine which comprises about one-quar-
ter of the total amino acid residues, except for that of
renal glomeruli where this amino acid made up one-fifth of
the total. The basement membranes were also all uniformly
high in their content of proline, hydroxyproline and glutamic
acid, as well as containing appreciable amounts of hydroxylysine. In general, the amino acid profiles of the basement
membranes were remarkably similar, although the content of
certain amino acids varied appreciably from one membrane to
the other, providing compositional features which made each
chemically distinct.

TABLE I

Amino Acid and Carbohydrate Composition of Isolated Basement Membranes

Residues/1000 Residues of Amino Acids

Residue	Sonicated Bovine Lens Capsules[1]	Bovine Lens Capsules	Sonicated Rat Renal Glomeruli[2]	Rat Renal Glomeruli	Rabbit Renal Tubules	Bovine Retinal Vessels	Bovine Brain Vessels
Hydroxyproline	110.7	106.1	52.0	70.2	92.0	98.5	88.4
Aspartic Acid	53.8	58.2	74.2	79.8	64.2	58.9	55.4
Threonine	28.4	30.5	46.3	38.4	32.5	31.6	25.3
Serine	44.9	50.5	62.8	58.9	52.4	38.0	43.2
Glutamic Acid	87.8	88.4	100.0	98.3	92.7	87.0	80.0
Proline	71.2	71.0	62.0	66.3	70.0	84.3	94.1
Glycine	267.1	253.7	200.0	202.3	246.6	267.6	273.6
Alanine	39.6	41.2	67.0	56.0	54.2	69.9	84.0
Valine	29.7	25.7	43.0	36.0	31.3	24.7	25.6
Methionine	12.5	14.6	11.0	11.7	9.4	8.6	3.5
Isoleucine	30.8	26.5	32.0	26.8	29.0	21.5	19.6
Leucine	57.2	56.6	66.0	64.8	43.8	43.8	37.5
Tyrosine	11.1	4.1	9.0	17.0	6.6	5.1	2.1
Phenylalanine	30.9	30.1	19.0	34.1	46.5	25.7	21.5
Hydroxylysine	44.8	52.0	21.8	23.4	31.3	27.4	23.4
Lysine	12.1	12.2	40.1	31.6	18.5	18.3	22.9
Histidine	11.5	16.7	20.9	18.9	7.9	9.0	11.4
Arginine	35.8	40.4	52.6	41.0	42.9	55.6	69.3
Half-Cystine	17.0	21.8	20.4	24.5	15.9	14.6	19.2
Glucose	38.8	38.3	(2.3)[3]	18.6	22.7	21.5	8.2
Galactose	40.0	37.3	(1.9)	18.8	24.3	20.9	6.9
Mannose	3.7	3.6	(0.8)	4.0	3.1	3.5	1.5
Fucose	2.0	1.3	(0.3)	0.6	0.7	2.2	1.6
Hexosamines	5.3	4.7	(0.65)	6.5	5.3	5.6	4.9

[1] Taken from Fukushi and Spiro (4).

[2] Taken from Kefalides and Forsell-Knott (19).

[3] Expressed as g/100 g.

The carbohydrate analyses of the isolated basement membrane samples also showed marked similarities. All of the samples examined had equimolar amounts of glucose and galactose which are the carbohydrates present in the largest amounts. The quantity of these sugars varied widely, however, from highest amounts in lens capsule basement membrane, intermediate levels in retinal vessel and renal glomerular and tubular basement membrane and lowest amounts in material derived from brain vessels. The hexosamines were the sugars found in next highest abundance in all samples, followed by mannose and small amounts of fucose. The quantities of these sugars were relatively constant from one basement membrane to the next with the exception of mannose which was present in appreciably lower levels in brain vessel basement membrane.

DISCUSSION

The basement membrane preparations isolated by the procedure described in this paper were ultrastructurally intact, free of any significant contamination with cell membranes and other cellular debris, and maintained the histo-architecture which these boundary structures exhibit in intact tissue sections. The maintenance of basement membrane structural integrity during the isolation procedure is based on the insolubility of this extracellular matrix in the detergents Triton X-100 (9) and sodium deoxycholate (10,16,17), agents which completely solubilize the cell membrane and all intracellular material. Because Triton X-100 solubilizes the the cell membrane without disrupting the nuclear envelope a graded solubilization of the cellular contents was possible which coupled with limited digestion with deoxyribonuclease minimized the difficulties introduced by exposure of the basement membrane to large amounts of cellular DNA. Following Triton X-100 extraction, many nuclei were removed by washing of the crude basement membrane fraction on a nylon sieve. The remaining nuclei adhering to the basement membrane could then be disrupted in 1 M NaCl and the released DNA digested with deoxyribonuclease. A final extraction with sodium deoxycholate dissolved any remaining cellular material yielding a pure basement membrane fraction. Direct treatment of the pure organ subfractions with sodium deoxycholate was adequate for the preparation of basement membrane from a few milligrams of tissue, but with larger amounts of tissue graded solubilization and deoxyribonuclease treatment were necessary to avoid the formation of a gel-like aggregate of DNA with basement membrane which was then difficult to disperse. The affinity of DNA for isolated basement membrane has been recently described (18).

Ultrastructurally the isolated basement membranes were indistinguishable from their appearance in the tissues from which they were obtained. Neither the thickness nor the fibrillar appearance of the basement membranes were altered by the detergent treatment and the gradation of electron density from lamina densa to lamina rarae was observable in the isolated preparations from retinal vessels and renal glomeruli. Most significant and unique to this isolation procedure the isolated basement membranes retained the spatial arrangement which they occupied *in vivo* making these preparations superior to those isolated by techniques involving sonication for ultrastructural studies.

Chemical analyses of the isolated basement membranes revealed that all had in common the characteristic amino acid and carbohydrate composition which is typical of the basement membranes whose structures have been investigated (5). Comparison of the compositions of bovine lens capsule and rat renal glomerular basement membrane isolated by the Triton X-100, deoxycholate procedure with those reported for well characterized preparations isolated by the standard technique of sonication (4,19), demonstrated that both techniques gave preparations of comparable composition (Table I), indicating that our detergent method yields chemically identical as well as ultrastructurally intact basement membranes. While all the basement membranes had similar amino acid compositions, those from lens capsule and renal tubules both associated with epithelial cells resembled each other most closely, while those from retinal and brain vessels both associated with endothelial cells and intramural pericytes formed another pair with similar compositions. Kidney glomerular basement membrane uniquely associated with epithelial, endothelial and mesangial cells had a composition which was distinctly different from the others, having a lower glycine and hydroxyproline content and a higher content of aspartic and glutamic acids and lysine. All of the basement membranes had the equimolar content of glucose and galactose characteristic of the disaccharide moiety attached to hydroxylysine present in basement membrane collagen (20,21). Brain vessel basement membrane was distinguished by its markedly lower content of these two monosaccharides, which make it distinctly less hydroxylated than the other basement membranes as indicated by its different hydroxylysine to glucose ratio (Table II). Lens capsule basement membrane was remarkable for its high hydroxylysine to lysine ratio and its high glucose and galactose content, although its degree of glycosylation was similar to the other preparations, excepting that from brain vessels. It is notable that the basement membranes differ from each other most

TABLE II

Hydroxylation Patterns of Basement Membranes

Ratio	Sonicated Bovine Lens Capsules[1]	Bovine Lens Capsules	Sonicated Rat Renal Glomeruli[2]	Rat Renal Glomeruli	Rabbit Renal Tubules	Bovine Retinal Vessels	Bovine Brain Vessels
Hydroxyproline/proline	1.55	1.49	0.84	1.06	1.31	1.17	0.94
Hydroxylysine/lysine	3.70	4.26	0.54	0.74	1.69	1.50	1.02
Hydroxylysine/hydroxyproline	0.40	0.49	0.42	0.33	0.34	0.28	0.26
Hydroxylysine/glucose	1.15	1.36	-	1.26	1.38	1.27	2.85

[1]Taken from Fukushi and Spiro (4).
[2]Taken from Kefalides and Forsell-Knott (19).

significantly in their pattern of hydroxylated amino acids and in their carbohydrate content, both of which are dependent on postribosomal events which could be affected by environmental as well as genetic factors. In this regard the similarity between the hydroxylysine, glucose and galactose composition of retinal vessel and renal glomerular basement membranes is of interest in view of the principal involvement of these structures in the pathological complications of diabetes mellitus.

The ultrastructural purity of the isolated basement membrane preparations coupled with their chemical identity to comparable preparations obtained by sonication indicated that the technique of detergent isolation did not measurably alter the appearance or structure of the basement membranes. In addition, the maintenance of the boundary architecture of the tissue with the detergent method would appear to make it the technique of choice for ultrastructural and chemical investigation of isolated basement membranes.

ACKNOWLEDGEMENTS: We would like to thank Dr. Sai Chang for advice concerning the methodology of amino acid analyses by gas chromatography and Ms. Betsy Hurd, Jerrolyn Campbell and Margaret Krasovich for technical assistance. This work was supported by Grants AM-15394, AM-HL-14977 and HL-17421-03S1 from the National Institutes of Health and by Grant 76-944 from the American Heart Association. Dr. Meezan is a Research Career Development Awardee of the National Institute of Arthritis, Metabolism and Digestive Diseases.

REFERENCES:

1. Krakower, C.A. and Greenspon, S.A., *Arch. Pathol.* 51:629-639, 1951.
2. Kefalides, N.A. and Winzler, R.J., *Biochemistry* 5:702-713, 1966.
3. Spiro, R.G., *J. Biol. Chem.* 242:1915-1922, 1967.
4. Fukushi, S. and Spiro, R.G., *J. Biol. Chem.* 244:2041-2048, 1969.
5. Kefalides, N.A., *Int. Rev. of Exptl. Pathol.* 10:1-39, 1971.
6. Meezan, E., Brendel, K. and Carlson, E.C., *Nature* 251:65-67, 1974.
7. Brendel, K., Meezan, E. and Carlson, E.C., *Science* 185:953-955, 1974.
8. Mohos, S.C. and Skoza, L., *J. Cell. Biol.* 45:450-455, 1970.

9. Westberg, N.G. and Michael, A.F., *Biochemistry* 9:3837-3846, 1970.
10. Meezan, E., Hjelle, J.T., Brendel, K. and Carlson, E.C., *Life Sci.* 17:1721-1732, 1975.
11. Cook, W.F. and Pickering, G.W., *Nature* 182:1103-1104, 1958.
12. Meezan, E., Brendel, K., Ulreich, J. and Carlson, E.C., *J. Pharmacol. Exp. Ther.* 187:332-341, 1973.
13. Brendel, K. and Meezan, E., *Fed. Proc.* 34:803, 1975.
14. Carlson, E.C., Brendel, K. and Meezan, E., *(this volume)* 1977.
15. Gehrke, C.W., Roach, D., Zumwalt, R.W., Stalling, D.L. and Wall, L.L., "Quantitative Gas-Liquid Chromatography of Amino Acids in Proteins and Biological Substances", Analytical Biochemistry Laboratories, Columbia.
16. Welling, L.W. and Grantham, J.J., *J. Clin. Invest.* 51:1063-1075, 1972.
17. von Bruchhausen, F. and Merker, H.J., *Histochemie* 8:90-108, 1967.
18. Izui, S., Lambert, P.H. and Miescher, P.A., *J. Exptl. Med.* 144:428-443, 1976.
19. Kefalides, N.A. and Forsell-Knott, L., *Biochim. Biophys. Acta.* 203:62-66, 1970.
20. Spiro, R.G., *J. Biol. Chem.* 242:4813-4823, 1967.
21. Kefalides, N.A., *Int. Rev. Connect. Tiss. Res.* 6:63-104, 1973.

STRIATED COLLAGEN FIBRILS DERIVED FROM

ULTRASTRUCTURALLY PURE ISOLATED BASAL LAMINA

Edward C. Carlson, Klaus Brendel

and Elias Meezan

Arizona Health Sciences Center
Tucson, Arizona

SUMMARY: Electron microscopic studies of ultrastruc-
turally pure basal lamina (BL) isolated from a pre-
paration of rabbit renal tubules showed continuous
folded layers devoid of unit collagenous fibrils and
other cellular contaminants. When these BL were treat-
ed up to 42 hours with 0.1% pepsin at 4°C, a meshwork
of fibrillar material was released selectively from one
side of the BL, but no striated collagen fibrils were
observed. However, when pepsinization of isolated BL
was carried out at 22°C (30 min) or at 4°C (up to 42
hrs) followed by 30 min at 22°C and pH 7.0, small
striated collagen fibrils (s00 A in diameter) were
present associated with the partially digested BL.
When other isolated BL were treated long term (5 days)
with sodium deoxycholate and Triton X-100, BL were
thinner and less dense. Furthermore, numerous stri-
ated collagen fibrils were anchored within the sub-
stance of the BL and projected from one laminar surface.
When these latter preparations were further treated
2 hrs with SDS-mercaptoethanol-urea, striated fibrils
were reduced to short segments of intertwined fine fila-
ments. Data in the present study suggest that masked
striated collagen fibrils may be confined within BL.
Furthermore, this or other forms of BL collagen may be
released by pepsin and under appropriate conditions may
be reaggregated to form fibrils

The nature of basal lamina (BL) has been the subject of investigations from several points of view. Biochemical studies indicate that BL is a complex composite of protein subunits which contain considerable amounts of carbohydrate (1,2,3,4). It seems clearly established that one of the protein components is a collagen (2,5,6). This collagenous component presently is being characterized in various laboratories, primarily by biochemical methods.

Morphological studies of isolated BL have not kept pace with biochemical investigations primarily because non-disruptive isolation methods have not been available. Fine structural illustrations of isolated BL usually have been relegated to one or more micrographs within larger studies to demonstrate the morphology of starting materials for biochemical investigations. Nevertheless, many investigators agree that ultrastructural examination provides the best tool for evaluating the purity of preparations (7,8).

The standard techniques employing sonication are not satisfactory for preparing isolated BL for morphological studies. Basal laminae prepared by sonication are often contaminated with cellular debris and the procedure is not precisely repeatable. Indeed, some studies have shown that compositional changes detectable by biochemical methods are related directly to the duration of sonication (9). Moreover, sonication fragments BL to the extent that its normal histoarchitecture is not maintained.

One recently published (4) non-disruptive method for the isolation of ultrastructurally pure BL provides a reproducible approach to BL studies. Solubilization of isolated organ subfractions with detergents yields a preparation of BL which maintains its structural integrity and is present in sufficient quantities for simultaneous chemical and morphological investigations.

The present study was carried out in an effort to characterize isolated rabbit renal tubule BL by electron microscopic techniques. Tubular BL was treated with pepsin, detergents, and detergents followed by sodium dodecyl sulfate (SDS)-mercaptoethanol-urea, in an attempt to unmask or otherwise enable the visualization of striated collagen fibrils derived from ultrastructurally pure preparations of BL.

MATERIALS AND METHODS

    The procedure for isolating rabbit proximal renal tubules
is illustrated in Figure 1 and has been described previously
(10). New Zealand white rabbits used in this study were
killed with a shot to the head followed by direct cannulation
of the renal arteries. Kidneys were perfused with Earle's
balanced salt solution and bovine serum albumin to remove the
majority of blood elements from the kidney. This was followed
by infusion with a suspension of magnetic iron oxide (0.72 g/l)
in HEPES buffer with 4% polyvinylpyrolidone at pH 7.4. The
cortex of the kidney was removed by blunt dissection, minced,
homogenized in a loose-fitting Teflon tissue homogenizer and
poured onto a 210 µm pore-size nylon sieve. The material
passing through the sieve was then poured onto a 153 µm sieve,
washed through and placed on a 64 µm sieve. The majority of
the residue from the final sieve was comprised of renal
tubules and glomeruli. These were resuspended in a buffer and
placed in a plastic beaker over permanent bar magnets. Iron-
containing glomeruli were removed by decanting and the resultant suspension was composed almost entirely of proximal convoluted tubules.

1

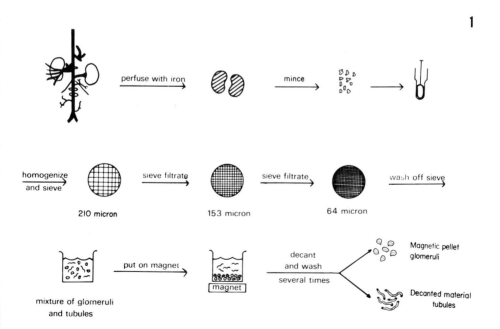

    Figure 1. Procedure for isolation of kidney glomeruli
and tubules. Details are described in the text.

Basal lamina was isolated from rabbit proximal convoluted tubules by the method described by us elsewhere in this volume (11). The procedure is basically an extraction of all cellular material with distilled water, 3% Triton X-100, DNA'se, and 4% sodium deoxycholate (DEOC). The resultant BL was treated with 0.1% pepsin in 0.05 N HCl for 30 minutes to 42 hours at 4°C or 22°C. This was followed by mild centrifugation and fixation in cold (4°C) Karnovsky's (12) fixative and post-fixation in 2% sodium cacodylate buffered OsO_4. Other BL were pepsinized for similar time periods, but prior to fixation were allowed to come to 22°C (at neutral pH) for 30 minutes prior to fixation in the cold. In a separate series of experiments, isolated BL were treated four days with 3% Triton X-100 followed by a one day treatment with 4% DEOC at room temperature. Some of these latter preparations were further treated two hours with 2% SDS, 2% mercaptoethanol and 8 M urea prior to fixation.

All BL were dehydrated in a graded series of ethanols. Propylene oxide was used as an intermediate solvent before final embedding in Epon-Araldite (13). Thin sections were mounted on naked 200-300 mesh copper grids, stained with uranyl acetate (5% in absolute ethanol) and lead citrate (14). Electron microscopic observations were carried out on Phillips EM 300 electron microscope at original magnifications of 1,800-33,000 diameters.

RESULTS

Scanning electron microscopic observations of isolated renal tubules (Figure 2) show an unusually pure preparation virtually free of non-tubular material. Tubules often exhibit long and tortuous contours, folding upon themselves or closely contacting adjacent tubules. They are usually smooth and show occasional wrinkles on their outside surfaces. This general smooth contour and lack of surface blebs suggest that the tubular BL has not been removed (15). Transmission electron microscopic studies of these preparations confirm that the BL is intact (16) and is ultrastructurally indistinguishable from its *in vivo* counterpart (17).

Following isolation by detergents, renal tubule BL consists of long continuous sheets of electron dense amorphous material that are in all morphological respects similar to BL surrounding isolated tubules. Pellets of this material (Figure 3) show that the lamina is of uniform thickness (1,200-1,500 A) and electron density. Both surfaces of tubular BL are sharply demarcated and rarely, if ever, fuse with adjacent laminar structures. This may indicate that electron lucent laminae rara are present, but not visualizable by usual electron microscopic methods. Significantly, striated collagen fibrils are excluded from these preparations.

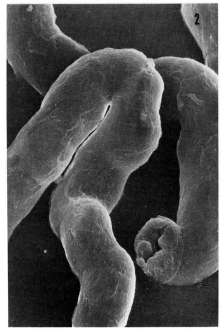

Fig. 2. Scanning electron micrograph of isolated rabbit renal tubules. The preparation contains almost no nontubular material. Individual tubules are long and tortuous, frequently folding upon themselves. Tubular surfaces are slightly wrinkled but show no gross blebbing. Broken ends occasionally exhibit microvilli. X 585.

Fig. 3. Transmission electron micrograph of a pellet of ultrastructurally "pure" basal lamina isolated from rabbit renal tubules with solubilization with DEOC and Triton X-100. Lipoprotein membranes and other cellular contaminants are not present. Layers of basal lamina (ca. 1,200-1,500 A thick) typically do not fuse but remain separated by a clear area which may correspond to the in vivo lamina rara. Unit collagenous fibrils are excluded from the preparation. X 19,500.

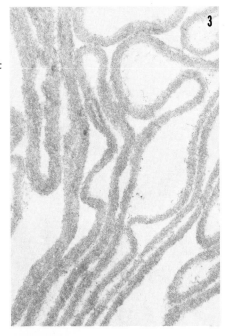

When isolated BL are treated with 0.1% pepsin at 4°C, the enzyme appears to attack selectively one side of the BL and a meshwork of interconnected fibrils is released into the adjacent space (Figure 4). The opposite side remains free

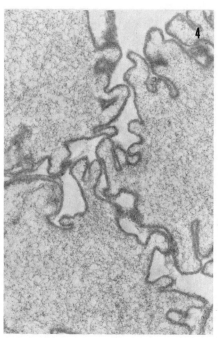

Fig. 4. Transmission electron micrograph of isolated rabbit renal tubule basal lamina treated 42 hours with 1% pepsin at 4°C. A meshwork of fibrillar materials is released selectively from one side of the basal lamina. The opposite side appears clear except for a few fine fibrils which may be entrapped within pockets of the basal lamina during pepsinization. No striated collagenous fibrils are observed in these preparations. X 24,600.

of digested substances with the exception of occasional fibrillar materials which appear to be entrapped within pockets of the pelleted lamina. This sidedness does not seem to be a function of enzyme availability since most isolated laminae exist as short "tubes". This finding suggests a previously undemonstrated structural heterogeneity which may indicate that the parenchymal surface of BL may be compositionally different from its connective tissue counterpart. When BL is pepsinized at room temperature (22°C), partial digestion occurs within 30 minutes (Figure 5). The most striking feature of these preparations, however, is the appearance of numerous small striated collagen fibrils which are released or otherwise derived from the BL. Fibrils average about 200 A in diameter and exhibit macroperiods of approximately 260 A. They do not show tapered ends typical of sectioned unit collagenous fibrils *in vivo*, but rather exhibit unwound terminal tufts. Such fibrils are not seen in untreated control BL, in BL treated with 0.05 N HCl only, or in BL treated up to 42 hours with pepsin at 4°C.

Fig. 5. Transmission electron micrograph of isolated rabbit renal tubule basal lamina treated 30 minutes with 1% pepsin at room temperature. The basal lamina is partially digested. Numerous small striated collagen fibrils (ca. 200 A in diameter) are associated with the lamina. These fibrils exhibit macroperiods of about 360 A.   X 12,800 (inset X 72,500)

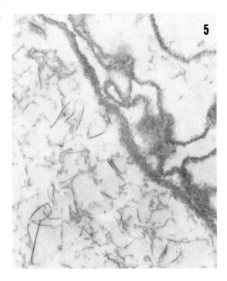

When BL is pepsinized at 4°C and allowed to come to room temperature for 30 minutes at neutral pH prior to fixation, striated collagen fibrils similar to those shown in Figure 5, are observed. These fibrils are not present, however, in preparations treated less than 24 hours in pepsin at 4°C. Longer digestion times result in greater number of fibrils. This suggests that under proper conditions collagen molecules solubilized by pepsin may reaggregate to form fibrils. The possibility that already formed fibrils may be present within the substance of the lamina and released by pepsin, however, cannot be eliminated.

In an effort to determine whether "masked" unit collagen-out fibrils were present within BL material, we subjected isolated tubular BL to Triton X-100 and DEOC for periods up to five days at room temperature. Following long-term treatment with detergents, isolated BL begin to lose structural integrity. The most striking feature of these specimens, however, is the appearance of individual unit collagenous fibrils anchored within the substance of some of the BL (Figure 6). Such fibrils are observed frequently in specimens treated with detergents for up to five days, but are not seen in untreated controls. These striated collagen fibrils average 300-400 A in diameter and are associated with BL which were approximately

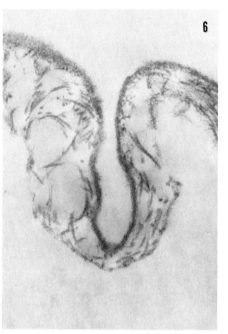

*Fig. 6. Transmission
electron micrograph of isolated
basal lamina treated four days
with Triton X-100 followed by a
one day treatment with DEOC.
Numerous striated collagen
fibrils (ca. 320 A in diameter)
appear anchored within the substance of the basal lamina.
Such fibrils may be unmasked by
long-term treatment with detergents. They are associated almost exclusively with one surface of the basal lamina.
These basal laminae are thinner
(ca. 1,000 A) than untreated
controls. X 13,000.*

30% thinner than controls. When long-term detergent treatment
is followed by a two hour exposure to SDS-mercaptoethanol-urea,
BL appear to become less granular and more fibrous. Striated
collagenous fibrils are never present in these preparations.
Instead, short segments of intertwined fine filaments (40-50 A
in diameter) are closely associated with the BL (Figure 7).

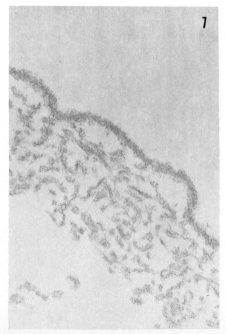

*Fig. 7. Transmission
electron micrograph of an
isolated basal lamina preparation similar to that seen in
Figure 6 but treated an additional two hours in SDS-mercaptoethanol-urea. No striated
collagen fibrils are present
in these preparations. Numerous bundles of intertwining
filaments (40-50 A) are continuous with similar filaments
within the substance of the
basal lamina. X 25,800.*

These filaments coalesce to form short blocks of fibrils continuous with similar structures within the substance of the BL. It seems possible that striated collagen fibrils unmasked by long-term treatment with Triton X-100 and DEOC may be reduced and/or denatured by the SDS-mercaptoethanol-urea to a more filamentous form. Such activity could result in a loss of morphological integrity of striated collagen fibrils.

DISCUSSION

    Data in the present study suggest that isolated BL from rabbit renal tubules contains a pepsin-digestible collagenous component which, under appropriate conditions, may reaggregate to form native striated fibrils. It is likely that most of the collagen molecules released by pepsinization become soluble and are lost in the supernatant. Nevertheless, some of the collagen molecules may form fibrils within the confines of the laminar matrix. This mechanism does not seem to occur, however, unless the preparation is allowed to come to room temperature prior to fixation.
    Previous studies have shown that pepsin exerts its solubilizing effect on collagen by cleaving non-helical terminal cross-linked peptides (18,19). Since it has been suggested (20) that chains of BL collagen are deposited in the extracellular space as a procollagen-like precursor and may not be converted to the respective collagen form (21), it is possible that pepsinization of BL collagen may alter the molecule to the extent that fibrous aggregation may occur. This interpretation is consistent with studies of vitreous collagen, the fibrillar morphology of which is altered follow-ing pepsinization (22). It seems reasonable, therefore, that the structures cleaved by pepsin may be important factors in the determination of the morphological characteristics of the final native fiber in vivo.
    It is interesting to note that other attempts to form fibrils from the putative type IV collagen of BL have not been successful. Only SLS-type aggregates have been formed follow-ing BL digestion and these require the addition of ATP to the solution (6).
    The possibility that pepsinized BL may contain collagen molecules that are capable of fibrillogenesis does not pre-clude the possibility that fibrous striated collagen may be present, but masked by other molecules which render them unidentifiable by electron microscopy. This hypothesis is consistent with our studies which suggest that long-term

treatment with detergents such as Triton X-100 and DEOC may partially remove masking substances and reveal striated fibrils similar in size and shape to interstitial collagen seen in most connective tissues *in vivo*. When these preparations are treated with SDS-mercaptoethanol-urea, the striated fibrils are not present. It seems possible that these agents may cleave disulfide bonds or interrupt intermolecular inter-actions which maintain the integrity of striated fibrils derived from BL. This could result in the appearance of short segments of intertwined fine filaments which may represent some elemental form of collagen.

In summary, our studies provide evidence for the presence of masked, striated collagen fibrils within the confines of ultrastructurally pure BL. Such fibrils may be unmasked by treatment with detergents. In addition, our data suggest that this or other forms of BL collagen may be released by pepsin and reaggregated to form fibrils under appropriate conditions.

ACKNOWLEDGEMENTS

We would like to acknowledge the technical assistance of Ms. Martina Bell in the isolation of basal lamina samples and Ms. Jerrolynn Campbell and Mr. Armando Romero for electron microscopic preparations.

This investigation was supported by research grants HL 17421-03S1 (E.C.), AM HL-14977 (K.B.) and AM-15394 (E.M.) from the National Institutes of Health and in part by grant 76-944 (E.C.) from the American Heart Association. Dr. Meezan is a Research Career Development Awardee of the National Institutes of Arthritis, Metabolism and Digestive Diseases.

REFERENCES

1. Beisswenger, P.J. and Spiro, R.G., *Science* 168:596-598, 1970.
2. Kefalides, N.A., *Int. Rev. of Conn. Res.* 6:63-104, 1973.
3. Mahieu, P. and Winand, R.J., *Eur. J. Biochem.* 12:410-418, 1970.
4. Meezan, E., Hjelle, J.T., Brendel, K., and Carlson, E.C., *Life Sci.* 17:1721-1732, 1975.
5. Daniels, J.R. and Chu, G.H., *J. Biol. Chem.* 250:3531-3537, 1975.
6. Kefalides, N.A., *Biochem.* 7:3103-3112, 1968.

7. Kefalides, N.A. and Winzler, R.J., *Biochem.* 5:702-713, 1966.
8. Spiro, R.G., *In:* "Diabetes Mellitus: Theory and Practice" (McGraw-Hill, New York), pp. 210-229, 1970.
9. Kefalides, N.A., *J. Clin. Invest.* 53:403-407, 1974.
10. Brendel, K. and Meezan, E., *Fed. Proc.* 34:803, 1975.
11. Meezan, E., Brendel, K., Hjelle, J.T., and Carlson, E.C., *In* "First International Symposium on the Biology and Chemistry of Basement Membranes" (Academic Press, New York), pp.
12. Karnovsky, M.J., *J. Cell Biol.* 27:137a-138a, 1965.
13. Anderson, W.A. and Ellis, R.A., *J. Protozool.* 12:483-499, 1965.
14. Venable, J.H. and Coggeshall, R., *J. Cell Biol.* 25:407-408, 1965.
15. Hay, E.D. and Dodson, J.W., *J. Cell Biol.* 57:190-213, 1973.
16. Carlson, E.C., Brendel, K., Hjelle, J.T. and Meezan, E., submitted, 1977.
17. Maunsbach, A.B., *In:* "Handbook of Physiology, Section 8, Renal Physiology", (Williams and Wilkins, Baltimore, pp. 31-79, 1973.
18. Kühn, K., Brauner, K., Zimmerman, B. and Pikkarainen, J., *In:* "Chemistry and Molecular Biology of the Intercellular Matrix" (Academic Press, New York) Vol. 1, pp. 251-273, 1972.
19. Start, M., Miller, E.J., and Kühn, K., *Eur. J. Biochem.* 27:192-196, 1972.
20. Grant, M.E., Kefalides, N.A. and Prockop, D.J., *J. Biol. Chem.* 247:3539-3544, 1972.
21. Kefalides, N.A., *Biochem. Biophys. Res. Commun.* 45:226-234, 1971.
22. Swann, D.A., Caulfield, J.B. and Broadhurts, J.B., *Biochim. Biophys. Acta* 427:365-370, 1971.

B. Morphology and Ultrastructure

STRUCTURE AND FUNCTION

IN

GLOMERULAR CAPILLARIES

Role of the Basement
Membrane in Glomerular Filtration[1]

Marilyn G. Farquhar

Yale University School of Medicine
New Haven, Connecticut

INTRODUCTION

Two major developments both of which occurred during the 1950's made possible a more penetrating analysis of basement membranes than had been possible before and led to work which established our current level of understanding of these structures: one was the application of the electron micro- scope to biomedical research, and the other was the introduc- tion of methods for preparation of purified basement membrane fractions. The latter development, which was marked by the publication of the technique for isolation of renal glomeruli and their basement membranes (1), made possible the biochemi- cal analysis of basement membranes. The information obtained from this approach is the focus of many other contributions to this symposium, whereas the purpose of this presentation is to review and summarize the results of the former develop- ment, i.e., introduction of the electron microscope, to our

[1]*This research was supported by Public Health Service Grant AM-17724 from the National Institutes of Health.*

knowledge of basement membranes and indicate how research
with the electron microscope has advanced our knowledge of
the structure and function of basement membranes, with
particular attention to the glomerular basement membrane
and its role in glomerular filtration.

BACKGROUND

What Are Basement Membranes and Where Are They Found?

To begin with, one should define what the term "basement
membrane" means to morphologists. Prior to the introduction
of the electron microscope this term was understood to refer
to a PAS-staining layer found under epithelia and around
muscle fibers which was visible in the light microscope.
Since the introduction of the electron microscope, this term
is generally used[2] to refer to the continuous sheet of
material of moderate electron density found at the base of
epithelia and endothelia, around muscle cells and in a
variety of other locations. What has come to light as a
result of studies with the electron microscope is that basement membranes are found wherever cells (other than connective cells)[3] meet connective tissue matrix (ground substance).
This means that basement membranes are found, for example, at
the dermal-epidermal junction of the skin (Figs. 1 and 2), at
the base of all lumen-lining epithelia throughout the digestive, respiratory, reproductive, and urinary tracts (Fig. 3),
underlying endothelia of capillaries and venules (Figs. 4 and
5), around Schwann cells, adipocytes, skeletal and cardiac
muscle cells, and at the base of parenchymatous cells of
exocrine (pancreas, salivary glands) and endocrine (pituitary,
thyroid, adrenal) glands wherever they face perivascular
connective tissue. In effect, basement membranes serve to delimit the domain of connective tissue and provide a barrier
between it and the domain of non-connective tissue elements.
In all the locations mentioned, the basement membrane
consists of a continuous sheet, 200-500 A in thickness, which

[2]*The term "basal lamina" or "basement lamina" is also
sometimes used to refer to this layer (2).*

[3]*The only cell lines which face the connective tissue
ground substance without an intervening basement membrane
layer are connective tissue cells (fibroblasts, mast cells,
macrophages, plasma cells) and cells of hematogenous origin
(granulocytes, monocytes, lymphocytes).*

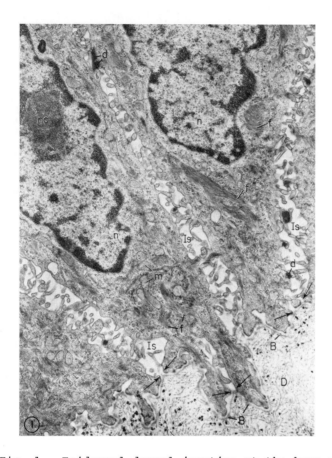

Fig. 1. Epidermal-dermal junction at the base of the skin (toad). Several epithelial cells of the s. germinativum are seen on the upper left, and part of the dermis (D) is present on the lower right. The basement membrane (B) forms a continuous (∿300 A) layer which closely parallels the basal cell membranes of the epithelial cells and separates dermal and epidermal elements. The epidermal-dermal junction is highly indented with the dermis forming a series of micro-papillae containing collagen fibrils. The intercellular spaces (Is) are dilated and open (i.e., not sealed by junctions) toward the dermis (arrows) so that only the basement membrane separates these spaces from the dermal ground substance. f - tonofilaments; d - desmosome; m - mitochondria; n - nucleus; nc - nucleolus. From Farquhar, M.G. and Palade, G.E., J. Cell Biol. 26:263, 1965. X 14,000.

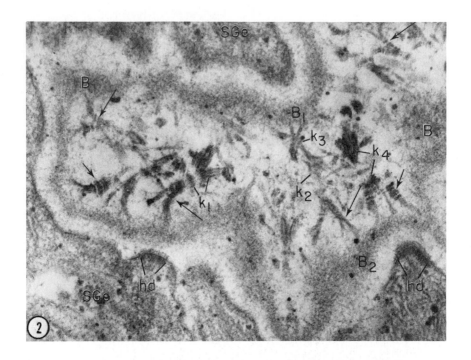

Fig. 2. Higher magnification of view of the dermal-epidermal junction (toad skin). A dermal micropapillae is cut obliquely, and parts of epidermal cells (SGe) are also present. Numerous anchoring filaments (diameter = 200-750 A) can be seen in the dermis. Most of them have stems with a characteristic banding pattern (short arrows). One end of the filament is split into two or more branches (long arrows) which are anchored in the basement membrane (B). The opposite end converges with others into a series of knots (k_1 to k_4) along the middle of the micropapilla. hd = hemidesmosome. From Palade and Farquhar (14). X 76,000.

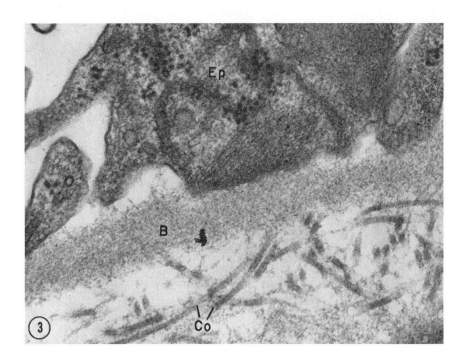

Fig. 3. Base of the proximal convoluted tubule of the
rat kidney showing a portion of the tubule basement membrane
(B) which is closely associated with the basal cell membrane
of the tubule epithelium (Ep) on its upper surface and faces
the ground substance of the connective tissue on its lower
surface. Banded collagen fibrils (Co) are intimately related
to this surface. The basement membrane is seen to be composed
of a tight feltwork of randomly oriented fine (30-40 A)
fibrils. (Courtesy of Dr. John P. Caulfield). X 95,000.

runs parallel to the basal cell membranes of the cell layer
in question (epithelia, endothelia, etc.) and separated from
the latter by a lighter ∿ 100 A layer. Thus the basement
membrane regularly faces cell membranes on one surface and
connective tissue ground substance on the other. There are,
however, a few locations which come to mind (smooth muscle,
glomeruli) where basement membranes face cell layers on both
surfaces. The best known example occurs in the renal glomerulus of mammals where the basement membrane faces endothelium
on one surface and epithelium on the other. This situation
in which the basement membrane does not face the connective
tissue ground substance makes the glomerulus a particularly
favorable object for biochemical analyses of basement membranes since it eliminates the problems introduced by contamination of basement membrane fractions with interstitial
fractions with interstitial collagens, a situation which is
unavoidable with basement membranes derived from most other
sources.[4]

Structure of Basement Membranes

     It is frequently stated that basement membranes are
"amorphous", but this is simply *not* the case unless one looks
at them under conditions of poor resolution in low power
electron micrographs. They are "amorphous" only in the sense
that they lack discernible order in the arrangement of their
fibrillar components. As shown many years ago for glomerular
basement membranes (3) and those underlying endothelia of
other capillaries (4), they are composed of tightly matted
arrays of fine (30-40 A) fibrils embedded in a finely granular
matrix (Figs. 3 to 6). In addition, larger (∿ 110 A) tubular
fibrils are frequently seen in association with basement
membranes. In glomerular capillaries such tubular fibrils
are typically found along the luminal surface of the basement
membrane between the endothelial or mesangial cell membrane
and the basement membrane (Fig. 7). In peripheral capillaries
fibrils of this description are seen on the outer surface of
the basement membrane (Figs. 4 and 5). They resemble the
proteinaceous "microfibrils" found in association with elastic
fibers (5). Whether or not all these ∿ 100 A fibrils are of

[4]*Another exception is the lens capsule which faces epithelium on one surface and the vitreous humor on the other,
and hence it also can be obtained without connective tissue
contamination.*

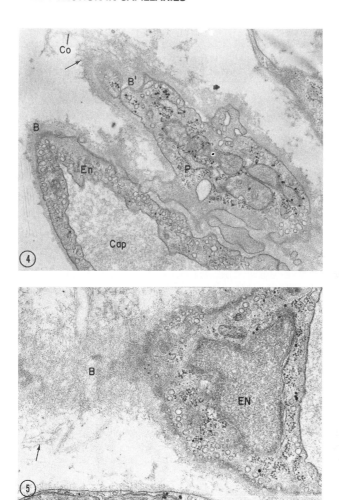

Figs. 4 and 5. Capillaries from skeletal muscle (rat diaphragm) to show the location and structure of the capillary basement membrane. Figure 4 shows a capillary that is cut in normal section with its basement membrane (B) which faces the endothelium (En) on one surface and the perivascular connective tissue on the other. The outer surface of the basement membrane is seen to be associated with collagen fibrils (Co) and smaller (100 A) fibrils (arrow). A portion of a pericyte surrounded by its basement membrane (B') is also present above. In Figure 5, which is an oblique section through a similar capillary, the basement membrane (B) can be seen to consist of a fine mat of 30-40 A fibrils which is closely associated with larger, ∿ 100 A fibrils (arrow) on its connective tissue surface. (Courtesy of Dr. G. E. Palade). Fig. 4 - X 33,000; Fig. 5 - X 43,000.

similar or different composition as well as their relationship to "microfibrils" found in association with elastic
fibers remain to be established.

Origin of Basement Membranes

    Work carried out during the last few years has made it
clear that basement membranes are not produced by cells of
the connective tissue (fibroblasts), but they are made by the
overlying cell layer with which they are in contact, i.e.,
epithelium, endothelium, muscle cell, etc. The origin of
basement membranes -- whether epithelial, fibroblastic, or
both -- was debated until relatively recently, although the
fact that epithelia can make basement membranes was already
suggested by work carried out in the early 60's which
demonstrated a) that the endoplasmic reticulum of the glomerular epithelium contains material which is morphologically (6,
7) and antigenically (8) similar to basement membrane, and
b) that a pure line of carcinoma cells (i.e., of epithelial
origin) was capable of synthesizing a basement membrane-like
material *in vitro* (9, 10). It has now been unequivocally
demonstrated that not only epithelia (11), but also vascular
endothelia (12) and smooth muscle cells (13) make morphologically recognizable basement membranes. Probably the most
interesting and definitive experiments bearing on the epithelial origin of basement membranes were those of Hay and
Dodson (11) who, working with isolated corneal epithelium
cultured on frozen-killed lens, showed by a combination of
morphologic studies and autoradiography at the electron
microscope level that epithelia can and do make basement membranes as well as interstitial collagen. On the other hand,
it should be added that there is no evidence to suggest that
fibroblasts or other connective tissue elements make basement
membranes.

Functions of Basement Membranes

    There are three main functions of basement membranes
which have been established so far and which have been understood only after work carried out with the electron microscope. It was previously mentioned above that one general
function of basement membranes is to delimit the connective
tissue-nonconnective tissue boundary. Thus it separates and
maintains separate the corresponding cell populations and
thereby maintains the orderly organization of organs and tissues.[3]

Another likely function of basement membranes is attachment -- attachment of cell layers to their associated connective tissue elements. This tight attachment is indicated
by the fact that to separate epithelia from their basement
membranes requires drastic procedures (e.g., digestion with
collagenase, or detergent treatment). In sites such as the
skin where the epithelial layers are subjected to severe
mechanical stress, and unusually tight epithelial-connective
tissue attachment is required, additional elements known as
"anchoring filaments" (14) with a unique banding pattern
occur at the epidermal-dermal junction along the dermal surface of the basement membrane (Fig. 2). These anchoring
filaments have one end anchored in the basement membrane and
the other in the dermal connective tissue.

A third function of basement membranes is that of
filtration. This function has been established by tracer
studies carried out with the electron microscope largely on
renal glomerular capillaries and to a lesser extent on other
capillaries.[5] The main purposes of this presentation are
a) to review what is known about glomerular filtration and
the role of the basement membrane and other layers of the
glomerular capillary wall therein, and b) to outline some of
the areas of agreement as well as some of the still unresolved problems which remain for the future from the perspec-
tive of a cell biologist and experimental pathologist interested in the relationship between structure and function.
The intention will be to give an updated synopsis and overview of this topic since a thorough review was published not
long ago (16) which can be consulted for details, especially
of earlier work (prior to 1975). Such an updated review is
justified because progress, particularly in reaching a con-
sensus on the primary role of the basement membrane in
filtration, has been rapid.

[5]There is no direct information concerning the size limit
of permeant molecules for basement membranes of epithelial or
smooth muscle cells, but in the intestinal epithelium it must
be unusually high since the local basement membrane can be
permeated by chylomicra and VLDL's. In the case of vascular
endothelia there is direct evidence to this effect primarily
on capillary endothelia (cf. 15) which indicates that the
size range (500-700 A) is considerably larger than that of
glomerular capillaries (< 100 A).

STUDIES ON GLOMERULAR CAPILLARIES

Glomerular Physiology

The function of the glomerulus is to filter the blood
plasma as the first step in urine formation (cf. 17, 18).
Since the time of Ludwig (1844) it has been appreciated that
this is done by a process of passive filtration: most of
the proteins and all of the cellular elements of the blood
are retained in the circulation and a plasma filtrate containing only small amounts of protein passes through the
glomerular capillary wall and down the tubule where re-
absorption of the major part of the filtrate (all of the
salts, protein, glucose and amino acids and most of the
water) takes place. Clearance studies carried out by
physiologists in the 1940's and 1950's using probe molecules
(proteins and dextrans (17-20) established that the glomerulus behaves like a sieve in that there is increasing restriction to passage with increasing molecular weight and
effective diameter up to molecular weights of ∿70,000, the
size of albumin, the major plasma protein. Molecules of this
molecular weight or greater are effectively retained and are
not filtered. It was also recognized some time ago (17, 19)
that molecular shape and charge could affect the filtration
properties of substances of similar molecular weights. Based
on such data, it was postulated (20) that the walls of
glomerular capillaries, like those of other capillaries, are
composed of pores of ∿90 A diameter. The higher filtration
rate of glomerular capillaries (compared to muscle capillaries) was explained by assuming the pores were more frequent
(1-2% as opposed to 0.1-0.2% of the capillary surface) than
in muscle capillaries.

Glomerular Morphology

Prior to the introduction of the electron microscope, it
had already been established by light microscopy that there
are four components of glomerular capillaries -- three cell
types (endothelium, epithelium and mesangium) and the acellular basement membrane which, like other basement membranes,
is intensely PAS-positive. However, the detailed organization
of these layers and their relationship one to another were
beyond the level of resolution afforded by the light microscope. With the introduction of the electron microscope,
the glomerulus was among the first histologic structures to
be studied in detail and whose fine structure was clarified

since its main features are evident in low power electron micrographs (See Fig. 6). It has, therefore, been established for some time that: 1) the glomerular capillaries are divided

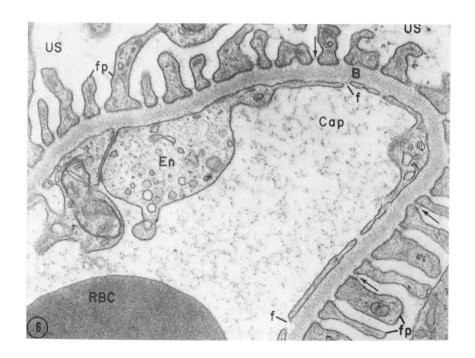

Fig. 6. Peripheral area of a glomerular capillary from a normal rat. The capillary wall is composed of three distinct layers: the endothelium (En) with its periodic interruptions or fenestrae (f); the basement membrane (B) which is a continuous layer, 100 to 150 nm in thickness; and the foot processes (fp) of the epithelium. The basement membrane consists of a broad middle layer of moderate density, the lamina densa, and two outer layers of lower density, the lamina rara interna and externa, which adjoin the endothelium and epithelium, respectively. In a number of places (arrows) a thin line or "slit membrane" can be seen bridging the narrow (∿250 A) gap between the foot processes. The epithelial filtration slits are defined as that portion of the space between foot processes extending from the level of the basement membrane to the level of the slit membrane. Cap - capillary lumen; US - urinary spaces; RBC - red blood cell. From Farquhar (7). X 40,000.

into peripheral and axial regions with the filtration surface
being limited to the peripheral regions and the cell bodies
of endothelial and mesangial cells concentrated in axial
regions; 2) the filtration surface contains three layers --
endothelium, basement membrane and foot processes of the
epithelium (the mesangium being incomplete and restricted to
axial regions); and 3) the basement membrane in these capillaries as well as in other capillaries[6] is the only continuous layer in the capillary wall, since the continuity of the
endothelium is interrupted by the fenestrae and that of the
epithelium by the filtration slits between the foot processes.
Although no pores of the size postulated by the physiologists
could be discerned, most of the morphologists of that time
assumed that the basement membrane, as the only continuous
layer, must represent the barrier which serves to prevent
passage of plasma proteins and therefore must contain the
hypothetical pores. It should be emphasized that the basement membrane of the glomerular capillary is unusual, not
only because it faces cell layers on both its inner and outer
surfaces, as already mentioned, but also in that it is much
thicker (up to 250 nm in humans) than basement membranes
found in other locations.

Tracer Studies

     To test the hypothesis that the basement membrane
represents the principal glomerular filter electron micros-
copists turned to the use of electron-opaque tracers. The
approach used was to inject a tracer of known dimensions into
the blood stream, to fix the tissue at selected intervals
thereafter, and to identify directly the barrier that prevented its passage.

Particulate Tracers

     The first tracers used were particulate tracers such as
ferritin, thorotrast, colloidal gold, and saccharated iron
oxide, all of which have inherent electron opacity and can
be visualized as individual particles. Of these, ferritin
was by far the most satisfactory and is still widely used as
a tracer. In 1961 we (3) obtained evidence which clearly
demonstrated that the basement membrane serves as the barrier

[6]*The only exceptions are the sinusoids of the liver and
bone marrow in which the basement membranes are discontinuous
in some species.*

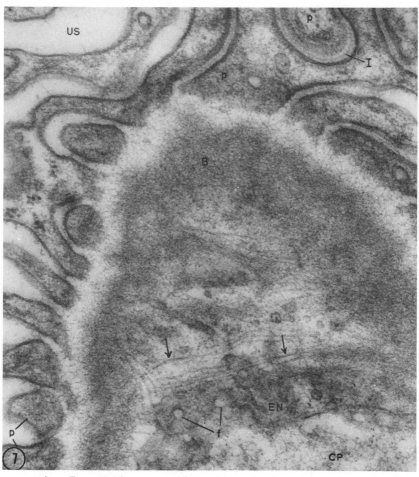

Fig. 7. *Oblique section through the glomerular capillary wall. The endothelium (EN) with its circular fenestrae (f) are seen "face on" below, and the epithelial foot processes (p) are cut obliquely at the level of the slit membrane on the lower left (p). The section grazes broadly through the base- ment membrane. The central dense portion (lamina densa) of the basement membrane (B) can be seen to be composed of a mat of 30-40 A fibrils. Some of the fibrils extend from the dense middle portion of the basement membrane across the outer lighter layer of the basement membrane (lamina rara externa) to the cell membrane at the base of the foot processes. Bundles of larger ($\sim$ 110 A) tubular fibrils are seen (arrows) in the subendothelial portion of the basement membrane (lamina rara interna). Cap - capillary lumen; US - urinary spaces. From Farquhar, Wissig and Palade (3). X 62,000.*

to the passage of ferritin: when ferritin was introduced into
the circulation there was a sharp drop in the concentration
of this tracer along the inner or subendothelial surface of
the lamina densa of the basement membrane, and with time
there was accumulation of ferritin against the basement
membrane particularly in the mesangial regions (Fig. 8).

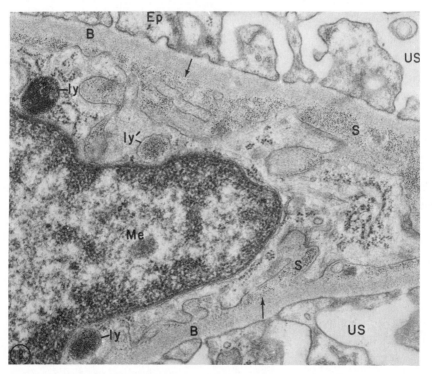

*Fig. 8. Mesangial region from a rat given ferritin 24 hr
prior to sacrifice. Ferritin molecules fill the spongy areas
(S) between the basement membrane (B) and the mesangial cell
(Me). Note, however, that their concentration falls off
sharply at the inner subendothelial surface of the basement
membrane (arrows), and very few molecules are found in the
dense portions of the basement membrane or lamina densa. The
spongy areas are apparently "cleared" by the activity of the
mesangial cells which incorporate the filtration residues in
pockets of varied sizes. Where the residues are massive as
in this case, incorporation takes place in large pockets. The
latter eventually become pinched off to form intracytoplasmic
vacuoles (ly') which become lysosomes (ly). Ep - epithelium;
US - urinary spaces. From Farquhar and Palade (24).
X 43,000.*

Eventually, the residues which accumulated against the basement membrane were taken up and disposed of by mesangial cells.
Thus these findings clearly pinpointed the basement membrane
as the main glomerular filter restricting the passage of
ferritin. The results obtained with native (anionic) ferritin
have been repeatedly confirmed by others (21-23). Identical
results were obtained with the other particulate tracers
mentioned (cf. 16). Based on the findings with ferritin
we proposed (3, 7, 24) a functional model for the glomerulus
and defined a role for each of its components in the filtration process:

> -- the basement membrane as the main filter;
>
> -- the endothelium as a valve, which by the
>   number and size of its fenestrae, controls
>   access to the filter;
>
> -- the epithelium as a monitor which partially
>   recovers (by pinocytosis) proteins that
>   leak through the filter; and
>
> -- the mesangium which serves to recondition
>   and unclog the filter by phagocytizing and
>   disposing of filtration residues which
>   accumulate against it.

Thus the problem of the identification of the filter appeared
to be solved, but this situation lasted only a few years.

Peroxidatic Tracers

In 1966, Graham and Karnovsky (25) introduced several
additional tracers -- horseradish peroxidase and myeloperoxidase -- which were not themselves electron dense, but could
be demonstrated by a secondary histochemical reaction involving peroxidatic activity. When these proteins were introduced
into the blood stream and the distribution of reaction product
determined, these workers observed a concentration of reaction
product between the epithelium and foot processes (in the
lamina rara externa) and in the epithelial slits up to the
level of the filtration slit membrane. Based on these findings, the authors proposed that in the glomerulus there are
two barriers in series: the basement membrane which serves
as a coarse filter to exclude very large molecules (> 100 A)
and the epithelial slits which were believed to contain "slit-
pores" and act as a fine filter by retaining molecules the
size of albumin. Subsequent findings using other peroxidatic
tracers such as lactoperoxidase (26), catalase (27), and
tyrosinase (28), were interpreted as consistent with this

hypothesis. The demonstration by Rodewald and Karnovsky (29) of a periodic structure interpreted as "slit-pores" in the slits using fixatives containing tannic acid lent credence to this hypothesis. This "two-barriers-in-series" model gained wide acceptance among physiologists (cf. 17) and morphologists (cf. 30) in spite of the fact that there was a clear discrepancy between the data obtained with peroxidatic tracers and the findings obtained with a variety of particulate tracers of varying sizes which pointed to the basement membrane as the main filter.

Use of Dextrans as Tracers

In 1972, in an attempt to resolve this issue, Caulfield and I began to do experiments using dextrans as tracers. Dextrans have the advantage that they are available in a wide range (19). Physiologic studies have established that dextrans behave like proteins in that there is increasing restriction to their glomerular passage with increasing molecular weight and effective diameter (17, 19). Hence we prepared three dextran fractions -- one corresponding to molecules approximately the same size as albumin (MW = 62,000) another considerably larger (MW = 125,000), and a third considerably smaller (MW = 32,000). If the two-barriers-in-series model were valid, one would expect to see accumulation against the basement membrane with the two larger fractions, and accumulation against the epithelial slits with the smaller one. However, such was not the case (35).[7] When these fractions were introduced into the circulation a sharp drop in concentration along the inner surface of the lamina densa was seen with all three fractions (Figs. 9 and 10). With the two larger fractions accumulation of dextran particles occurred against the lamina densa in the mesangial regions. The only difference in the findings with the three fractions was in the amount of dextran filtered, the smaller molecular weight fraction being extensively filtered and lost from the circulation after 2-3 hr. No accumulation of dextran was seen in the epithelial slits or against the slit-membranes with any of the fractions. Hence the findings were the same

[7]*Similarly if two barriers were present one would expect to see a sieving affect with other particulate tracers which are polydisperse, (colloidal gold, thorotrast, and saccharated iron oxide) resulting in accumulation of large particles against the basement membrane and smaller particles against the slits. However, accumulation against the basement membrane was found with all these tracers, but accumulation against the slits was <u>not</u> observed (24, 31-34).*

as those obtained with ferritin and all the other particulate tracers which pointed inescapably to the basement membrane as the filter.

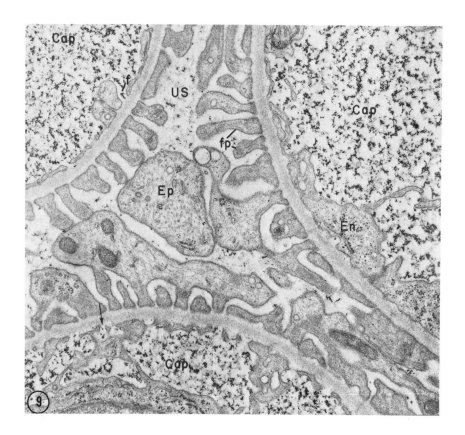

Fig. 9. Portions of a glomerular capillary from an animal sacrificed 3.5 hr after the injection of 125,000 MW dextran. Three capillary lumina filled with dextran are present in the field. The luminal concentration of tracer is still quite high, indicating the tracer has been effectively retained in the circulation. Dextran penetrates the endothelial fenestrae and is present in the subendothelial portions (lamina rara interna) of the basement membrane but the lamina densa and epithelial slits are free of dextran. This sharp drop in the dextran concentration which occurs in the subendothelial area is particularly evident in the lower left (arrow). From Caulfield and Farquhar (35). X 32,000.

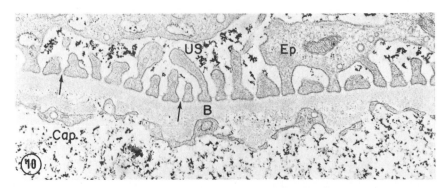

Fig. 10. Glomerular capillary 5 min after the injection of 32,000 MW dextran. Section is from the edge of a mesangial area. The tracer is present in high concentration in both the capillary lumina (Cap) and the urinary spaces (US). Even with this relatively small MW fraction, a sharp drop in the concentration of dextran is seen along the subendothelial portions of the basement membrane (B), but there is no accumulation of tracer in the slits (arrows). From Caulfield and Farquhar (35). X 23,000.

We next decided to check the identify of the filtration barrier in aminonucleoside nephrotic rats (36). This is an interesting situation in which to check the identity of the filter because there is increased permeability to albumin along with striking changes in the glomerular capillary wall (cf. 6 and 37). It has been known for over 20 years from studies on renal biopsy specimens from children (38) that in the nephrotic syndrome there is a loss of the usual foot process organization along with a striking reduction in the number of spithelial slits (Fig. 11). Moreover, the slits are not only reduced in number, but they are also reduced in width since many are replaced by occluding junctions (6, 37). In addition, the epithelium becomes filled with protein absorption droplets or lysosomes (36, 37, 39). At the same time -- a fact which is <u>not</u> usually appreciated -- the basement membrane shows thinning; the lamina densa appears thinner than usual, and the lamina rara externa is correspondingly widened (37) (Fig. 12).

This combination of changes -- i.e., increased leakage of albumin, together with a reduction in the number and narrowing of the epithelial slits, and thinning of the lamina densa is difficult to reconcile with the idea that the filtration barrier resides at the level of the slits, but it is easy to reconcile with the concept of the basement membrane as the principal glomerular filter.

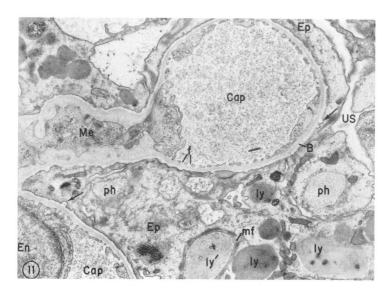

Fig. 11. Glomerular capillary from a nephrotic rat given 10 daily injections of aminonucleoside. The endothelium and its fenestrae (f) and the basement membrane (B) appear essentially normal whereas the epithelium shows the striking changes associated with nephrosis -- loss of foot processes and increased numbers of protein absorption droplets or lysosomes (ly). The outer surface of the basement membrane is covered by large sheets of epithelial cytoplasm (Ep). Numerous pinocytotic invaginations are seen along the portions of the epithelial cell membranes facing the basement membrane (arrows). Cap - capillary lumen; Me - mesangial cell; US - urinary spaces; ph - phagosome; mf - microfilaments. From Caulfield, Reid and Farquhar (37). X 18,000.

To study dextran permeability in the nephrotic rat we used the two largest dextran fractions only. The physiologic data (17, 40) tell us that the defect is fine and widespread since most of the albumin is retained, but increased quantities leak through and appear in the filtrate. Therefore, one would expect most of the dextran particles to be retained in the lumen. Indeed such was the case, and the main findings in the nephrotic were the same as in the normal: most dextran molecules were retained in the lumen, there was a sharp concentration drop along the inner surface of the basement membrane with accumulation against the basement membrane in the mesangial regions, and no piling in the slits (Fig. 14).

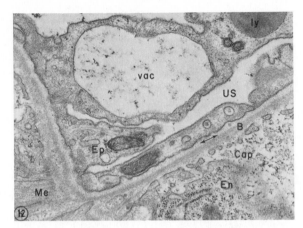

Fig. 12. Portions of glomerular capillary from another 10 day nephrotic animal given dextran 17 min prior to sacrifice. The dense portion of the basement membrane (B) appears thinner than usual, especially to the right; the subepithelial, lighter layer of the basement membrane (lamina rara externa) appears correspondingly wider (↔). A large phagocytic vacuole (vac) and a protein absorption droplet or lysosome (ly) are present in the epithelial cell (Ep). The phagosome contains some dextran particles. Cap - capillary lumen; US - urinary spaces; Me - mesangial cell; and En - endothelial cell. From Caulfield and Farquhar (36). X 126,000.

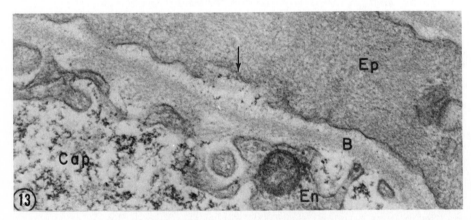

Fig. 13. Glomerular capillary from a 10 day nephrotic rat, 17 min after the injection of 125,000 MW dextran showing an area in which the lamina densa (B) appears thin and dextran particles are present in the lamina rara externa between the epithelial cell (Ep) and the basement membrane (B) (arrow). Cap - capillary lumen; En - endothelium. From Caulfield and Farquhar (36). X 126,000.

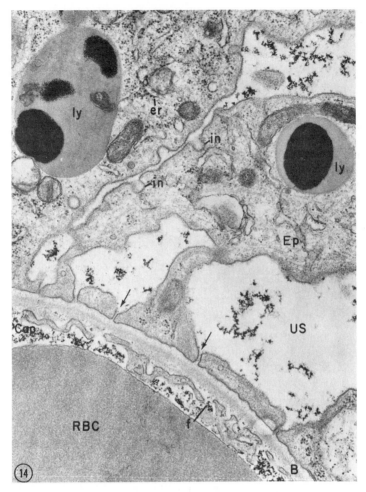

Fig. 14. *Glomerular capillary from a 10 day nephrotic rat sacrificed 2 hr after the injection of 125,000 MW dextran. Most of the dextran is retained in the lumen by the basement membrane as shown by the high concentration of dextran present in the capillary lumen (Cap) and in the endothelial fenestrae (f), the sharp drop in concentration that occurs at the level of the inner (subendothelial) surface of the basement membrane (B), and the lack of accumulation of dextran in the residual slits (short arrows). Yet there is evidence of increased leakage of dextran through the basement membrane because more dextran is present on the epithelial side of the capillary in the urinary spaces (US) and in the protein absorption droplets or lysosomes (ly) of the epithelial cell (Ep). The dextran is segregated into large, dense masses within the lysosomes. RBC - red blood cell; er - endoplasmic reticulum. From Caulfield and Farquhar (36). X 38,000.*

However, there was also evidence that the basement membrane was more leaky than is normally the case in that more
dextran was seen on the epithelial side of the basement membrane, 1) in the lamina rara externa (LRE) (Fig. 13);
2) in the urinary spaces, and 3) within the epithelium (Fig.
14). As in the case of ferritin (6), uptake of dextrans by
the epithelium occured by the usual endocytic pathway --
incorporation by pinocytosis, followed by fusion of pinocytic
vesicles with phagosomes, which in turn fuse with protein
absorption droplets or lysosomes (cf. 37). In the nephrotic
there was increased pinocytotic activity of the epithelium,
and with time large accumulations of dextran were seen in
lysosomes or absorption droplets in the epithelial cells
(Fig. 14).

To summarize, the findings indicate that in nehprotics,
as in the normal, the basement membrane represents the main
filter, though it is a leaky filter, and accordingly, the
monitoring activities of the epithelium are enhanced.

*How Do We Then Explain the Findings of Karnovsky and Co-work-
ers With Histochemically-Demonstrable Tracers?*

There were several problems with this work (cf. 16)
including the method used to fix the tissue which involved
a fixation delay. A piece of tissue was removed and immersed
in fixative rather than being fixed *in situ* by injection (3,
35) or by dripping the fixative on surface glomeruli (41).
In retrospect, however, it is clear that the greatest problem
in that work is the fact that most of the tracers used were
basic proteins (cf. 42) with isoelectric points above 7.0;
the most basic was myeloperoxidase which has an isoelectric
point of 10. This being the case they might be expected to
bind to the highly-negatively charged cell coat of the epithelial cell, and the images obtained could be explained on
the basis of such binding. It has been known for some time
that the glomerular epithelial cell bodies and the foot
processes are covered with a thin layer of sialoglycoprotein,
referred to as "glomerular polyanion", which stains with the
colloidal iron technique (30, 43-45).

In order to test the idea that basic proteins can bind
to the epithelial cell surface, we carried out some experiments (16, 46) in which we infused solutions of lysozyme, a
basic protein (pI = 11) of small size (MW = 14,000) which is
readily filtered. It proved to bind avidly to the epithelial
cell membrane which becomes completely outlined by a thick
(300-400 nm) layer (Figs. 15-17).

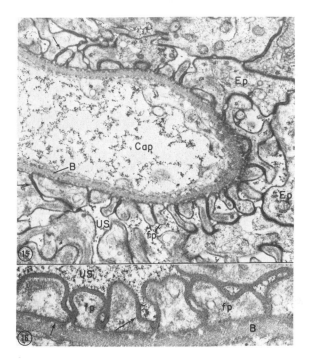

Figs. 15 and 16. Portions of glomerular capillaries from
an animal perfused with a mixture of lysozyme (3%) and dextran
(5%). In Fig. 15, a few clumps of dextran particles are seen
both in the capillary lumen (Cap) and the urinary spaces (US).
The epithelial cell membrane appears outlined in its entirety
by a thick (∿ 40 nm), dense-staining layer; both those portions of the cell membrane surrounding the foot processes
(arrows) and those portions surrounding the epithelial cell
bodies (Ep) are outlined. This layer is presumed to result
from the adsorption of lysozyme, a basic protein (pI = 11),
to the highly negatively charged surface of the epithelial
cell. In Fig. 16, the dense layer of lysozyme adhering to the
membrane of the foot processes is seen to better advantage
(long arrow). Numerous dextran particles are found free in
the urinary spaces, but none adheres to the epithelium. Both
the dextran (which is uncharged) and lysozyme are apparently
filtered, but only the lysozyme (a cationic protein) sticks
to the anionic epithelial cell membrane. Material which is
presumed to represent lysozyme (since it is of the same density and texture as the layer along the cell membrane) is also
present in all the layers of the basement membrane and is
especially concentrated in the subepithelial areas (short
arrow) between the basement membrane and the base of the foot
processes (lamina rara externa). The distribution of lysozyme is similar to that seen with other basic proteins used
as tracers (see text for a discussion of this point). From
Farquhar (16). "Reprinted from Kidney International (vol. 8,
p. 197, 1975) with permission." Fig. 15 = X 28,000; Fig. 16 =
X 65,000.

Thus it is clear that basic proteins can and do bind to the epithelial cell coat, as well as to the basement membrane (see below). Therefore, when basic proteins are used as tracers one cannot distinguish between retention due to ionic interaction and retention due to their size being larger than the slit pores. It appears that the findings with all the peroxidatic tracers except catalase, an acidic protein (pI = 5.7)[8], depended on the former, rather than the latter.

Work published quite recently from Karnovsky's laboratory (41, 47) indicates that these workers no longer believe that the slits represent the site of the primary filtration barrier to albumin. This conclusion is based on work carried out by Ryan and Karnovsky (41) on the immunocytochemical localization of endogenous albumin. These workers fixed glomeruli *in situ* and incubated the tissue sections in Fab fragments of antibodies prepared against rat albumin (or IgG) which had been coupled to peroxidase. The coupled antibodies bound to albumin, and their localization was demonstrated by carrying out a cytochemical reaction for peroxidase.

In such preparations they found albumin concentrated in the vessel lumen and penetrating the endothelial fenestrae. Some staining for albumin was found in the lamina rara interna, but there was a sharp drop in its concentration along the inner surface of the lamina densa, and none was seen deeper in the basement membrane. In short, the findings were very similar to those obtained with ferritin and dextrans and other particulate tracers. They have further shown that if one ligates the renal artery and vein for 5 min, albumin can be seen in the basement membrane and urinary spaces, indicating, not too surprisingly, that with such drastic alteration in the flow conditions in the glomerulus albumin can penetrate the basement membrane and the slits to reach the urinary

---

[8]*The initial tracer studies with catalase carried out by Venkatachalam et al. (27) on specimens fixed by immersion showed reaction product for catalase throughout the basement membrane and filling the epithelial slits, but not in the urinary spaces. These findings were interpreted as consistent with the two-barriers-in-series model. More recently, Ryan and Karnovsky (47) have repeated those experiments with catalase and have found that when specimens are fixed in situ (by drip fixation), reaction product for catalase is confined to the capillary lumina and the lamina rara interna, indicating that catalase, like ferritin, dextrans and other particulate tracers, does not penetrate the basement membrane under normal flow conditions.*

spaces. If the vessels are unclamped the findings are
reversed. Very similar findings were obtained with IgG (47).
     What is very clear from these findings is a) that there
is no evidence either under normal or altered flow conditions
for the role of the slits in restriction of albumin's passage, and b) that the filtration barrier exists along the
inner surface of the basement membrane. Indeed the authors
conclude that the filtration barrier is localized along the
inner surface of the endothelium and basement membrane. Thus,
as reviewed elsewhere (16), there is at present no evidence
that the slits serve as the critical filtration barrier, but
there is considerable evidence that the basement membrane does.

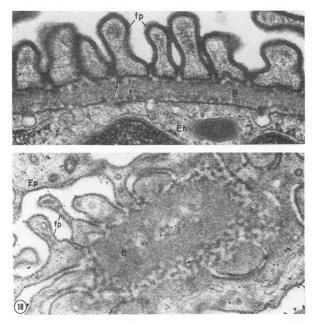

*Figs. 17 and 18. Portions of glomerular capillaries from
a rat perfused with lysozyme to show binding of lysozyme to
components of the basement membrane as well as the epithelium
(Ep). In Fig. 17, a normal section, discrete densities due to
lysozyme binding are seen in the lamina rara interna and externa (arrows). The lamina densa (B) also shows a slight but
definite increase in density. Note also that the binding
occurs all along the membranes of the epithelial cell coat of
the foot processes (fp), but it is considerably less at their
base. In Fig. 18, a grazing section, the densities in the
lamina rara externa can be seen to form a regular reticular
pattern (arrows). Due to its cationic nature lysozyme is presumed to bind to negatively-charged structures and in effect
acts as a stain for anionic sites. From Caulfield and Farquhar (46). Fig. 17 = X 43,000; Fig. 18 = X 57,000.*

Localization of the Charge Barrier in the Glomerulus.

Clearance Studies

Recently emphasis has shifted to locating the charge barrier in the glomerulus. As mentioned already, it has been known for some time (cf. 17) that charged molecules, mainly proteins, behave differently from uncharged molecules such as dextran or polyvinylpyrrolidone (PVP) in that the uncharged molecules are more freely permeable than charged molecules of the same effective molecular radius. This is of importance since albumin, the major plasma protein which one is interested in retaining, is a polyanion in physiological solution. Recently Brenner's group (48, 49) has carried out a series of studies in which glomerular transport of neutral vs. charged (sulfated) dextrans was compared in order to investigate the role of charge in glomerular filtration. They found in normal rats that the clearance of sulfated dextrans is reduced in comparison to neutral dextrans: restriction begins at a lower molecular weight and complete exclusion is reached at a much lower effective molecular radius than with neutral dextrans (see Fig. 19)[9]. Based on this observation, Brenner

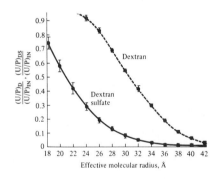

Fig. 19. Fractional clearances of dextran sulfate and neutral dextran in the normal rat glomerulus plotted as a function of effective molecular radius. The clearance of the positively charged dextran sulfate is greatly retarded over the neutral dextran. "Reprinted from Kidney International (vol. 8, p. 212, 1975) with permission."

[9]*A similar situation pertains to other capillaries since Areekul (50) had shown earlier that the passage of dextran sulfate is considerably more restricted than that of neutral dextran in the isolated perfused rabbit ear.*

and co-workers postulated that there are fixed negatively charged components in the glomerular capillary wall which serve to electrostatically repulse circulating anionic macromolecules. Because of the existence of the highly negatively charged epithelial cell coat which extends into and fills the slits (30, 43-45), they postulated (48) that the charge barrier might be located along the epithelial slits.

Anionic Sites in the Basement Membrane

Rennke et al. (22,23,52) and Caulfield and I (46), have recently obtained evidence that the basement membrane also contains anionic binding sites.[10] Such evidence was obtained from our studies with lysozyme which because of its basic nature binds to anionic sites and in effect acts as a stain for anionic groups. We found anionic sites in all layers of the basement membrane, i.e., the lamina rara interna (LRI), lamina densa, and the lamina rara externa (LRE), as well as on the epithelial (Figs. 16-18) and endothelial membranes. In the LRE, and to a lesser extent the LRI, there is concentration of anionic sites which in grazing section take on a reticular pattern (Fig. 18). Similar extraglomerular sites were also seen elsewhere between epithelia and endothelia and their basement membranes. More recently we have demonstrated anionic sites in the same locations using alcian blue (53).

The experiments of Rennke et al. involved the investigation of the behavior of native anionic ferritin (pI = 4.6) vs. cationized ferritin (pI = 5.9). They found cationized ferritin accumulates in all layers of the basement membrane far exceeding the native ferritin, and the higher the isoelectric point the greater the accumulation. They concluded that the charge of a molecule affects its filtration behavior and that the glomerular basement membrane and endothelium are negatively-charged structures which form a barrier to anionic proteins.

Thus, from two different sources there is evidence that identifies the basement membrane as the charge barrier as well as the size barrier, a situation which is infinitely more practical than localization of the charge barrier at the level of the slits or distal to the size barrier.

[10]*Some evidence for weak binding of cationic substances to basement membrane components had already been obtained by others (43-45,51), but in all these cases attention was focused on the stronger binding to the epithelium.*

Loss of Anionic Sites and Proteinuria

The presence of anionic sites in the basement membrane
may turn out to be very important in glomerular disease
because Brenner and co-workers (49) have recently shown that
the restrictive clearance of dextran sulfate over neutral
dextran is lost in several experimental glomerular diseases
associated with proteinuria -- nephrotoxic serum nephritis
(54,55) and aminonucleoside nephrosis (56). The clearance of
neutral dextrans is similar to normal or decreased, but the
clearance of dextran sulfate is increased, suggesting that
there is a loss of fixed negative charges from the glomerulus.
It is known from the work of Michael, Blau and Vernier
(57) that anionic sites are lost from the epithelium in
aminonucleoside nephrosis since staining for sialoproteins
(as demonstrated by the colloidal iron method) is reduced.
Brenner and associates (55) have shown that this is the case
in nephrotoxic serum nephritis as well.
In order to determine whether or not the anionic groups
re also lost from the basement membrane in nephrosis we
carried out studies in aminonucleoside nephrotic rats similar
to those described above in normals in which we investigated
lysozyme binding to glomerular components (58). We found
that at early stages in the disease process (8-10 days) when
the epithelium has reduced numbers of foot processes and the
LRE is widened, lysozyme deposits still occurred on the epithelial cells, but the number of discrete sites in the LRE
of the basement membrane were reduced. Later on (at 11-15
days) when the epithelium is partially detached from the
basement membrane (cf. 37), lysozyme no longer bound to the
epithelium or any of the basement membrane components. Thus
the findings indicate anionic sites are lost from both the
epithelium and basement membrane in aminonucleoside nephrosis
coinciding with the onset of proteinuria and epithelial
detachment. These observations therefore raise the possibility that loss of fixed negative charges from the basement
membrane may contribute to proteinuria and/or epithelial
detachment.

What is the Nature of the Anionic Sites in the Glomerulus?

As already discussed, the anionic sites on the membrane
of the epithelial cell stainable with colloidal iron are most
probably due to the local presence of sialoglycoproteins since
they are removed by neuraminidase treatment (43,44). If the
situation is comparable to that of the red blood cell (59,60),
the sialoproteins may represent an intrinsic component of the

cell membrane of the glomerular epithelium, i.e., an intrinsic
membrane protein instead of an extraneous cell coat, with the
portion of the protein containing the carbohydrate protruding
from the outer membrane surface.

The nature of the anionic sites in the basement membrane
is less certain. They could be carboxyl groups of the collagenous or noncollagenous peptides, sialyl groups of glycoproteins or sulfated groups of glycosaminoglycans (GAG's).
The first two have been detected in isolated glomerular basement membrane fractions (61-63), whereas sulfated GAG's have
not. There has been some debate (64,65) concerning whether
the sialoglycoprotein present in glomerular basement membrane
preparations represents a true basement membrane component or
contamination from residual epithelial cell membranes; however, at present most workers believe that sialoglycoproteins
are present in small amounts in basement membranes.

As mentioned, sulfated GAG's have not been detected in
isolated glomerular basement membranes. However, Mohos and
Skoza (44) obtained histochemical evidence (based on diamine
iron staining), suggesting the presence of sulfate esters in
the basement membrane. It should be mentioned that sulfated
GAG's have been demonstrated in association with basement
membranes in other locations, primarily in embryonic tissues
(basement membranes of the developing lens, neural tube, and
notochord (66) and developing corneal epithelium (67)).
Thus the possibility that sulfated GAG's could be present in
small amounts in the glomerular basement membrane cannot be
ruled out at present. Clearly, however, more work is needed
to clarify the nature and identity of the anionic sites in
the glomerular basement membrane.

Functional Model

Figure 20 gives an updated version of the dynamic model,
which we proposed in 1961, for the role of each of the components of the glomerular capillary wall in filtration.

Basement Membrane

According to our model, the basement membrane was identified as the main filter serving to retain plasma proteins.
This concept was contested following the work of Graham and
Karnovsky in 1966. However, as reviewed above, currently
there is general agreement that the basement membrane acts as
the main filter -- acting as both the size and the charge
barrier. It remains only to ask how this fits with the physiologic data. Renkin and Gilmore (17) have pointed out that

FUNCTION OF GLOMERULAR STRUCTURES

FILTRATION   OTHER

ENDOTHELIUM: "Valve" which controls - SYNTHESIS OF GBM?
              access of filtrate
              to the GBM

BASEMENT MEMBRANE: Main filter
                    (size and
                     charge)

EPITHELIUM: Monitors filtrate and   - SYNTHESIS OF GBM
             recovers lost protein
             (by endocytosis)
             Regulates hydraulic flux
             by providing "porous
             support" for GBM

MESANGIUM: Phagocytosis of filtration- REMOVAL OF GBM(?)
            residues (particles,
            immune complexes, etc.)- CONTRACTION (?)

*Fig. 20. Summary of functions proposed for each of
the structural elements in the glomerular capillary wall.
On the left the proposed role for each of the glomerular
components in filtration is given, and on the right, other
functions are listed. (See text for discussion)*

the glomerular filtration data could be equally well explained
by the presence of circular pores with a radius of 36 A, slits
with a half-width of 36 A, or fiber networks with fiber half-
interspaces of 26 A. Of these, the fiber model appears to
fit in best with what is known about the morphology and biochemistry of the basement membrane. It follows that alterations in either the fiber spacing or distribution of charged
groups could affect filtration processes. The basement membrane is already known to undergo thickening as well as
changes in composition in a variety of glomerular diseases.

Endothelium

     According to the model, the endothelium is seen to repre-
sent a valve which controls access to the basement membrane
by virtue of the number and arrangement of its fenestrae as
well as possibly by density of charged groups present on its
cell membranes. Alterations in either of these factors could
affect filtration. It is known, for example, in some glomerular diseases (e.g., preclampsia) that the endothelium is
swollen and the number of fenestrae is reduced at a time when
filtration is reduced.

Epithelium

     According to the model, the epithelium was seen as a
monitor which served to take up protein which passed through
the basement membrane. Normally this monitoring activity
appears to go on at a relatively modest level, but it is
highly amplified under circumstances (e.g., nephrosis) when
the basement membrane is leaky. Subsequent work with both
particulate tracers, particularly that with dextrans, as well
as work with peroxidatic tracers has confirmed the monitoring
function of the epithelium. In addition, the epithelium can
be assumed to act as a supporting grid for the basement membrane with the slits allowing passage of the filtrate into
the urinary spaces. The situation is analogous to that in
an artificial ultrafiltration membrane composed of a thin
membrane which accounts for selective permeability and a
"porous support" on which the thin membrane rests. In such
systems (cf. 16,36) the porous support serves to strengthen
the thinner membrane, to increase the effective path length
across it and to limit hydraulic fluxes across it either by
reducing the frequency or the radius of the pores in the
support. By analogy the epithelium may limit hydraulic
fluxes across the glomerular capillary wall as a result of
the distribution and width of the filtration slits. There is

some evidence that this is the case in aminonucleoside nephrosis since at a time (7-10 days) when the collective slit area is reduced (from ∿20% to ∿2% of the outer surface of the peripheral capillary loop (6)) glomerular filtration is greatly reduced (56).

Mesangium

As far as the mesangium is concerned, it was seen to function in unclogging and reconditioning the filter by incorporating and disposing of filtration residues. Mesangial cells, which are undoubtedly part of the mononuclear phagocyte system, have been shown to take up a variety of tracers, as well as aggregated serum proteins and immune complexes. In addition, the mesangial cells, which are rich in contractile proteins and resemble smooth muscle cells in some respects, (i.e., abundance of microfilaments and presence of adhesion plaques) may also have a contractile function (cf. 30).

In addition to their role in glomerular filtration, the epithelium, endothelium, and mesangium probably also function in the biosynthesis and turnover of basement membrane constituents. It has already been established that the epithelium plays a role in biosynthesis of one or more basement membrane components. Since endothelia in other locations can make basement membranes, the possibility exists that it may also contribute one or more components to the glomerular basement membrane, a point which is discussed in other papers at this conference. The mesangium being phagocytic, may participate at the removal end of the process, by incorporating and disposing of superficial, worn-out basement membrane layers.

In summary, in 1961 we proposed a dynamic model for the function of glomerular constituents. The model has been put to the test and appears still to be valid.

SUMMARY AND CONCLUSIONS

As a result of electron microscopic studies carried out over the past 20 years, we now have a much more comprehensive view of the structure and function of basement membranes in general and of the glomerular basement membrane in particular than we did before. We know that basement membranes are ubiquitous structures which are found wherever cells of any type meet connective tissue ground substance. Structurally they consist of a continuous sheet of moderately-dense material made up of a fine feltwork of 30-40 A fibrils.

With a few exceptions (glomerular basement membrane, smooth muscle) basement membranes typically face a cell layer on one surface and the connective tissue ground substance on the other. They are products of the overlying cellular layer (epithelium, endothelium, smooth muscle), rather than of connective tissue elements (fibroblasts). Their known functions are: 1) to delimit the connective tissue-non-connective tissue boundary, thereby keeping the corresponding cell populations separate and maintaining the orderly arrangement of cells and tissues; 2) attachment of cell layers to their associated connective tissue elements; and 3) filtration of macromolecules, a function most highly developed and most thoroughly studied in the capillaries of the renal glomerulus.

The main function of glomerular capillaries is to filter the blood plasma and to produce a protein-free plasma filtrate, and they are structurally specialized to carry out this function. The filtration surface of the capillaries consists of three elements -- endothelium, basement membrane and epithelium; however, only the basement membrane is continuous. The glomerular filtrate passes through the endothelial fenestrae, the basement membrane and through the filtration slits between the epithelial foot processes to cross the glomerular capillary wall. There has been considerable debate concerning which layer -- the basement membrane or epithelial slits -- represents the main barrier which retains plasma proteins. Electron microscope studies carried out in the early 60's with particulate tracers pointed to the basement membrane as the filter. Results carried out with peroxidatic tracers in the late 60's were interpreted as indicating that the epithelial slits were the main barrier, and the basement membrane was seen to represent only a coarse prefilter to retain molecules > 100 A. Recent work from several different laboratories using different approaches [use of dextrans as tracers (35,36), localization of endogenous albumin (41) and IgG (47), work with cationized ferritin (22, 23) and with lysozyme (16,46)] have all clearly pointed to the basement membrane as the main barrier based on both size and charge. Thus at present there is general agreement that the barrier lies along the endothelial surface of the basement membrane and that the slits do not serve this function. The results obtained with peroxidatic tracers can be explained in part due to the fact that the majority of such tracers are cationic proteins which bind to glomerular components. Present work is concentrated on investigating the nature and properties of the basement membrane which affect its filtration functions and on defining the role of the other layers in the filtration process and in the synthesis and turnover of the filter. With the main barrier agreed upon one can take a more rational approach to the interpretation of physiological and pathological data. One can also take advantage of the large volume of information available on the biochemical composition of the basement membrane, including that generated by this conference.

Our own work at present is centered on attempts to identify and characterize further the anionic sites in the glomerular basement membrane. We have shown that all three layers of the basement membrane (as well as the epithelial cell surface) contain anionic binding sites which can be demonstrated by lysozyme binding. These sites are particularly concentrated in the lamina rara externa where they form a continuous reticular pattern. We have also shown that the lysozyme binding sites in the basement membrane, as well as those in the epithelium, are lost in aminonucleoside nephrosis. Since physiologic studies by Brenner and his associates indicate that: a) the permeability of the glomerulus to anionic macromolecules is less than that to neutral macromolecules of similar molecular radius, and b) this selectivity is lost in certain experimental glomerular diseases associated with proteinuria, the loss of charged sites from the basement membrane (as well as the epithelium) can be assumed to play a role in the abnormal glomerular permeability found in these diseases. As pointed out elsewhere (46), the existence of fixed negative charges in the basement membrane may also be important in the trapping of antigen-antibody complexes which occurs in immune complex disease. In this regard it is of interest to note that in certain types of immune complex-induced glomerulonephritis occurring in man or induced experimentally in animals which are characterized by the deposition of antigen-antibody complexes in glomeruli, the deposits typically occur in the lamina rara externa (68-71) where the anionic sites are most concentrated.

What remains for the future is to understand the morphologic basis for alterations in glomerular permeability that occur in glomerular diseases. The situation at present is that we have considerable information on the pathologic changes which occur in various glomerular components in a variety of experimental and human glomerular diseases without understanding the functional implications of such findings. With the available framework and understanding of the structural basis of filtration in the normal glomerulus, future work can now focus on attempts to explain the molecular and cellular changes that underlie the alterations in glomerular permeability found in various diseased states. This particular endeavor is likely to keep many of us busy for some time to come.

ACKNOWLEDGEMENT

The author would like to express her thanks to Pamela Stenard for preparation of the photographs and to Lynne Wootton for typing and preparation of the manuscript.

REFERENCES

1. Krakower, C.A. and Greenspon, S.A., *Arch. Pathol.* 52:629, 1951.
2. Bloom, W. and Fawcett, D.W., *A Textbook of Histology*, W. B. Saunders Co., Phila., Pa., p. 97, 1975.
3. Farquhar, M.G., Wissig, S.L. and Palade, G.E., *J. Exp. Med.* 113:47, 1961.
4. Bruns, R. and Palade, G.E., *J. Cell Biol.* 37:244, 1965.
5. Ross, R. and Bornstein, P., *J. Cell Biol.* 40:366, 1969.
6. Farquhar, M.G. and Palade, G.E., *J. Exp. Med.* 114:699, 1961.
7. Farquhar, M.G., *In:* "Small Blood Vessel Involvement in Diabetes Mellitus", edited by Siperstein, M.D., Colwell, A.R. and Meyer, K., Washington, D.C., American Institute Biological Sciences, p. 31, 1964.
8. Andres, G.A., Morgan, C., Hsu, K.D., Rifkind, R.A. and Seegal, B.C., *J. Exp. Med.* 115:929, 1962.
9. Pierce, G.B., Midgley, A.M., Jr., and Rana, J.S., *J. Exp. Med.* 117:339, 1963.
10. Pierce, G.B. and Nakane, P.K., *Lab. Invest.* 17:499, 1967.
11. Hay, E.D. and Dodson, J.W., *J. Cell Biol.* 57:190, 1973.
12. Jaffe, E.A., Minnick, R.C.R., Adelman, B., Becker, C.G. and Nachman, R., *J. Exp. Med.* 144:209, 1976.
13. Ross, R., *J. Cell Biol.* 50:172, 1971.
14. Palade, G.E. and Farquhar, M.G., *J. Cell Biol.* 50:172, 1971.
15. Simionescu, N., Simionescu, M. and Palade, G.E., *J. Cell Biol.* 53:365, 1972.
16. Farquhar, M.G., *Kidney Int'l.* 8:197, 1975.
17. Renkin, E.M. and Gilmore, J.P., *In:* "Handbook of Physiology", edited by Orloff, J. and Berliner, R.W., Wash., D.C., American Physiological Society, Vol. 8, p. 185, 1973.
18. Renkin, E.M. and Robinson, R.R., *N. Eng. J. Med.* 299: 785, 1974.
19. Wallenius, G., *Acta Soc. Med. Upsal.* 59:1, 1954.
20. Pappenheimer, J.R., *Physiol. Rev.* 33:387, 1953.
21. Schneeberger, E.E., Leber, P.D., Karnovsky, M.J. and McCluskey, R.T., *J. Exp. Med.* 139:1283, 1974.

22. Rennke, H.G., Cotran, R.S. and Venkatachalam, M.A.,
     J. Cell Biol. 67:638, 1975.
23. Rennke, H.G. and Venkatachalam, M.A., *Kidney Int'l.*
     11:44, 1977.
24. Farquhar, M.G. and Palade, G.E., *J. Cell Biol.* 13:55,
     1962.
25. Graham, R.C. and Karnovsky, M.J., *J. Exp. Med.* 124:1123,
     1966.
26. Graham, R.C. and Kellermeyer, R.W., *J. Histochem.*
     Cytochem. 16:275, 1968.
27. Venkatachalam, M.A., Karnovsky, M.J., Fahimi, H.D. and
     Cotran, R.S., *J. Exp. Med.* 132:1153, 1970.
28. Oliver, C. and Essner, E., *J. Exp. Med.* 136:291, 1972.
29. Rodewald, R. and Karnovsky, M.J., *J. Cell Biol.* 60:423,
     1974.
30. Latta, H., *In:* "Handbook of Physiology", edited by
     Orloff, J. and Berliner, R.W., Washington, D.C.,
     American Physiological Society, Vol. 8, p. 1, 1973.
31. Farquhar, M.G. and Palade, G.E., *Anat. Rec.* 133:378,
     1959.
32. Latta, H., Maunsbach, A.B. and Madden, S.C., *J. Ultra-*
     struc. Res. 4:455, 1960.
33. Pessina, A.C., Hulme, B. and Peart, W.S., *Proc. R. Soc.*
     Lond. (Biol.), 180:61, 1972.
34. Deodhar, S.D., Cuppage, F.E. and Gableman, E., *J. Exp.*
     Med. 120:677, 1964.
35. Caulfield, J.P. and Farquhar, M.G., *J. Cell Biol.* 63:883,
     1974.
36. Caulfield, J.P. and Farquhar, M.G., *J. Exp. Med.* 142:61,
     1975.
37. Caulfield, J.P., Reid, J. and Farquhar, M.G., *Lab. Invest.*
     34:43, 1976.
38. Farquhar, M.G., Vernier, R.L. and Good, R.A., *J. Exp.*
     Med. 106:649, 1957.
39. Farquhar, M.G. and Palade, G.E., *J. Biophys. Biochem.*
     Cytol. 7:297, 1960.
40. Oken, D.E. and Flamenbaum, W., *Clin. Invest.* 50:1498,
     1971.
41. Ryan, G.B. and Karnovsky, M.J., *Kidney Int'l.* 9:36, 1976.
42. Karnovsky, M.J. and Ainsworth, S.K., *Adv. Nephrol.* 2:35,
     1973.
43. Nolte, A. and Ohkuma, M., *Histochemie* 17:170, 1969.
44. Mohos, S.C. and Skoza, L., *Exp. Mol. Pathol.* 12:316,
     1970.
45. Jones, D.B., *Lab. Invest.* 21:119, 1969.
46. Caulfield, J.P. and Farquhar, M.G., *Proc. Nat. Acad. Sci.*
     U.S.A., 73:1646, 1976.

47. Ryan, G.B. and Karnovsky, M.J., *Lab. Invest.* 34:415, 1976.
48. Chang, R.L.S., Deen, W.M., Robertson, C. R. and Brenner, B.M., *Kidney Int'l.* 8:212, 1975.
49. Brenner, B.M., Baylis, C. and Deen, W.M., *Physiol. Revs.* 56:502, 1976.
50. Areekul, S., *Acta Soc. Med. Upsalien* 74:129, 1969.
51. Latta, H., Johnston, W.H. and Stanley, T.M., *J. Ultra. Res.* 51:354, 1975.
52. Venkatachalam, M.A. and Rennke, H.G., *In:* "Proceedings of the First International Symposium on the Biology and Chemistry of Basement Membranes" (Kefalides, N.A., Ed.) Academic Press, New York, p.   , 1977.
53. Caulfield, J.P. and Farquhar, M.G., *In:* "Proceedings of the First International Symposium on the Biology and Chemistry of Basement Membranes" (Kefalides, N.A., Ed.) Academic Press, New York, p.´   , 1977.
54. Bennett, C.M., Glassock, R.J., Chang, R.L.S., Deen, W.M., Robertson, C.R. and Brenner, B.M., *J. Clin. Invest.* 57: 1287, 1976.
55. Chang, R.L.S., Deen, W.M., Robertson, C.R., Bennett, C.M., Glassock, R.J. and Brenner, B.M., *J. Clin. Invest.* 57: 1272, 1976.
56. Baylis, C., Bohrer, M.P., Troy, J.L., Robertson, C.R. and Brenner, B.M., *Kidney Int'l.* 10:554, 1976.
57. Michael, A.F., Blau, E. and Vernier, R.L., *Lab. Invest.* 23:649, 1970.
58. Caulfield, J.P. and Farquhar, M.G., *J. Cell Biol.* 70: No. 2, Part 2, 92a, 1976.
59. Marchesi, V.T., Tillack, T.W., Jackson, R.L., Segrest, J.P. and Scott, R.E., *Proc. Natl. Acad. Sci., U.S.A.,* 69:1445, 1972.
60. Steck, T., *J. Cell Biol.* 62:1, 1974.
61. Spiro, R.G., *In:* "Chemistry and Molecular Biology of the Intercellular Matrix" (Balazs, E.A., Ed.) Academic Press, New York, p. 511, 1970.
62. Kefalides, N.A., *Intern. Rev. Conn. Tissue Res.* 6:63, 1973.
63. Westberg, N.G. and Michael, A.F., *Biochemistry* 9:3837, 1970.
64. Mohos, S.C. and Skoza, L., *J. Cell Biol.* 45:450, 1970.
65. Nicholes, B.K., Krakower, C.A. and Greenspon, S.A., *Proc. Soc. Exp. Biol. and Med.* 142:1316, 1973.
66. Hay, E.D. and Meier, S., *J. Cell Biol.* 62:889, 1974.
67. Trelstad, R.L., Hayashi, K. and Toole, B.P., *J. Cell Biol.* 62:815, 1974.
68. Cochrane, C.G. and Kottler, D., *Adv. Immunol.* 16:185, 1973.

69. Wilson, C.B. and Dixon, F.J., *Ann. Rev. Med.* 25:83, 1974.
70. Davies, D.R. and Clarke, A.E., *Br. J. Exp. Pathol.* 56: 28, 1975.
71. Feenstra, K., Lee, R., Greben, H.A., Arends, A. and Hoedemaeker, Ph. J., *Lab. Invest.* 32:235, 1975.

THE DISTRIBUTION OF ANIONIC SITES

IN THE GLOMERULAR BASEMENT MEMBRANE

OF NORMAL AND NEPHROTIC RATS

John P. Caulfield

Yale University School of Medicine
New Haven, Connecticut

SUMMARY: Fixed negative charges have been demonstrated in normal and nephrotic rat glomeruli using lysozyme (MW = 14,000; pI = 11) and alcian blue (MW = 1,342) as cationic probes. The results are: First, all three layers of the capillary wall, endothelium, basement membrane and epithelium, are negatively charged. Second, the strongest charge is on the epithelial cell coat and in the lamina rara externa of the basement membrane where the probes bind to the anionic sites in a reticular pattern in tangential section. Third, alcian blue stains a double set of fibrils in the lamina rara externa, one of which runs parallel to the long axis of the basement membrane between the lamina densa and lamina rara externa and the other set runs perpendicular to the first and extends from (or near to) the first set across the lamina rara externa to the plasma membrane of the foot processes. Fourth, in the late stages of aminonucleoside-nephrosis, the epithelium and basement membrane will no longer bind lysozyme. Fifth, in nephrosis when the epithelium detaches, the anionic sites in the lamina rara externa remain attached to the basement membrane suggesting they are a part of the basement membrane rather than protrusions of the epithelial cell coat into it.

INTRODUCTION

Lysozyme (MW = 14,000; pI = 11) has been used recently to examine the distribution of anionic sites in the glomerular capillary wall. Lysozyme bound to all layers of the capillary wall -- endothelium, basement membrane and epithelium (6). The strongest binding sites were located in the lamina rara externa and on the epithelial cell coat (6).

In this study alcian blue (MW = 1,342 (25), a cationic derivative of copper phthalocyanin has been used to study the distribution of anionic sites and these results were compared with those obtained using lysozyme. Alcian blue was chosen because it is approximately ten times smaller than lysozyme yet carries almost as much charge as the larger molecule -- i.e., 2-4 positive charges (25,35) vs. 6.5, respectively. In addition, others (2) have shown that alcian blue binds to the epithelial cell coat and to fibrillar structures in the lamina rara externa. Therefore, alcian blue appears to give a better definition of structure than lysozyme.

In addition to these studies in normal animals, the glomerular capillary wall during aminonucleoside nephrosis has been examined using lysozyme in order to determine what changes in fixed negative charges on the basement membrane accompany proteinuria and/or the detachment of the epithelium late in the disease.

MATERIALS AND METHODS

Dextrans T-70 (MW - 70,000) and T-40 (MW = 40,000) were obtained from Pharmacia Fine Chemicals, Piscataway, N.J. Egg white lysozyme (Grade I) was obtained from Sigma Chemical Co., St. Louis, Mo. Alcian blue (C.I. No. 74240) was obtained from Harleco, Philadelphia, Pa.

Animals

Sprague-Dawley male rats, 150-200 gms, were used for all nephrotic experiments. Ten normal animals were used as lysozyme controls; 5 animals were used as alcian blue controls; and 11 nephrotic animals (4 early term and 7 late) were used.

Production of Nephrosis

The daily injection protocol was used in which rats were given 1.67 mg/100 gm of body weight of aminonucleoside (Lederle Laboratories, Pearl River, N.Y.) in a 0.5% solution of normal saline subcutaneously for 5 to 10 days. Animals were sacrificed 7 to 11 days after the start of injections.

Lysozyme Perfusion Experiments

There experiments were done the same as described previously (6). In brief, the vasculature of the left kidney was isolated from the systemic circulation with clamps as previously described (5), and the kidney was perfused retrograde through the aorta with a buffer followed by the lysozyme test solution. Perfusion was carried out with a Harvard Infusion pump at rates varying from 0.97 to 7 ml/min. Rats were perfused with 2-3 ml of Krebs-Ringer's bicarbonate, pH 7.4 (KRB) followed by 20 ml of a solution consisting of 1-3% lysozyme and 4% dextran. Dextran was used to test the adequacy of perfusion and fixation, since dextran is seen in the capillary lumen and urinary space only if lysozyme is fixed around the dextran and holds it in place.

Alcian Blue Perfusion Experiments

Alcian blue was either perfused with the fixative or in an acid buffer after the kidney has been fixed by perfusion. In the first case the left kidney was perfused as described above with 2-3 ml of minimal essential medium (Eagle's) followed by 20 cc of 0.5 percent Alcian blue in Karnovsky's aldehyde fixative (17), which consists of 1 percent formaldehyde and 3 percent glutaraldehyde in 0.1 M cacodylate buffer, pH 7.4. In the second case the left kidney was perfused by 2-3 cc of medium, 7-10 cc of Karnovsky's fixative (17) and 20 cc of 0.1 percent Alcian blue in 0.05 M Na Acetate, pH 5.0.

Tissue Processing

Most kidneys perfused with lysozyme-dextran were fixed by injection *in situ* as previously described (3) using the fixative mixture of Simionescu et al (38) required for the demonstration of dextrans (which contains 1.5% formaldehyde, 2.5% glutaraldehyde, 0.66% OsO_4, and 2-3 mg per 100 ml of lead citrate in 0.1 M arsenate buffer, pH 7.4). Blocks of these tissues were dehydrated and embedded in Epon without further treatment. Two of the kidneys perfused with lysozyme

were perfused fixed *in situ* with Karnovsky's aldehyde fixative (17). Blocks from these kidneys as well as from the
kidneys perfused with Alcian blue were post-fixed in 1% OsO_4
in acetate veronal buffer (pH 7.2) and stained in block with
uranyl acetate (12) before dehydration. Techniques for
sectioning and microscopy were the same as given previously.

RESULTS

General Morphology

     The structure of the peripheral capillary wall is de-
scribed elsewhere in this volume (see papers by Farquhar and
Venkatachalam).
     The sequence of changes produced in the capillary wall
during the daily administration protocol of the aminonucleoside of puromycin can be divided into two phases (5). Early
in the disease (days 6-9) shortly after proteinuria has
begun, there is a reduction in the number of foot processes
and filtration slits and occluding junctions are seen in many
of the residual slits. The basement membrane is no longer
uniform, but in some areas there is a thinning of the lamina
densa which is accompanied by a concomitant widening of the
lamina rara externa. In the late stages of the disease (days
10-15), when proteinuria is maximal and when the animals begin
to die, the changes are more severe. The slits remain
essentially the same, but the epithelial cytoplasm is filled
with large (4-5 μ) vacuoles. In addition, denuded regions
of basement membrane are seen where there is initially partial,
and eventually complete detachment of the epithelium from the
basement membrane.

Lysozyme Experiments

Controls

     As reported previously (6) and as illustrated in the
paper by Farquhar in this volume, lysozyme is seen in the
normal capillary wall as an electron dense deposit which out-
lines the epithelial cell coat as a 300-400 Å layer except
at the base of the foot processes where the layer is thinner
and at the level of the slits where the two layers were thick
enough to fill the slits. In the lamina rara externa there
are dense deposits which in normal section through the capillary wall extend from the lamina densa to the epithelial
plasma membrane and are typically located at the base of the

foot processes, but are absent along the filtration slits.
When cut in grazing section these densities are seen forming
an apparent reticular pattern. The lamina densa shows a
slight, but definite increase in electron density. Some
deposits are seen in other components of the capillary wall,
but in these locations the binding of lysozyme is less
striking. Areas of increased density are seen in the lamina
rara interna and mesangium, but the pattern is less regular.
The luminal surface of the endothelium is also covered by a
continuous dense layer in some capillary loops, but not in
others.

Nephrotic, early (days 6-8)

The binding of lysozyme to the epithelial cell coat in
early nephrotics is usually decreased (Fig. 2), but this is
not an invariable finding. In the lamina rara externa the
dense deposits are still seen, but are decreased in number
(Figs. 1 and 2) and are often absent under pinocytic invaginations (Figs. 1 and 2). In some areas the dense
deposits do not extend across the lamina rara externa from
lamina densa to epithelial plasma membrane as in the normal,
but instead a lucency is seen between the deposits and the
epithelium (Fig. 2). Where the epithelium is detached,

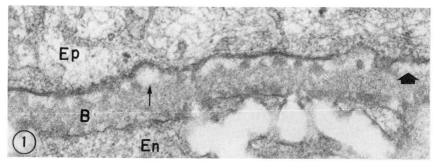

Fig. 1. *Normal section of glomerular capillary from
an 8 day nephrotic rat perfused with lysozyme. Note that
the number of dense deposits in the lamina rara externa,
between the lamina densa (B) and the epithelium (Ep), is
decreased and occasionally absent (thick arrow). Note
the somewhat mottled appearance of the lamina densa compared to the normal (See Fig. 17 in paper by Farquhar)
and continued presence of densities between the lamina
densa and endothelium (En). Tissue fixed in situ with
aldehydes, postfixed in OsO_4, and stained in block with
uranyl acetate. X 75,000.*

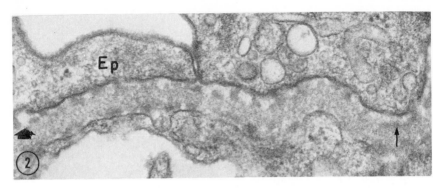

Fig. 2. Normal section of glomerular capillary
from an 8 day nephrotic rat perfused with lysozyme.
Note the reduced thickness of the epithelial cell coat
(Ep), the absence of density under the pinocytic in-
vagination (thick arrow), the appearance of lucency
between the epithelium and the binding sites in the
lamina rara externa (thin arrow) and the overall
decrease in binding sites in the lamina rara externa.
Tissue preparation as in Fig. 1. X 75,000.

the deposits are still seen adjoining the lamina densa and
the epithelial plasma membrane is free of them (Fig. 4).
In tangential section in early nephrotics the deposits in
the lamina rara externa form a reticular pattern similar to
that seen in the normal. The lamina densa is still increased
in density, but occasionally appears mottled (Fig. 1) when
compared to the normal. In other areas of the capillary wall
the pattern of lysozyme binding is the same as in the normal.

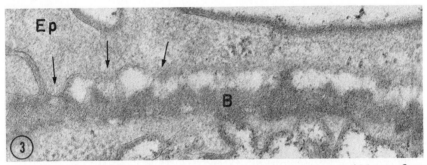

Fig. 3. Normal section of glomerular capillary from
a 10 day nephrotic rat perfused with lysozyme and dextran,
showing the epithelium (Ep) pulling away from the basement
membrane (B). Note that the densities in the lamina rara
externa appear to stay with the basement membrane and that
the epithelium extends towards them (↓). Tissue fixed in
situ with a special fixative to demonstrate dextrans.
X 75,000.

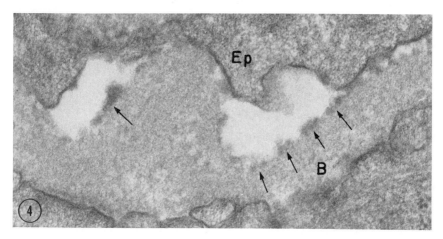

Fig. 4. Normal section of glomerular capillary
from an 8 day nephrotic rat perfused with lysozyme showing early separation of the epithelium (Ep) from the
basement membrane (B). Note that the densities in the
lamina rara appear to stay with the basement membrane
(arrows). Tissue preparation as in Fig. 1. X 75,000.

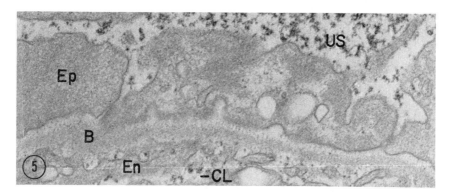

Fig. 5. Normal section of glomerular capillary from
a 10 day nephrotic rat perfused with lysozyme and dextran.
Note the absence of binding of lysozyme to all three layers
of the capillary wall and that dextran is present in the
urinary space (US) and capillary lumen (CL). Ep = epithelium, B = basement membrane, En = endothelium. Tissue
preparation as in Fig. 3. X 43,000.

Nephrotic, late (days 9-11)

In the later stages of the disease lysozyme no longer
binds to any component of the capillary wall (Fig. 5) in most
(70%), but not all animals. However, dextran which was

included in the perfusate is seen in the capillary lumen and
urinary space, indicating the perfusate had reached the
glomerulus.

In those late term animals when lysozyme does bind, the
thickness of the layer on the epithelial cell coat is usually
decreased, but may be normal (Fig. 3). In normal section the
dense deposits in the lamina rara are similar to those seen in
the early nephrotic except there are many more and wider
areas of separation between the deposits and the epithelium
(Fig. 3). In these areas the epithelium extends across the
widened lamina rara to the dense deposits (Fig. 3). Tangential section through the deposits in the lamina rara externa
shows that many of the deposits are unconnected to one
another unlike the normal where they formed a reticular
pattern. The lamina densa is still increased in density over
non-lysozyme containing tissue and may even appear homogenous (Fig. 3) as in the normal. The binding in the remainder
of the capillary wall is the same as in the normal.

Extraglomerular sites of lysozyme binding

In normal and all nephrotic animals the pattern of
extraglomerular lysozyme binding is the same. Discrete dense
areas of lysozyme binding similar to those seen in the laminae
rarae are seen between the tubular epithelium and its basement
membrane, between Bowman's capsule and the parietal epithelium
and between the endothelium of peritubular capillaries and its
basement membrane. A continuous dense layer is seen along the
luminal plasma membrane of parts of the tubular epithelium and
on the blood front of arterial and capillary endothelia.
Lysozyme also binds to collagen fibers in a 660 Å period.

Alcian Blue Experiments

Alcian blue is seen as an electron dense deposit in all
three layers of the peripheral capillary wall in a distribution similar to that seen with lysozyme. The dye binds to
the epithelial cell coat as a uniform dense layer 100-200 Å
thick (Fig. 6). The slit diaphragms are stained by the dye
and are easily seen (Fig. 6). In the lamina externa there are
densely staining < 100 Å fibrils which run in two directions--
one set runs perpendicular to the long axis of the basement
membrane to (or near to) a second set running parallel to the
long axis between the lamina rara externa and the lamina densa
(Fig. 6). These parallel fibrils are seen intermittently and
are apparently discontinuous. In the epithelial slits the
parallel fibrils may be present, but the perpendicular fibrils

are generally absent (Figure 6). Some electron dense fibrils
are seen apparently randomly distributed in the lamina densa.

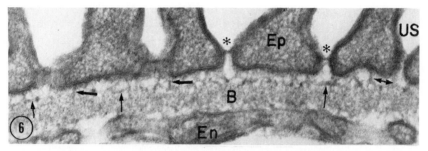

Fig. 6. Section of normal glomerular capillary
from an animal perfused with alcian blue. Note the
increased density of the epithelial (Ep) cell coat
and the slit diaphragm (*). Note also the double
set of fibrils in the lamina rara externa, one set
(thin arrows) runs parallel to the long axis of the
basement membrane (B) and the other (thick arrows)
runs perpendicular to the first, from them to the
plasma membrane of the epithelial cell. Note also
that fibrils can be seen beneath the endothelium
(En) and in the lamina densa. US = urinary space.
Tissue prepared by perfusing 0.5% alcian blue in
aldehydes, postfixing in OsO_4 and staining in block
with uranyl acetate. X 75,000.

In the lamina rara interna fibrils are seen which are similar
to the ones seen in the externa, but are less well defined in
orientation. In tangential section the fibrils in both
lamina rarae form a reticular pattern (Fig. 7). Alcian blue
also forms an electron dense deposit on the endothelial plasma
membrane in most capillary loops.

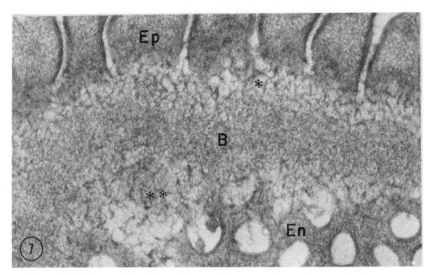

*Fig. 7. Tangential section of normal glomerulus
perfused with alcian blue showing that the fibrils in
the lamina rara interna (**) and externa (*) are arranged
in a reticular pattern. Note that fibrils are also seen
in the lamina densa (B). Ep = epithelium, En = endothelium. Tissue prepared by perfusing 1% alcian blue in
aldehydes, postfixing in OsO_4 and staining in block with
uranyl acetate. X 75,000.*

The extraglomerular localization of alcian blue sites is
similar to that seen with lysozyme. Fibrils similar to the
fibrils seen in the laminae rarae are seen between the parietal epithelium and Bowman's capsule, between proximal and
distal tubules and their basement membranes and between the
endothelia of various blood vessels and their basal laminae.
In addition, deposits are present at 660 Å intervals along
collagen fibers. The luminal surface of the endothelial cell
plasma membrane of peritubular capillaries is also stained.

DISCUSSION

Lysozyme (MW = 14,000; pI = 11) and alcian blue (MW = 1,342) have been used as cationic probes to examine the distribution of fixed negative charges on the glomerular capillary wall. It is clear that both probes are binding to the glomerulus as a consequence of their high positive charge because lysozyme will no longer bind if the charge is made negative (pI = 4.5) by succinylation (6) and alcian blue binding is inhibited by high salt and by cetylpyridinium chloride (26, 27, 34, 35). However, the two probes are different in that lysozyme is approximately ten times the size of alcian blue even though they both carry approximately the same number of positive charges at the pH of the buffers used in these experiments, i.e., 6.5 positive charges on lysozyme and 2-4 on alcian blue. The results with both tracers are similar in that they bind to the epithelial and endothelial plasma membranes, weakly in the lamina densa and to discrete reticular sites in the laminae rarae. The sites in the laminae rara externa have a somewhat different appearance with the two probes, but probably each probe is binding to the same structures and is providing a slightly different type of information. First, let us consider the similarity in lysozyme and alcian blue binding. Both bind from the epithelial cell plasma membrane to the juncture of the lamina rara externa and the lamina densa and both are absent in the slits. In addition, tangential or grazing sections reveal a reticular pattern. The probes bind differently in that the lysozyme is seen as large ($\sim$ 300 Å) densities whereas alcian blue stains a double set of fibrils one of which runs parallel to the long axis of the basement membrane between the lamina densa and the lamina rara externa and one of which runs perpendicular to the parallel fibrils across the lamina rara to the plasma membrane of the foot processes. However, these appearances are not mutually exclusive. It is clear that anions appear larger after staining with lysozyme than with alcian blue--e.g., lysozyme binds to the epithelial cell coat in a layer two or three times as thick as alcian blue does. In fact, the lysozyme layer is so thick that it overlaps in the slit and the slit diaphragm is not seen whereas with alcian blue the diaphragm is seen and appears to bind the dye. A similar phenomenon could occur in the lamina rara externa where lysozyme, binding to fibrils that are close together, overlaps and is seen as a large density except where the fibrils are too widely spaced for the overlap to occur such as in the slits or under pinocytic invaginations. Therefore, alcian blue by virtue of its small size and high charge density would appear to give a better definition of the final structure of the lamina rara externa. On the other hand, lysozyme because of its lesser charge density provides a more sensitive assay of the quantity of charge.

Localization of Negative Charges in the Normal Glomerulus

The results with both tracers indicate that there are fixed negative charges on all three layers of the peripheral capillary wall. There is general agreement among the results of various investigators using different cationic probes that there is strong binding of the probes to the epithelial plasma membrane and slit diaphragm and binding in both laminae rarae as well as on the endothelial plasma membrane, albeit weaker than that seen on the epithelium (Table I). However, there is disagreement on whether the lamina densa carries any negative charge. The failure to observe binding of cationic probes in the lamina densa may be due to the large size of some of the probes used (e.g., colloidal iron, colloidal thorium (15,16,21,22,24,28) or to the use of diffusion as a method to deliver the probe to the glomerulus (15,16,24). Two other questionable areas remain, both concerning the lamina rara externa. First, what is the morphology of the sites in the lamina rara externa? Alcian blue apparently provides the best picture of what the sites truly look like because of its small size and high charge density. With alcian blue the anionic sites in the lamina rara externa appear as a double set of fibrils. Perpendicular fibrils are only seen under areas where the plasma membrane is at its usual distance from the lamina densa--i.e., in the slits or under pinocytic invaginations, where the epithelium is absent or elevated, the fibrils are absent or pulled to one side. The layer of parallel fibrils seen in the plane between the lamina densa and lamina rara externa possibly serves to anchor the perpendicular fibrils to the lamina densa. This is suggested by the many Y or T shaped branches seen on tangential section and the inverted T shaped images in normal section. However, these images can only be suggestive because the relative thickness of the section (500 Å) compared to the thickness of the fibrils (< 100 Å) could produce an overlapping image of fibrils on different planes of the section that would be indistinguishable from truly branching fibrils. We have attempted to see whether the fibrils are truly branched by examining them with tilting stages and have found that some fibrils remain branched over a range of 40° of tilt.

TABLE I

BINDING OF CATIONIC PROBES TO GLOMERULUS
LOCATION IN CAPILLARY WALL

Probe	Endo	LRI	LD	LRE	Ep	SD	Ref
CI	+	−	−	+	+	+	22
"	+	+	−	+	+	+	16
"	−	DS	−	+	+	DS	21
"	+	+	−	+	+	+	15
"	+	+		+	+	DS	24
C.T.	+	DS	−	+	+	DS	28
"	+	DS	DS	DS	+	DS	15
R.R.	+	+	DS	+	+	DS	14
"	+	+	−	+	+	DS	16
"	+	DS	+	+	+	+	15
"	+	+	+	+	+	+	19
AB	DS	DS	+	+	+	+	2
AB	+	+	+	+	+	+	
AB-La	+	+	+	+	+	+	37
Lysozyme	+	+	+	+	+	+	6

Abbreviations: Endo = endothelial cell plasma membrane
LRI = lamina rara interna
LD = lamina densa
LRE = lamina rara externa
Ep = epithelial cell plasma membrane
SD = slit diaphragm
Ref = reference
C.I. = colloidal iron
C.T. = colloidal thorium
R.R. = ruthenium red
AB = alcian blue
AB-La = alcian blue lanthanum
+ means author described some binding
− means author described absence of binding
DS means author did not say whether there was or
was not binding

It is not clear that this is sufficient evidence for true
branching, but it suggests at least extremely close apposition.
In addition to the possibility of the perpendicular fibrils
branching from the parallel fibrils which is favored by the
normal section appearance, one must consider the possibility
of the parallel fibrils being a true reticulum of branching
fibrils. This reticulum must cover the entire surface of
the lamina densa since on tangential section the pattern of
the reticulum is seen even under the slits.

    The second question concerning the sites in the lamina
rara externa is are they part of the basement membrane or are
they extensions of the epithelial cell coat into the basement
membrane? We believe the sites are indeed part of the basement membrane because they are morphologically different from
the epithelial cell coat when examined with alcian blue and
because they remain with the basement membrane when the
epithelium sloughs off in nephrosis.

Influence of Fixed Negative Charges on Glomerular Filtration
in Normal and Nephrotic Glomeruli

    From physiological data in normal animals it is clear
that negatively charged molecules are hindered from passing
into the urine more than neutral molecules of the same
molecular weight (9). In addition, macromolecules, charged
or uncharged, appear in the urine inversely proportional to
their molecular weight (9,10). From these observations it
is suggested that the glomerulus has both a size and charge
barrier which stereologically restricts plasma proteins from
crossing the capillary wall as well as electrochemically
opposing the negative charges of plasma proteins with fixed
negative charges on the capillary wall (9). From studies
using particulate tracers such as ferritin (13) and dextrans
(3) as well as studies in which endogenous plasma proteins
were localized with immunocytochemical techniques (31,33), it
is clear that macromolecules are restricted to the endothelial
side of the basement membrane in the normal capillary under
normal flow conditions. These experiments appear to reasonably establish the basement membrane as the stereological and
consequently as the charge barrier, because it is illogical
to postulate a charge barrier distal to the size barrier.
Other experiments agree with this interpretation. Experiments
which attempt to localize fixed negative charge in the capillary find them in the lamina rara interna and on the endothelium as well as on the epithelium (Table I). In addition,
cationic ferritin has been shown to penetrate further into the
basement membrane than the native anionic model (29,30).

If the basement membrane is the size and charge barrier to plasma proteins, what is the function of the negative charge on the epithelial cell coat and in the slits? Two answers can be suggested. First, the podocyte will pinocytose macromolecules that leak through the basement membrane just like the cells of the proximal tubule (3,13). Perhaps the cell coat aids in pinocytosis. This is suggested by the fact that other cells with strongly negatively charged coats are extremely pinocytic, e.g., proximal tubule, small intestine (28), and by the fact that pinocytic activity diminishes in nephrosis when the cell coat loses its charge (5,21). Second, recent experiments have shown that neutralization of cell surface charge results in a reduction in slit number and formation of occluding junctions similar to changes seen in aminonucleoside-nephrosis (36). This experiment suggests that charge on the epithelium helps maintain foot process architecture, and slit arrangement. This is an extremely important function because the epithelium probably controls hydraulic fluxes across the capillary as is shown by the reduced glomerular filtration rates early in aminonucleoside-nephrosis when the number of slits is reduced (4).

From the above, loss of negative charge on the basement membrane and epithelium observed in aminonucleoside-nephrosis should produce severe changes in glomerular function. First, loss of charge on the basement membrane should result in a more permeable structure particularly to anionic macromolecules. It has been shown previously by a number of investigators that the basement membrane is more permeable in this disease to various particulate and mass tracers (4,11,41,42). In addition, it has been shown that the clearance of sulfated dextran increases over normal values where the clearance of neutral dextran decreases in the early stages of nephrosis (1). Secondly, the neutralization of negative charge on the epithelial cell coat should result in the reduction of slit number, which is seen (5,11). In addition, the time of the loss of lysozyme binding to the epithelium correlates with the appearance of vacuoles in the epithelial cell cytoplasm and detachment of the epithelium from the basement membrane (5). It is not clear whether this relationship is causal or coincidental.

Possible Role for Anionic Sites in Lamina Rara Externa as Attachment Devices

It has been postulated that the anionic sites in the lamina rara externa might represent attachment devices between the epithelial cell and the basement membrane because

similar sites are widely distributed between many basement
membranes and epithelia and therefore probably do not have a
function related only to glomerular filtration (6). The
present study adds to this theory the following: First, the
role of attachment device is favored by the distribution of
the alcian blue staining perpendicular fibrils which are
seen only in areas where the epithelial plasma membrane is
the normal distance from the lamina densa, i.e., they are
not randomly distributed, but depend on close apposition of
the epithelial cell. Second, at the time when lysozyme
binding is lost, the epithelium begins to detach. However,
areas are seen in which lysozyme is bound and the epithelium
is detached and areas are seen in which lysozyme is not
bound and the epithelium is still attached. These observations
suggest that the quantity of negative charge necessary to bind
lysozyme is irrelevant to the attachment of the epithelium to
the basement membrane, but negative charge in general can
provide a useful marker of these sites. If indeed these sites
or fibrils are attachment devices, the suggestion is that the
epithelium is becoming detached from them at or near to where
the fibrils insert into the epithelial plasma membrane, because normal size lysozyme binding sites remain attached to
the basement membrane during detachment. The suggestion that
detachment occurs at the level of the cell-fibril interaction
is consistent with a previous description of defective cell
function at this time of the disease (5).

Chemical Composition of Anionic Sites

      It is generally accepted that the anionic material which
binds cationic dyes on the epithelial surface is primarily
sialic acid (16,21). What anions are present in the lamina
rara interna and externa is a matter of controversy. Histochemical studies in which enzymes are used to remove the
anion have claimed that it is removed with neuraminidase (21)
and with hyaluronidase (16). Biochemical analyses of isolated basement membranes find variable amounts of sialic acid,
which is possibly contamination from the epithelial cell coat
material (23), carboxyl side chains of amino acids, but no
detectable levels of uronic acid or sulfate (18,39,43). In
analogous systems, structures similar to those seen in the
laminae rarae have been described as chondroitin sulfate in
the notochort (20) and cornea (40), but have defied characterization in the aorta (44). At present, the evidence is
too confusing to assign a chemical definition to the anionic
sites within the glomerular basement membrane.

This work was reported in part before the First International Congress of Cell Biology (7), and the First International Congress on Basement Membranes (8).

ACKNOWLEDGEMENT

I wish to thank Dr. Marilyn G. Farquhar for allowing me to do this work in her laboratory, for the guidance she has provided during the course of the work and for the financial support derived from her grant AM-17724. I also wish to thank Ms. JoAnne J. Reid and Ms. L. Liaw for their excellent technical assistance, Ms. Lynne Wootton for her superb secretarial help and Ms. Pamela Stenard for her fine help in preparation of the illustrations.

REFERENCES

1. Baylis, C., Bohrer, M.P., Troy, J.L., Robertson, C.R., and Brenner, B.M., *Abst. Am. Soc. of Nephrol.* p. 68, 1976.
2. Behnke, O. and Zelander, T., *J. Ultrastruct. Res.* 31:424, 1970.
3. Caulfield, J.P. and Farquhar, M.G., *J. Cell Biol.* 63:883, 1974.
4. Caulfield, J.P. and Farquhar, M.G., *J. Cell Biol.* 142: 61, 1975.
5. Caulfield, J.P., Reid, J.A., and Farquhar, M.G., *Lab. Invest.* 34:43, 1976.
6. Caulfield, J.P. and Farquhar, M.G., *Proc. Natl. Acad. Sci., USA,* 73:1646, 1976.
7. Caulfield, J.P. and Farquhar, M.G., *J. Cell Biol.* 70:92a, (abstr.), 1976.
8. Caulfield, J.P. and Farquhar, M.G., *First International Congress on Basement Membranes,* (abstr.), 1976.
9. Chang, R.L.S., Deen, W.M., Robertson, C.R., and Brenner, B.M., *Kidney Int.* 8:212, 1975.
10. Farquhar, M.G., *Kidney Int.* 8:197, 1975.
11. Farquhar, M.G. and Palade, G.E., *J. Exp. Med.* 114:699, 1961.
12. Farquhar, M.G. and Palade, G.E., *J. Cell Biol.* 26:263, 1965.
13. Farquhar, M.G., Wissig, S.L. and Palade, G.E., *J. Exp. Med.* 113:47, 1961.
14. Fowlwer, B.A., *Histochemie* 22:155, 1970.
15. Groniowski, J., Biczyskowa, W., and Walski, M., *J. Cell Biol.* 40:585, 1969.

16. Jones, D.B., *Lab. Invest.* 21:119, 1969.
17. Karnovsky, M.J., *J. Cell Biol.* 27:137A (abstr), 1965.
18. Kefalides, N.A. and Winzler, R.J., *Biochemistry* 5:702, 1966.
19. Latta, H., Johnston, W.H. and Stanley, T.M., *J. Ultrastruct. Res.* 51:354, 1975.
20. Hay, E.D. and Meier, S., *J. Cell Biol.* 62:889, 1974.
21. Michael, A.F., Blau, E. and Vernier, R.L., *Lab. Invest.* 23:649, 1970.
22. Mohos, S.C. and Skoza, L., *Exp. Mol. Pathol.* 12:316, 1970.
23. Mohos, S.C. and Skoza, L., *J. Cell Biol.* 45:450, 1970.
24. Nolte, A. and Ohkuma, M., *Histochemie* 17:170, 1969.
25. Pearse, A.G.E., *In:* "Histochemistry, Theoretical and Applied", Vol. 1, p. 344, Little, Brown and Company, Boston, 1968.
26. Quintarelli, G., Scott, J.E. and Dellovo, M.C., *Histochemie* 4:86, 1964.
27. Quintarelli, G., Scott, J.E. and Dellovo, M.C., *Histochemie* 4:99, 1964.
28. Rambourg, A. and LeBlond, C.P., *J. Cell Biol.* 32:27, 1967.
29. Rennke, H.G., Cotran, R.S. and Venkatachalam, M.A., *J. Cell Biol.* 67:638, 1975.
30. Rennke, H.G. and Venkatachalam, M.A., *Kidney Int.* 11:44, 1977.
31. Ryan, G.B., Hein, S.J. and Karnovksy, M.J., *Lab. Invest.* 34:415, 1976.
32. Ryan, G.B. and Karnovsky, M.J., *Kidney Int.* 8:219, 1975.
33. Ryan, G.B. and Karnovsky, M.J. *Kidney Int.* 9:36, 1976.
34. Scott, J.E. and Dorling, J., *Histochemie* 5:221, 1976.
35. Scott, J.E., Quintarelli, G. and Dellovo, M.D., *Histochemie* 4:73, 1964.
36. Seiler, M.W., Venkatachalam, M.A. and Cotran, R.S., *Science* 189:390, 1975.
37. Shea, S.M., *J. Cell Biol.* 51:611, 1971.
38. Simionescu, N., Simionescu, M. and Palade, G.E., *J. Cell Biol.* 53:365, 1972.
39. Spiro, R.G., *J. Biol. Chem.* 242:1915, 1967.
40. Trelstad, R.L., Hayashi, K. and Toole, B.P., *J. Cell Biol.* 62:815, 1974.
41. Venkatachalam, M.A., Cotran, R.S. and Karnovsky, M.J., *J. Exp. Med.* 132:1168, 1970.
42. Venkatachalam, M.A., Karnovsky, M.J. and Cotran, R.S., *J. Exp. Med.* 130:381, 1969.
43. Westberg, N.G. and Michael, A.G., *Biochemistry* 9:3837, 1970.
44. Wight, T.N. and Ross, R., *J. Cell Biol.* 67:660, 1975.

THE BASEMENT MEMBRANE PORES

Antonio Martinez-Hernandez

University of Colorado Medical Center
Denver, Colorado

SUMMARY: Although filtration has been postulated to be the main function of glomerular basement membrane, numerous ultrastructural studies failed to demonstrate pores in glomerular basement membrane. We studied two purified basement membranes, glomerular and parietal yolk sac carcinoma, both in suspension and by thin sections, using negative staining, immunohistochemistry and positive staining. With these methods pores were demonstrated in both basement membranes. The radius calculated for these pores is 70 ± 10 Å. This radius is slightly larger than that calculated from physiological studies but it is in agreement with the data obtained from morphological tracer studies. From our morphological findings a structural model for basement membrane is proposed.

INTRODUCTION

Although the function of basement membranes (BM) is not clear, a filtering capacity has been postulated, particularly for glomerular BM (GBM) (1-6).

Physiological studies indicate that a filtering effect could be mediated by GBM if it had pores of 35 to 60 Å in radius (3,6-13) but morphological studies, using ultrathin tissue sections, have never demonstrated the existence of pores in GBM (1,14,23).

In approaching this problem, we used negative staining of a BM synthesized by the ascites form of parietal yolk sac carcinoma (PYS Ca). This basement membrane (NBM) has been used as a model to study BM origin, structure, function, biosynthesis, and degradation (24-30).

Negative staining of NBM demonstrated pores. These observations were then confirmed by a variety of metods in GBM.

MATERIALS AND METHODS

Neoplastic basement membrane (NBM) was obtained from the ascites form of the murine (PYS) carcinoma (28,30). Ascites fluid was centrifuged at 10,000 g for 30 min; the pellet was resuspended in 0.5% deoxycholate (DOC) in Tris-HCl buffer (0.01 M; pH 7.4), and sonicated in an ice water bath at 30 sec intervals for 10 min. This suspension was centrifuged at 10,000 g for one hour, and the precipitate was resuspended, sonicated, washed in Tris-Hcl buffer, and centrifuged five more times. The pellet was then resus-pended in Tris-HCl buffer containing 10% NaCl, sonicated and centrifuged. This final pellet was extensively dialyzed against distilled water and lyophilized.

Since these rigorous methods might artifactually alter the appearance of NBM, other samples were prepared as follows: The pellet obtained from the ascites was suspended in 0.5% DOC in the Tris-HCl buffer and extracted overnight at 4°C with continuous stirring. The suspension was centrifuged at 10,000 g for one hour and the procedure repeated until no protein could be detected in the supernatant by the Lowry method (31). The pellet was dialyzed against distilled water and used for the morphological studies.

Glomeruli were obtained from mouse kidneys according to the method of Krakower and Greenspon (32) using a stainless steel mesh #170 and processed by their methods and those listed above.

Suspensions of purified NBM or GBM were negatively stained with 1% phosphotungstic acid at pH 6.5, 6.8, and 7.0 (33). For routine electron microscopy, purified BM was fixed in 3% glutaraldehyde, post-fixed in 1% OsO_4, dehydrated in progressive grades of ethanol and embedded in Epon-Araldite via propylene oxide. Ultrathin sections were stained with uranyl acetate (34) and lead citrate (35).

For immunohistochemical studies, affinity purified anti-bodies against PYS carcinoma BM (anti-NBM) were prepared as previously described (36-38). Fifty mg of purified NBM or

GBM were suspended in phosphate buffered saline (PBS) at room temperature for 30 minutes, reacted with peroxidase-labeled anti-NBM (24-27,29,36) for 60 minutes, incubated in diaminobenzidine solution (23) for 10 minutes and reacted with 1% OsO_4 for one hour. Aliquots of these preparations were either processed for ultrathin sectioning or spread on copper grids and examined without further staining. All samples were examined in a JEOL-100B at 100 Kv.

RESULTS

Parietal Yolk Sac Carcinoma Basement Membrane

By negative staining, NBM fragments appeared as irregular polygons of variable size. All the fragments contained multiple intercommunicating pores with a radius of 70 ± 10 Å, imparting to the NBM a sponge-like appearance (Figs. 1, 2).

 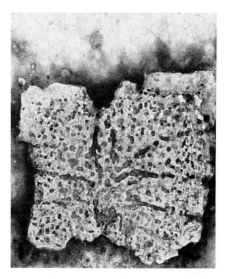

Fig. 1. Negatively stained suspension of NBM. The NBM fragment appears electron lucent, while the edges and pores within the NBM are gray to black. The relative thickness of the NBM fragment can be appreciated at the edges (PTA 1% sol., pH 6.8). Magnification, X 110,000.

Fig. 2. Negatively stained suspension of NBM. Due to the plane of fracture, the intercommunication between pores, forming a channel network is apparent (PTA 1% sol., pH 6.8). Magnification, X 110,000.

Suspensions of NBM stained with peroxidase-labeled anti-NBM produced a reverse image ("positive staining") (Fig. 3). The average radius of the pores as measured by this technique was 190 ± 10 Å, more than double the radius obtained by negative staining.

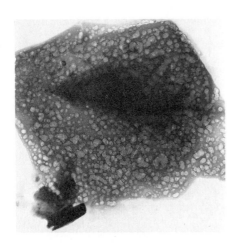

Fig. 3. Suspension of purified NBM stained with enzyme-labeled antibody (see Materials and Methods for details). The NBM appears electron dense, while the pores are electron lucent. The relative thickness of the fragment is obvious from the double and triple images at the edges. Magnification, X 60,742.

NBM examined by thin sectioning and routine staining appeared as a dense network of fibers of different diameters permeated by poorly defined round to oval empty spaces (Fig. 4).

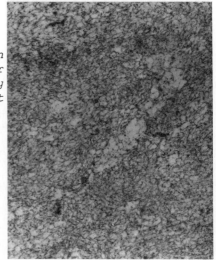

Fig. 4. Ultrathin section of purified NBM. The fibrillar network is apparent. The empty space at center could represent a tangential section through a pore. However, pores are difficult to identify in thin sections. Uranyl acetate and lead citrate. Magnification, X 35,300.

Immunohistochemically stained purified NBM examined by thin sectioning was also fibrillar with slightly better defined empty spaces (Fig. 5).

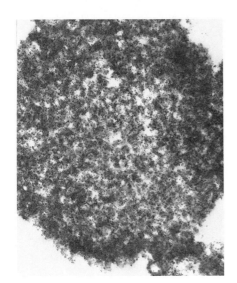

Fig. 5. Ultrathin section of purified NBM stained with enzyme-labeled antibody. The fibrillar network is not obvious, but empty spaces, probably representing pores, are more prominent than those in routine sections (Fig. 9). Magnification, X 98,000.

To exclude the possibility that the pores demonstrated by the preceding methods represented artifacts, NBM purified without sonication or freezing was examined by the same staining techniques. The appearance of these preparations of NBM and the pore dimensions were identical to those of sonicated and frozen NBM (Fig. 6). The only difference was that the preparation obtained without sonication contained plasma membranes and other cell debris.

Fig. 6. Suspension of NBM, purified without sonication or freezing. Although the preparations had significant contamination by cell debris, and the NBM fragments were larger the porous appearance and dimensions were identical to that of the sonicated and frozen preparations. Enzyme-labeled antibody staining. Magnification, X 55,500.

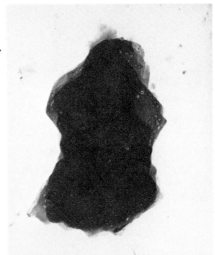

Glomerular Basement Membrane

Since sonication did not introduce any detectable change
in the structure of NBM, all GBM was prepared by sonication.
GBM preparations contained interstitial collagen, occasional
elastic fibers and cell debris (Fig. 7). The average thickness of purified GBM examined by thin sections of plastic
embedded material was 450 ± 20 Å.

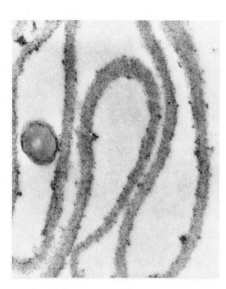

*Fig. 7. Ultrathin section
of a pellet of purified GBM.
The folded, convoluted appearance is characteristic of GBM.
Anti-NBM staining, magnification, X 30,000.*

Suspensions of GBM examined either by negative staining
or by immunohistochemistry had an average thickness of
550 ± 20 Å. By negative staining, GBM was heteroporous with
an average pore radius of 70 ± 10 Å. GBM had an appearance
similar to that of NBM (Fig. 8), with multiple pores forming
a channel network running obliquely through the width of GBM.
The pore radius was 130 ± 10 Å for GBM. GBM reacted with
phosphotungstic acid at pH 7.0 was "positively stained" and
appeared electron dense and the pores electron lucent. The
average radius of the pores with this technique was 200 ± 10 Å
(Fig. 9).

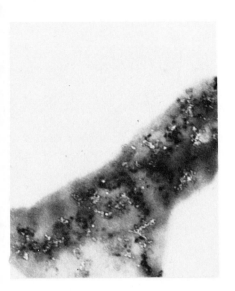 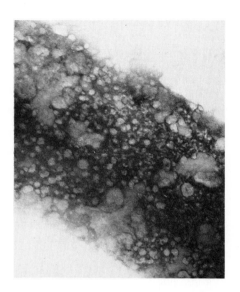

Fig. 8. Suspension of purified GBM stained with enzyme-labeled antibody. The stain density is significantly less than that obtained in NBM (there is only partial cross-reactivity with GBM). The pores are electron lucent. Unidentified areas in GBM have high antigenicity (black stain). Magnification, X 90,000.

Fig. 9. Positively stain-ed suspension of GBM. The porous structure is obvious, occasional communications be-tween the pores are apparent (PTA 1%, pH 7.0). Magnifi-cation, X 80,000.

DISCUSSION

The total thickness of the three layers of murine GBM is 1100-1600 Å (5,19,22). Each layer forms one-third of this total thickness. The purified GBM used in our studies had a thickness between 450 to 550 Å and the staining character-istics of lamina densa. These observations suggest that both lamina rara were solubilized during the purification procedure. In addition, any other soluble components present in the lamina densa of GBM (25,26) could have been removed in the purification procedure. In other words, our findings apply only to the insoluble backbone of BM.

With these reservations in mind, our studies confirm
(17,22) the fibrillar structure of lamina densa (Figs. 4, 5).
More importantly, they establish that within this dense
fibrillar network there is a system of channels or pores
obliquely traversing the GBM. The fact that the pore radius
demonstrated with antibody staining is twice that demonstrated
by negative staining can be explained if we assume that the
inner wall of the pore is non-antigenic (Fig. 10). The
dimensions given by negative staining include this non-antigenic portion. We propose, therefore, that the average radius
of the pores is 70 Å in murine GBM.

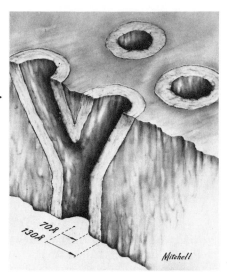

Fig. 10. Diagrammatic
interpretation of purified GBM
structure based on the results
of negative and "positive"
staining. The pore radius as
seen by negative staining 70 Å.
The pore's radius by immuno-
histochemical staining is
130 Å, thus there is a non-
antigenic 60 Å layer surround-
ing the pore (light area in
diagram), surrounded by anti-
genic areas (darker areas in
diagram).

There are two reasons why these relatively large pores
have not been clearly visualized before. A perfectly ortho-
gonal section would be required to visualize them in ultrathin sections. Another factor that may play a role is the
demonstration by Westberg and Michael (39) that human GBM
contains albumin, IgG and fibrinogen. If these (and maybe
other) proteins are trapped within the GBM pores, it seems
probable that when tissues are fixed for electron microscopy,
these proteins may be cross-linked to GBM and consequently
obliterate the pores.

The pores we are describing should permit the passage of molecules as large as ferritin (MW = 482,000, molecular radius, 61 Å) (40,41) through the basement membrane. In fact, studies using ferritin as a tracer showed that this molecule, although retarded within the basement membrane, reaches the urinary space (5,15). Furthermore, recent studies by Rennke *et al.* (18) indicate that cationized ferritin (pI = 8.8) readily passes through GBM. Therefore, the dimensions obtained from our morphological studies are not in disagreement with the data obtained from the tracer studies.

In summary, these studies demonstrated the presence of pores in the insoluble backbone of glomerular and neoplastic basement membrane, that fall in the range required to explain glomerular filtration. Although these studies do not settle the question about the location of the glomerular filter, they provide a structural basis for filtration by basement membrane.

ACKNOWLEDGEMENTS

This work was supported in part by grants AM-15663, CA-15823, and CA-13419 from the National Institutes of Health and grant PDT-23Q from the American Cancer Society.

REFERENCES

1. Meneffee, M.G. and Mueller, C.B., *IN* "Ultrastructure of the Kidney", Academic Press, New York, 1967.
2. Moffat, D.B., *The Mammalian Kidney,* pp. 70-100, Cambridge University Press, Cambridge, England, 1975.
3. Arthurson, G., Groth, T. and Grotte, G., *Clin. Sci.* 40: 137-158, 1971.
4. Caulfield, J.P. and Farquhar, M.G., *J. Cell Biol.* 63:883-903, 1960.
5. Farquhar, M.G. and Palade, G.E., *J. Biochem. Biophys. Cytol.* 7:297-304, 1960.
6. Gekle, D., V. Bruchhausen, F. and Fuchs, G., *Pflügers Arch.* 289:180-190, 1966.
7. Davson, H., *A Textbook of General Physiology,* pp. 845-848 Williams & Wilkins, Baltimore, 1970.
8. Giebisch, G., Lauson, H.D. and Pitts, R.F., *Amer. J. Physiol.* 178:168-176.
9. Huhme, B. and Hordwicke, J., *Clin. Sci.* 34:515-529, 1968.
10. Pappenheimer, J.R., *Physiol. Rev.* 33:387-423, 1953.

11. Pappenheimer, J.R. and Soto-Rivera, A., *Amer. J. Physiol.* 152:471-491, 1948.
12. Wallenius, G., *Acta Soc. Med. Opsali* 59, Suppl. 4, 1954.
13. Winne, D., *Pflü. Arch. ges. Phys.* 283:119-136, 1965.
14. Hall, V., *Amer. Heart J.* 54:1-9, 1957.
15. Karnovsky, M.J. and Ainsworth, S.K., *Adv. Nephrol.* 2:35-60, 1973.
16. Kurtz, S.M. and McManus, J.F.A., *Amer. Heart J.* 58:357-371, 1959.
17. Latta, H., *J. Ultrastruct. Res.* 32:526-543, 1970.
18. Rennke, H.G., Cotran, R.S. and Venkatachalam, M.A., *J. Cell Biol.* 67:638-646, 1975.
19. Rhodin, J.A.G., *Exp. Cell Res.* 8:572-574, 1955.
20. Venkatachalam, M.A., Karnovsky, M1J., Fahimi, H.D. and Cotran, R.S., *J. Exp. Med.* 132:1153-1167, 1970.
21. Venkatachalam, M.A., Cotran, R.S. and Karnovsky, M.J., *J. Exp. Med.* 132:1168-1180, 1970.
22. Yamada, E., *J. Biophys. Biochem. Cytol.* 1:551-566, 1955.
23. Graham, C.R. and Karnovsky, M.J., *J. Exp. Med.* 124:1123-1133, 1966.
24. Martinez-Hernandez, A., Nakane, P.K. and Pierce, G.B., *Amer. J. Path.* 76:389-399, 1974.
25. Martinez-Hernandez, A. and Pierce, G.B., *J. Cell Biol. Suppl. 67,* 263a, 1975.
26. Martinez-Hernandez, A. and Pierce, G.B., *Fed. Proc.* 35: 2803a, 1976.
27. Martinez-Hernandez, A., Fink, L.M. and Pierce, G.B., *Lab. Invest.* 34:455-462, 1976.
28. Pierce, G.B., Midgley, A.R., Sri Ram, J. and Feldman, J.D., *Amer. J. Path.* 41:549-566, 1962.
29. Pierce, G.B. and Nakane, P.K., *Lab. Invest.* 17:499-514, 1967.
30. Pierce, G.B. and Dixon, F.J., *Cancer* 12:584-589, 1957.
31. Lowry, O.H., Rosebrough, N.J., Farr, A.L. and Randall, R.J., *J. Biol. Chem.* 193:265-275, 1951.
32. Krakower, C.A. and Greenspon, S.A., *Arch. Path.* 51:629-639, 1951.
33. Heschemeyer, R.H. and Meyers, R.J., *IN:* "Principles and Techniques of Electron Microscopy", Van Nostrand Reinhold Co., New York, pp. 101-147, 1972.
34. Locke, M. and Krishman, N., *J. Cell Biol.* 50:550-557, 1971.
35. Reynolds, E.S., *J. Cell Biol.* 17:208-213, 1963.
36. Martinez-Hernandez, A., Merrill, D.A., Naughton, M.A. and Gecsy, C., *J. Histochem. Cytochem.* 23:146-148, 1975.
37. Nakane, P.K. and Kawaoi, A., *J. Histochem. Cytochem.* 22: 1084-1091, 1974.

38. Nakane, P.K. and Pierce, G.B., *J. Histochem. Cytochem.* 14:929-931, 1967.
39. Westberg, N.G. and Michael, A.F., *Biochemistry* 9:3837-3846, 1970.
40. Farrant, J.L., *Biochim. Biophys. Acta.* 13:569-576, 1954.
41. Haggis, G.H., *J. Mol. Biol.* 14:598-602, 1965.

ELASTIC SYSTEM FIBERS AND BASEMENT LAMINA

Gerson Cotta-Pereira

State University of Rio de Janeiro
Rio de Janeiro, Brazil

F. Guerra Rodrigo

Gulbenkian Institute of Science
Oeiras, Portugal

SUMMARY: The use of tannic acid-glutaraldehyde fixed
specimens allowed the demonstration of the ultrastruc-
tural patterns of the elastic system fibers. These
fibers (oxytalan, elaunin and elastic fibers) have the
same microfibrillar component differing only in the
amount of elastin. In human dermis, the elastic system
fibers are distributed in three layers from the surface
to the depth. The superficial fibers (oxytalan) are
oriented perpendicularly to the dermo-epidermal junc-
tion and anchor the basement lamina. They depart from
a plexus of elaunin fibers parallel to the dermo-
epidermal junction and connected with elastic fibers
in the deepest layers of the dermis. In ciliary zonule,
the zonular fibers were identified as oxytalan fibers
connecting the lens capsule to the ciliary body. In
both situations, the role of anchorage of oxytalan
fibers in basement lamina was suggested.

INTRODUCTION

In previous work (1), we proposed that the connective tissue fibers can be grouped in two main systems: the collagen and the elastic systems. The connective fibers formed by molecular aggregations of tropocollagen are concerned with the collagen system, while elastic system fibers have a common unit: the tubular microfibril 10 to 12 nm in diameter.

In the light microscope the elastic system fibers (oxytalan, elaunin and elastic fibers) are classified by their different tinctorial affinities and, by electron microscopy, their identification is possible after using the tannic acid-glutaraldehyde fixation. Ultrastructurally, with this technique, the elastic fibers are observed as being formed by bundles of tubular microfibrils peripherically situated around a homogeneous, abundant and dense amorphous material, identified as elastin. This protein is responsible for the elastic properties of the elastic fibers. Analogously, the elaunin fibers are formed by bundles of similar microfibrils intermingled with patches of dense amorphous material, whereas in the oxytalan fibers we visualize only the bundles of microfibrils (2,3).

In the present study we report the interactions between the elastic system fibers and the basement lamina observed in the dermo-epidermal junction and forming the lens capsule.

MATERIALS AND METHODS

Fragments of human skin biopsies and fragments of rat eyeballs were fixed in neutral 10% formaldehyde for light microscopical studies or in a solution containing 3% glutaraldehyde and 0.25% tannic acid in Millonig buffer (pH 7.3) for electron microscopical studies (2).

After formaldehyde fixation, the paraffin sectioned material was submitted to several staining methods in order to reveal the elastic system fibers: Verhoeff's iron hematoxylin, Weigert's resorcin-fuchsin, Gomori's aldehyde fuchsin and Unna-Tanzer's orcein with or without previous oxidation (4,5,6).

The glutaraldehyde fixed specimens were postfixed in 1% osmium tetroxide and embedded in Epon. Ultrathin sections were studied in an Elmiskop IA, at 80 kV, after staining with uranyl acetate and lead citrate.

RESULTS AND DISCUSSION

   In human skin the elastic system fibers are distributed
in three interconnected layers from the surface to the depth.
By light microscopy we observed that the most superficial
fibers, the oxytalan fibers, are oriented perpendicularly to
the dermo-epidermal junction and depart from a plexus of
elaunin fibers parallel to the surface. This plexus is
connected with the elastic fibers present in the deepest
layers of the dermis (Figure 1).

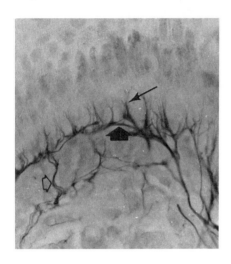

*Fig. 1. Human skin. Note
in this photomicrograph the
oxytalan fibers (thin arrow)
departing from elaunin fiber
plexus (thick arrow). Elastic
fibers (open arrow) are seen
deeply in the dermis and connected with the elaunin plexus.
Weigert's resorcin-fuchsin
after oxidation. (X 500).*

   Ultrastructurally, the oxytalan fibers appear to join
the basement lamina and the extremities of their microfibrils
are intermingled with the dense filamentous material of the
basement lamina (Figure 2).
   In previous work, a functional significance of the
architectural distribution of the elastic system fibers in
the human dermis was proposed (3). The oxytalan and the
elaunin fibers, the first lacking elastin and the second having small amounts of this protein, would provide resistance
to mechanical stress, thus contributing to the adhesion of
the epidermis to the dermis. On the other hand, the elastic
fibers, having elastin in abundance, would contribute to the
elasticity of the deepest layer of the dermal elastic system,
acting by absorbing shock due to stretch or compression.

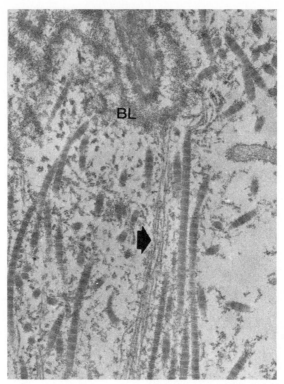

Fig. 2. Human skin. Electron
micrograph of dermo-epidermal junction
showing oxytalan fiber (arrow) binding
the basement lamina (BL). Tannic acid-
glutaraldehyde fixation. (X 40,800).

If this is correct, the dermal elastic system would
anchor the epidermis to the dermis using the basement lamina
as an intermediary structure. The dermal face of the basement lamina is joined to the dermal elastic system while its
epidermal face is known to be connected to the basal cells
of the epidermis by hemdesmosomes.

Consequently, the basement lamina may be representing,
despite other functions, a factor of anchorage between the
epidermis and the dermis. In fact, in the places where
oxytalan fibers bind the basement lamina it was observed that
it is pulled in the direction of the dermis. The cytoplasmic
processes of the basal cells accompany the basement lamina
(Figure 2).

In several tissues the oxytalan and elaunin fibers have been described as being structures that provide mechanical support, as in periodontium, epineurium, perineurium, adventitia of blood vessels (5,7), tendons, cartilage, etc. (6).

Additional evidence of this role suggested for the oxytalan fibers is the fact that they can be identified as forming the ciliary zonule in the eye, binding the lens to the ciliary body. By light microscopy, the zonular fibers have the same tinctorial characteristics of the oxytalan fibers (Figure 3). They are revealed by Weigert's resorcin-fuchsin, Gomori's aldehyde fuchsin or Unna-Tanzer's orcein staining methods only after oxidation. Biochemically, it has been demonstrated that zonular microfibrils and elastic system microfibrils have the same susceptibility to enzymatic digestion (8). Ultrastructurally, the zonular microfibrils are 10 to 12 nm in diameter, similar to elastic system microfibrils.

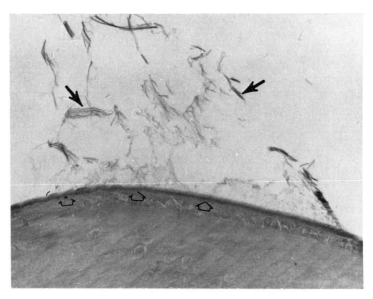

Fig. 3. Rat eyeball. Photomicrograph showing zonular fibers (arrows) anchoring the lens capsule (open arrows). Weigert's resorcin-fuchsin after oxidation. (X 500).

The zonular microfibrils are observed intermingled with the lens capsule filamentous material in the equatorial zone (Figure 4).

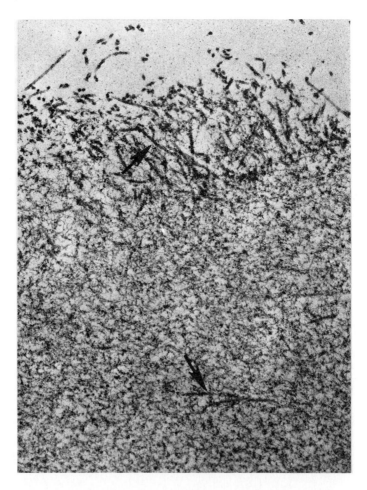

Fig. 4. Lens capsule. Electron micrograph showing zonular microfibrils (arrows) intermingled with lens capsule filamentous material. Tannic acid- glutaraldehyde fixation. (X 48,000).

Finally, these morphological evidences make us suggest that in the dermo-epidermal junction and in the lens capsule (basement lamina), by their connections with elastic system fibers, the basement lamina performs a function of anchorage.

ACKNOWLEDGEMENTS

The authors wish to thank Mr. S. T. Ribeiro Filho for his criticism and Mr. J. C. Lemos and Mrs. Maria Tereza Plantier for their technical assistance.

REFERENCES

1. Cotta-Pereira, G., Rodrigo, F. G. and David-Ferreira, J.F., *Proc. 1st Int. Conf. Elastin and Elastic Tissue,* Utah, 1976.
2. Cotta-Pereira, G., Rodrigo, F.G. and David-Ferreira, J.F., *Stain Technol.* 51:7, 1976.
3. Cotta-Pereira, G., Rodrigo, F.G. and Bittencourt-Sampaio, S., *J. Invest. Dermatol.* 66:143, 1976.
4. Pearse, A.G.E., *Histochemistry, Theoretical and Applied,* 3rd edition, 1, J.A.A. Churchill Ltd, London, p. 620, 1968.
5. Fullmer, H.M. and Lillie, R.D., *J. Histochem. Cytochem.* 6:425, 1958.
6. Gawlik, Z., *Folia Histochem. Cytochem.* 3:233, 1965.
7. Fullmer, H.M., *Science* 127:1240, 1958.
8. Raviola, G., *Invest. Ophthalmol.* 10:851, 1971.

C. Development and Differentiation

ROLE OF BASEMENT MEMBRANES

IN DEVELOPMENT AND DIFFERENTIATION

Elizabeth D. Hay

Harvard Medical School
Boston, Massachusetts

*SUMMARY: The basement membrane that lies under
embryonic epithelia and surrounds muscle is a cell
exoskeleton composed of a dense, collagenous mat,
which is probably attached to the plasmalemma;
this mat contains an inner and outer layer of glycosaminoglycan particles in all embryonic tissues in
which it has been studied by ruthenium red-osmium
fixation and appropriate enzyme digestion. Basement
membranes serve as substrata for the various embryonic mesenchymal cells that migrate under and around
the epithelial germ layers and epithelial organs.
Via their basement membranes, epithelial cells influence the polymerization of underlying collagen
fibrils. Basement membranes probably trap ions and
other molecules and, by this filtering function,
control intracellular environments. These exoskeleton sheaths undoubtedly serve many additional
mechanical and physical functions and it has recently
been shown that the ECM molecules which they contain
can interact with the cell surface and actually influence epithelial synthetic events. Basement membranes in their relation to embryonic cells and adjacent extracellular matrices show a flexibility and
adaptibility during morphogenesis that will challenge
our imagination for some time to come.*

119

The various cells of the vertebrate organism are cloaked
in glycoprotein surface coats which, in the case of epithelia,
vary in their composition from the apical to the lateral to
the basal cell surfaces. Termed the glycocalyx by Bennett
(1) and the greater membrane by Revel and Ito (2), such cell
coats stabilize and protect the outer leaflet of the plasmalemma of mesenchymal as well as epithelial cells. Recalling
the chitinous and collagenous exoskeletons of arthropods and
annelids (3), we suggest the name *cell exoskeleton* for the
glycoprotein cloak that surrounds the vertebrate cell.

The chemistry and structure of the basement membrane
(basal lamina) have been covered elsewhere in this volume.
I wish to emphasize here that in the embryo this cell exoskeleton contains a well developed sheet of glycosaminoglycan (GAG) particles (Figs. 1 and 2) in both its inner and
its outer layers (4,5). Since GAG does not stain conspicu-
ously with routine preparative methods for electron microscopy, GAG might be a principal component of the internal and
external lamina rara of the adult basement membrane as well,
but this question requires further study.

In the embryonic cornea, the GAG particles (which are
probably proteoglycan) of the basal lamina measure 100-200 A
in diameter following ruthenium red-osmium fixation (4,5).
It is possible that the component proteoglycan is aggregated
or the globular form exaggerated by this fixative-stain (6).
Overstaining obscures the particles (6,7). The corneal and
lens epithelia are producing chondroitin and heparan sulfates
(8) in the period illustrated here (Figs. 1-3), but it would
be impossible to explain the great regularity of the arrangement of the GAG particles (4) or their confinement to the
lamina rara by the theory that this GAG is merely in transit
through the basal lamina. Enzyme digestion establishes that
chondroitin sulfate is the major GAG component of the particles. The particles are sensitive to chondroitinase and
testicular, but not streptococcal, hyaluronidase (4,5).

After enzyme digestion to remove chondroitin sulfate,
proteinaceous filaments remain in the basement membrane which
may be collagenous; such filaments seem to attach to GAG-rich
plaques in the outer leaflet of plasmalemma (Fig. 3). The
plaques resist digestion by both neuraminidase (9) and
testicular hyaluronidase (5) and may be heparan sulfate (5,
10). On the cytoplasmic side of the basement membrane in
such regions, microfilaments of the cytoskeleton may be
attached to the plasmalemma (11). In older corneas, hemi-
desmosomes develop along the epithelial basal plasmalemma
which seem to serve as attachment points between cytoplasmic
filaments and the basement membrane (12).

The functions of the basement membrane in development
and differentiation are diverse. Some are probably similar
to those of the GAG and collagen of the general extracellular
matrix (ECM), while others undoubtedly reflect the unique
structure of the basement membrane as a continuous sheet or
boundary membrane (13) separating epithelium and muscle from
the connective tissue proper. Using selected examples of
embryonic material, let us consider possible morphogenetic
functions of the internal surface of the lamina, the lamina
proper and the external (juxtaepithelial) surface, in that
order.

The Basement Membrane as a Substratum for Cell Migration

The internal surfaces of basement membranes in the early
embryo may serve as substrata for the migration of the primary mesenchymal cells from the primitive streak (Fig. 4),
for macrophages in the developing eye (Fig. 5), and for
secondary mesenchymal cells derived from the somite (11) and
neural crest (14-20). Examples of migration of cells on
basement membranes in the adult are seen in epidermal wound
healing and in muscle regeneration (21).
Trelstad, Hay and Revel (17,18) suggested that the
mesenchymal cells migrating away from the primitive streak
contact patches of developing basal lamina on the basal
surfaces of the epithelial germ layers, the epiblast and
hypoblast; filopodia that adhere to basal lamina (Fig. 4,
inset) may contract and propel the cell forward, whereas
filopodia that contact bare epithelial plasmalemma form
junctions that may inhibit motility. Thus, the basement
lamina of the epiblast and hypoblast could provide a form of
contact guidance (19) for the migrating mesenchymal cells.
It would be interesting to determine whether or not the
epithelial exoskeleton specifically directs mesenchymal cells
to different locations in the definitive mesoderm.
The developing cornea provides two good examples of
migration of cells on a basal lamina (the lens capsule).
When the developing lens pinches off the overlying ectoderm,
macrophages migrate between the basal laminas of the lens
and the ectoderm to clean up debris (stage 18, Fig. 5). Subsequently, after the epithelium has secreted the primary
corneal stroma (stage 22, Fig. 5), macrophages migrate between
the corneal stroma and the lens capsule (12,20). They are
followed by mesenchymal cells (endothelium, stage 25, Fig. 5)
that seem to use both the internal surface of the corneal
stroma and the lens capsule as substrata (14,20).

Figs. 1-3. Electron micrographs illustrating the structure of the corneal epithelial basement membrane in 3 day old chick embryos fixed in ruthenium red-osmium (5). The basal lamina, including the external lamina rara, is a filamentous sheet about 1000 A wide which is closely applied to the cell surface. In contrast, the external coats of mesenchymal cells, viewed after a similar staining procedure, are thin and ill-defined (11). Particles 100-200 A in diameter containing glycosaminoglycan (GAG) are spaced 600-700 A apart (4) along the inner and outer surfaces of the central dense component of the basal lamina (open arrows, Figs. 1,2). These particles are seen to good advantage in the tengentially sectioned, thick basal lamina of the lens, the lens capsule (tan, Fig. 1). The particles are digested by testicular hyaluronidase (Fig. 3). Dense anion-rich plaques associated with the plasmalemma (curved arrows) resist enzymatic digestion. Between the plaques, the epithelial plasmalemma is poorly preserved (X, Fig. 2). GAG particles in the stroma are 200-400 A in diameter (short, solid arrows, Fig. 1). Figure 1, X 60,000. Figures 2 and 3, X 92,000. From Hay and Meier (5), copyright Rockefeller Press, N.Y.

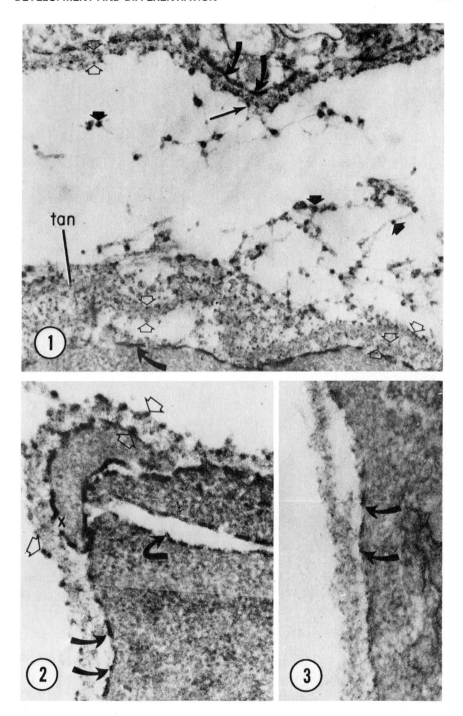

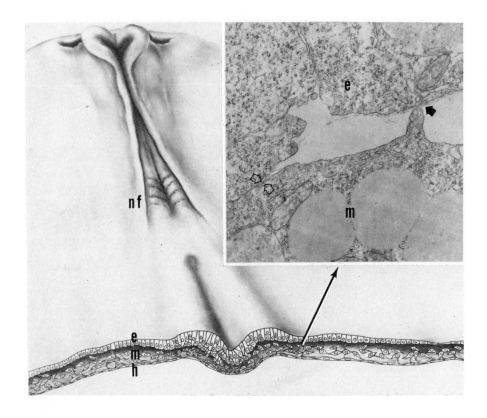

Fig. 4. Primary mesenchymal cells (m) migrating
from the primitive streak of a stage 8 chick embryo
contact the epiblast (e) and the hypoblast (h). Electron
micrographs of the contact region show that filopodia on
the leading edge of mesenchymal cells touch the developing basal lamina of the epiblast (solid arrow, inset).
Where no basal lamina is present, close apposition of
cell membranes (open arrows, inset) may impede motility
(17). In the drawing of the embryo, the neural folds
can be seen to be forming over the developing somites
cephalad to the primitive streak (nf). Drawing from
Hay (18), copyright Williams and Wilkins, Baltimore.
Inset, X 14,000, from Trelstad et al. (17), copyright
Academic Press, N.Y.

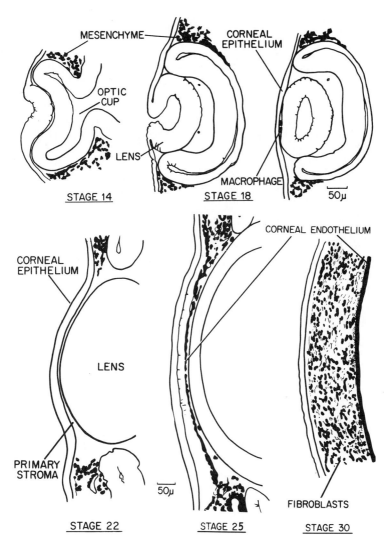

Fig. 5. *Camera lucida drawings showing formation of the lens placode (stage 14), lens (stage 18) and cornea (stages 22-30) of the chick embryo. The primary corneal stroma is seen in electron micrographs at stage 22 to consist of 25 orthogonally arranged layers of collagen fibrils. The endothelium migrates between the stroma and the lens capsule and then fibroblasts invade the stroma. From Hay and Revel (12), copyright S. Karger, Basel.*

The mesenchymal cells may show a stronger adhesion to the stroma (14) than to the lens capsule (Fig. 6), but experiments (20) have shown that the mesenchymal cells readily migrate on isolated lens capsule and that they give rise to a normal, monolayered endothelium only when lens is present.

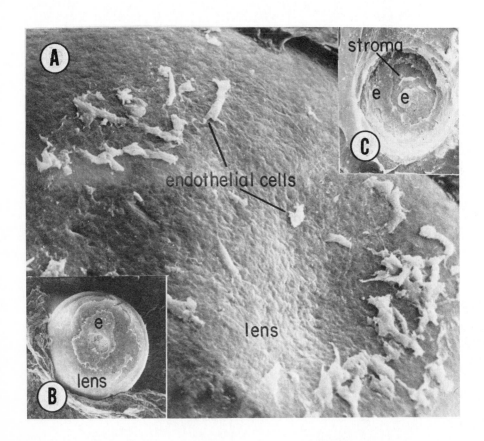

Fig. 6. Scanning electron micrographs showing migrating endothelial cells on the surface of the embryonic chick lens capsule. Endothelial cells (e) may stick to the underlying lens (inset B) or the overlying stroma (inset C) when the lens and stroma are pulled apart at this time (B,C, stage 25). The larger micrograph (A) shows migrating endothelial cells which are not yet confluent (stage 23). Micrograph A, X 450. From Bard et al. (20), copyright Academic Press, N.Y.

Role of the Basement Membrane in Polymerization of
    Underlying ECM

Tissues which lie on or are surrounded by basement membranes must continuously remodel these membranes and the adjacent ECM in order to grow. Bernfield (this volume) suggests that the surrounding mesenchyme lyses the basement lamina at the ends of the branches of growing glands and it is obvious that the skeletal muscle lamina has to be removed and replaced continuously for a fiber to increase in girth. In order for an outer epithelial layer and the adjacent connective tissue to expand in a growing embryo, the basement membrane must move outward with the epidermis and increase its lateral dimensions at the same time.

Let us examine a specific example of the latter type of growth. The epidermis of the amphibian larva rests on a basement lamella consisting of a basement lamina next to the epidermis and about 25 subjacent layers of collagen fibrils, orthogonally arranged. In transmission electron micrographs, the basal lamina can be seen to be connected to the epithelial plasmalemma and the underlying ECM by small filaments, which are probably collagenous in nature (open arrows, Fig. 7). Within the internal border of the basal lamina, small fibrils (curved and straight arrows, Fig. 7) occur that may represent new collagen in the process of polymerization, as suggested for chick cornea by Trelstad (22) and for fish skin by Nadol, Gibbons and Porter (23).

In autoradiographs, Hay and Revel (24) found that new ^{3}H-proline labeled protein (largely collagen) secreted by the amphibian epidermis and underlying fibroblasts is deposited mainly in or under the basement membrane (Fig. 8). However, some collagen is added to the deep layers of the dermis where the collagen fibrils continue to increase in size. In pulse experiments, it can be shown that new, unlabeled collagen is subsequently deposited under the epidermis. Thus, the basal lamina and epidermis are progressively pushed away from the collagen that was produced in the earlier period. Finally, the initially labeled band of collagen becomes the innermost surface of the lamella where it is removed, presumably by phagocytes (25).

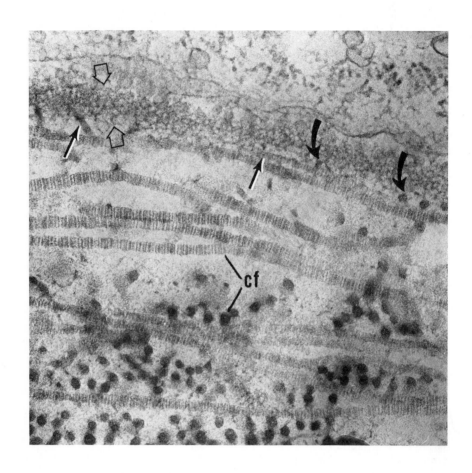

Fig. 7. Electron micrograph of the base of an epidermal cell (top) and part of the basal lamella of a salamander larva. The basal lamella, an embryonic dermis, consists of a basal lamina (open arrows) and about 25 orthogonally arranged layers of collagen fibrils, several of which appear here (cf). On the internal border of the lamina, new fibrils are seen in cross section (curved arrows) and in longitudinal or oblique section (straight arrows). X 100,000. From Hay (25), copyright Academic Press, N.Y.

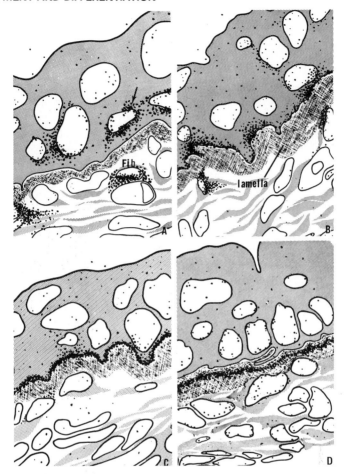

Fig. 8. Camera lucida drawings of autoradiographs showing amphibian larval skin fixed one hour
(A), 2 hours (B), 4 hours (C) and 4 days (D) after
a single (pulse) intraperitoneal injection of ^3H-
proline at time 0. At one hour (A), much of the
epidermal label (black dots) is juxtanuclear, associated with secretory organelles (24). Labeled protein
(largely collagenous) moves to the basal cytoplasm
(B) in small vesicles (24) where it is discharged into the basal lamina (C). Deposition of new, unlabeled
collagen after the initial radioactive pulse results
in a translocation inward of the labeled band of collagen (D). The fibroblasts also synthesize proline-
rich protein (Fib. A) which is secreted into the
lamella to accumulate largely in the juxtaepithelial
area. X 600. From Hay (25), copyright Academic Press,
New York.

Filtering Role of the Basement Membrane of
   Differentiating Cells

The filtering role of the basement membrane of the
glomerulus has received a good deal of emphasis at this
symposium. Konigsberg and Hauschka (26) mentioned the
possible effect of the extracellular matrix in controlling
access of nutrients or other substances to developing cells,
but not much attention has been paid to the manner in which
the basement membrane might filter incoming ions and otherwise monitor the composition of the intracellular environment
in embryonic cells. I would like to emphasize the possible
importance of such a function with a hypothetical example.

The neural tube of most vertebrate embryos arises from
the neural folds (nf, Fig. 4), which rise up, move to the
midline and fuse into a tube, largely as a result of the
interaction of cytoplasmic microtubules and microfilaments
(27). Prior to neural tube formation in regions of the
embryo occupied by the primitive streak, little sulfated
GAG is produced (28,29). The notochord forms anterior to
the primitive streak from cells that invaginated from the
regressing primitive node; the ECM produced by the notochord bisects the embryo, an event which probably stimulates
somite formation (30). GAG produced in part by the notochord and somites accumulates in the basement membrane under
the forming neural plate (Fig. 9). The intracellular environment in the neural place could be regulated by this anion-
rich basement membrane; changing intracellular ion concentrations might influence microtubular polymerization (31) and,
consequently, neural fold formation. Ions have, interestingly,
been implicated in neural induction by Barth and Barth (32).

While it must be stressed that this discussion of a
basement membrane "filtering" mechanism in the embryo is
highly speculative, I hope it will serve to attract attention
to possible morphogenetic functions of the anionic components
of cell exoskeletons.

Influence of the Cell Exoskeleton on ECM Synthesis

The morphology of the basement membrane is compatible
with the idea that its juxtacellular component interacts
directly with the plasmalemma (Figs. 1-3). As we have already
indicated, specializations occur along this surface which
probably attach the epidermis to underlying connective tissue
(11,24). In this final section, evidence will be reviewed
which suggests that, in addition to such mechanical functions,

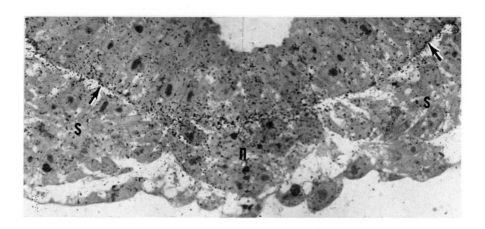

Fig. 9. An autoradiograph of the neural fold region of a stage 8 chick embryo (nf, Fig. 4) which received $^{35}SO_4$ 2 hours prior to fixation. The secreted product, which is probably largely chondroitin sulfate (5), accumulates (arrows) under the neural ectoderm in the newly formed basement membrane that separates it from the notochord (n) and developing somites (s). X 725. From Hay (29), copyright American Society of Zoologists.

interaction of the collagenous exoskeleton with the plasmalemma influences cell metabolism, particularly the synthesis and secretion of ECM. Collagen secreted by fibroblasts was shown by Konigsberg and Hauschka (26) to stabilize muscle differentiation *in vitro* and both collagen and GAG promote chondrogenesis (33-36). Do basement membranes influence the metabolism of the overlying epithelium in a similar fashion?

The corneal epithelium of the 5 day chick embryo (stage 25, Fig. 5) resides on a typical basement lamina containing the GAG particles described above and a filamentous component believed to represent collagen in large part (4,5). The epithelium is in the process of secreting GAG (8) and several types of collagen, including types I and II (37), into the underlying acellular corneal stroma. If the epithelium is divorced from its basal lamina by treatment with trypsin-collagenase or EDTA (ethylenediaminetetraacetic acid), the naked plasmalemma blebs angrily and continues to do so if the epithelium is placed on Millipore filter (38,39). Cultured on filter, glass or any variety of noncollagenous substrata, the epithelium fails to produce a new corneal stroma *in vitro*, although it synthesizes GAG and collagen at a low or "baseline" level (38-40).

When the isolated corneal epithelium is placed on a basal
lamina, the lens capsule, the basal plasmalemma smoothes out
(Fig. 10A) and the cells produce a facsimile of the original
corneal stroma, including patches of new basement lamina,
within 24 hours (Fig. 10B). Frozen-killed and NaOH-extracted
lens capsule is just as effective as living lens in promoting
the synthesis and secretion of corneal stroma (40). Indeed,
gels of pure type II collagen and rat tail tendon (type I)
collagen equally support epithelium production of corneal
stroma (40,41). All types of collagen tested stimulate a
2-3 fold increase in GAG and collagen synthesis over the baseline level (40). Pure GAG (chondroitin and heparan sulfate
synthesis over and above the stimulatory effect of pure collagen on GAG, but GAG has no effect on collagen synthesis
(42).

Since the lens capsule and other collagenous substrata
used in this study are not soluble under the conditions of
culture, Meier and Hay (40) concluded that the stimulatory
effect of collagen is transmitted by an interaction at the
epithelial cell surface. Proof of this hypothesis was sought
by growing the corneal epithelium on Millipore filter trans-
filter to lens capsule to prevent cell contact with the insoluble ECM; stroma production was prevented under these
circumstances (Fig. 10C). Nucleopore filter allows epithelial
cell processes to extend through its straight pores to contact
the underlying substratum; the epithelium thus in contact with
the stimulatory ECM produces stroma (Fig. 10D). Moreover,
radioactive collagen from such a substratum does not enter
the cells (41) and the stimulatory effect on collagen synthesis is directly proportional to the area of cell-ECM contact
(43).

Concluding Remarks

What emerges from this brief review, is the concept that
the basement lamina in embryonic tissues serves as a remarkable cell exoskeleton of multiple functions. On the one hand,
basement membranes can serve as substrata for the migration
of embryonic cells which recognize and respond to the GAG and
collagenous components without taking up permanent residence
thereon. On the other hand, more differentiated cells may
possess in their plasmalemma receptors for those components
of the ECM that enhance or stabilize their subsequent development in a particular direction. Corneal epithelium does
migrate over a lens capsule *in vitro* (44), but even as the

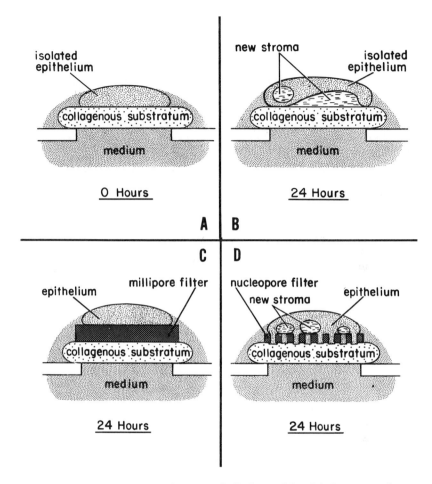

Fig. 10. When isolated 5 day old chick corneal epithelium is placed on a collagenous substratum, such as lens capsule, it smoothes out on the substratum (A) and within 24 hours produces a facsimile of the corneal stroma (B). If contact with the lens capsule is prevented by a Millipore filter, no stroma is produced (C). On the other hand, Nucleopore filter allows cell processes to traverse it to contact the underlying collagenous substratum and in this case the corneal epithelium produces stroma in vitro (D). From Hay (11), copyright Raven Press, New York.

peripheral cells continue to migrate, the relatively immobile
central epithelium attached to the lens capsule produces a
new corneal stroma under the influence of the collagen in the
capsule (38-41). Yet other cells, such as those of the
presumptive endothelium, can migrate on lens capsule without
forming a permanent attachment (14,20). Since the structure
of the lens capsule is the same, it is the cells that are
following different programs and their response is undoubtedly mediated in part via their distinctive cell surfaces.

The way the basement membrane serves as the platform
supporting the epidermis of the amphibian tadpole while
itself being constantly remodeled recalls a sign often seen
on downtown Boston stores: "Open for business during
renovation". Autoradiographic studies suggest that both new
basement membrane and new collagen fibrils are formed under
the epidermis which is itself constantly moving away from
the very same underlying stroma that gives it support (24,25).
Assuming there are receptors for collagen and other molecules
of the ECM in the epithelial plasmalemma (11,41), one must
postulate continual disruption and reforming of the cell-ECM
bonds as the epithelium grows. We need to explain both the
assembly (22-25) and the disassembly of basement membranes
which take place during embryonic business hours without
altering the structure and function of the lamina.

Finally, one must consider the role of the embryonic
basement membrane as a skeletal sheath that totally separates
the epithelium and muscle of the body from the connective
tissue proper. We discussed the obvious possibility that the
basement lamina may trap ions and filter other molecules,
thus influencing the epithelial intracellular environment,
but why are the glycoprotein coats of mesenchymal cells,
which may have a similar function, so puny by comparison?
All mesenchymal cells arise from epithelial tissues; epithelium, in a sense, is the more primitive tissue (18,25). There
is much left to be learned about the structure and function
of the embryonic extracellular matrices and it is probably
important that we begin to compartmentalize our thinking
about these cell exoskeletans. This conference, by focusing
our attention on a single class of ECM, the basement membrane,
has done much to start us in that direction.

REFERENCES

1. Bennett, H.S., *J. Histochem. Cytochem.* 11:14-23, 1963.
2. Revel, J.P. and Ito, S., *In:* "The Specificity of Cell
    Surfaces" (Davis, B. and Warren, K.B., eds.) Prentice-
    Hall, New Jersey, pp. 211-234, 1967.
3. Humphreys, S. and Porter, K.R., *J. Morph.* 149:33-51,
    1976.
4. Trelstad, R.L., Hayashi, K. and Toole, B.P., *J. Cell
    Biol.* 62:815-830, 1974.
5. Hay, E.D. and Meier, S., *J. Cell Biol.* 62:889-898, 1974.
6. Luft, J.H., *Anat. Rec.* 171:369-415, 1971.
7. Bernfield, M.R. and Banerjee, S.D., *J. Cell Biol.* 52:
    664-673, 1972.
8. Meier, S. and Hay, E.D., *Dev. Biol.* 35:318-331, 1973.
9. Hay, E.D., Hasty, D.L. and Kiehnau, K., *First Munich
    Symposium on the Biology of Connective Tissues.* In Press.
10. Kraemer, P.M., *Biochem.* 10:1437-1445, 1971.
11. Hay, E.D., *In:* "Cell and Tissue Interactions" (Burger,
    M.M. and Lash, J.W., eds.), Raven Press, New York, In
    Press.
12. Hay, E.D. and Revel, J.P., *Fine Structure of the Develop-
    ing Cornea,* S. Karger, Basel, 144 pp., 1969.
13. Low, F.N., *Anat. Rec.* 160:93-108, 1968.
14. Nelson, G.A. and Revel, J.P., *Dev. Biol.* 42:315-333,
    1975.
15. Bard, J.B.L. and Hay, E.D., *J. Cell Biol.* 67:400-418,
    1975.
16. Cohen, A.M., *J. Exp. Zool.* 179:167-182, 1972.
17. Trelstad, R.L., Hay, E.D. and Revel, J.P., *Dev. Biol.*
    16:78-106, 1967.
18. Hay, E.D., *In:* "Epithelial-Mesenchymal Interactions"
    (Fleischmajer, R. and Billingham, R., eds.), Williams
    and Wilkins, Baltimore, pp. 31-55, 1968.
19. Weiss, P., *Exp. Cell Res., Suppl.* 8:260-281, 1961.
20. Bard, J.B.L., Hay, E.D. and Meller, S.M., *Dev. Biol.*
    42:334-361, 1975.
21. Mauro, A., Shafig, S.A. and Milhorat, A.T., *Regeneration
    of Striated Muscle and Myogenesis,* Excerpta Medica,
    Amsterdam, 299 pp., 1970.
22. Trelstad, R.L., *In:* "Extracellular-Matrix Influences
    on Gene Expression", (Slavkin, H.C. and Greulich, R.,
    eds.), Academic Press, New York, pp. 331-340, 1975.
23. Nadol, J.B., Jr., Gibbons, J.R. and Porter, K.R., *Dev.
    Biol.* 20:304-331, 1969.
24. Hay, E.D. and Revel, J.P., *Dev. Biol.* 7:152-168, 1963.

25. Hay E.D., *In:* "The Epidermis", (Montagna, W. and Lobitz, W.C., eds.), Academic Press, New York, pp. 97-116, 1964.

26. Konigsberg, I.R. and Hauschka, *In:* "Reproduction: Molecular, Subcellular and Cellular", (Locke, M., ed.) Academic Press, New York, pp. 243-290, 1965.

27. Burnside, B., *Am. Zool.* 13:989-1006, 1973.

28. Manasek, F.J., *Curr. Top. Dev. Biol.* 10:35-102, 1975.

29. Hay, E.D., *Am. Zool.* 13:1085-1107, 1973.

30. Lipton, B.H. and Jacobson, A.G., *Dev. Biol.* 38:91-103, 1974.

31. Olmsted, J.B., *In:* "Cell Motility", (Goldman, R., Pollard, T. and Rosenbaum, J., eds.) Cold Spring Harbor Laboratory, N.Y., pp. 1081-1092, 1976.

32. Barth, L.G. and Barth, L.J., *Dev. Bicl.* 28:18-34, 1972.

33. Kosher, R.A. and Church, R.L., *Nature* 258:327-330, 1975.

34. Kosher, R.A. and Lash, J.W., *Dev. Biol.* 42:362-378, 1975.

35. Kosher, R.A., Lash, J.W. and Minor, R.R., *Dev. Biol.* 35:210-220, 1973.

36. Nevo, A. and Dorfman, A., *Proc. Nat. Acad. Sci., U.S.A.,* 69:2069-2072, 1972.

37. Linsenmayer, T.F., Smith, G.N. and Hay, E.D., *Proc. Nat. Acad. Sci., U.S.A., In Press.*

38. Dodson, J.W. and Hay, E.D., *Exp. Cell Res.* 65:215-220, 1971.

39. Dodson, J.W. and Hay, E.D., *J. Exp. Zool.* 189:51-72, 1974.

40. Meier, S. and Hay, E.D., *Dev. Biol.* 38:249-270, 1974.

41. Hay, E.D. and Meier, S., *Dev. Biol.* 52:141-157, 1976.

42. Meier, S. and Hay, E.D., *Proc. Nat. Acad. Sci., U.S.A.,* 71:2310-2313, 1974.

43. Meier, S. and Hay, E.D., *J. Cell Biol.* 57:190-213, 1973.

44. Hay, E.D. and Dodson, J.W., *J. Cell Biol.* 57:190-213, 1973.

THE BASAL LAMINA IN EPITHELIAL-MESENCHYMAL

MORPHOGENETIC INTERACTIONS

Merton R. Bernfield and Shib D. Banerjee

Stanford University School of Medicine
Stanford, California

ABSTRACT: Mouse submandibular epithelia require a
basal lamina to maintain lobular morphology and require mesenchyme for continued branching morphogenesis.
The mesenchyme is not involved in the organization,
synthesis or deposition of the lamina, since the
epithelium alone produces a lamina with identical biochemical and ultrastructural characteristics as the
lamina in intact glands. Epithelia stripped of a
lamina with hyaluronidase lose lobular morphology
during culture with mesenchyme, but loss of morphology
is prevented by delaying combination with mesenchyme
for 2 hr, during which a new lamina is deposited.
Thus, the mesenchyme may have a property deleterious
to epithelial recovery from hyaluronidase, but normally
involved in morphogenesis. In intact glands, newly
synthesized laminar glycosaminoglycan (GAG) accumulates
more rapidly at the distal ends of lobules than at the
lateral aspects and within clefts, the sites of greatest
amounts of histochemically-identified GAG. Pulse-chase
studies with $[^3H]$glucosamine and histochemistry following
inhibition of GAG secretion indicate that the distal ends
also lose laminar GAG more rapidly. This pattern is not
due to lobular growth, nor altered by inhibition of
proliferation or by inhibitors of collagen secretion and
cross-linking. However, the pattern occurs only in the

*presence of mesenchyme, indicating that the mesenchyme
is involved in the differential loss of laminar GAG.
Unbranches epithelia do not show site-specific differences in laminar GAG, but during morphogenesis, loss
becomes greater at regions where branching occurs and
lower at areas which do not branch and these sites
accumulate more laminar GAG. Thus, the mesenchyme may
influence changes in epithelial morphology by selective
remodeling of GAG within the basal lamina.*

The generation of structural form of several embryonic
epithelial organs, such as the lung, kidney, mammary and
salivary glands, occurs by a common sequence. Each organ
arises as a rounded bud consisting of a layer of epithelial
cells, and during development, the bud undergoes a distinctive pattern of folding and branching which results in a
morphology that is characteristic of the organ. This morphogenesis is completely dependent upon the close association of
a highly cellular loose connective tissue, or mesenchyme. In
the absence of mesenchyme, the epithelia fail to develop and
slowly lose their shape.

Between the epithelium and mesenchyme lies an extracellular matrix consisting of fibrillar collagen, amorphous
materials and a well-defined epithelial basal lamina (Fig. 1).
In mouse embryonic lung and submandibular glands, these
materials are not distributed uniformly over the epithelial
surface. Fibrillar collagen is in greater amounts in the
clefts between the lobules and on the stalk than at the distal
aspects of the lobules, the sites where further branching will
take place (1,2). The basal lamina completely encompasses the
epithelium, but is thinner and may be interrupted at the distal aspects of the lobules, where intimate contacts between
epithelial and mesenchymal cells are seen (3,4).

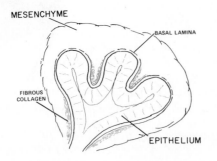

*Fig. 1. Distribution
of extracellular materials
on a 13-day mouse embryo
submandibular epithelium.
See text for description.*

The Embryonic Epithelial Basal Lamina

In addition to other components, the basal lamina of epithelia which undergo branching morphogenesis contains glycosaminoglycan (or GAG) (5). Ultrastructural studies of submandibular epithelia indicate that the GAG within the basal lamina is highly ordered and is intimately associated with the plasmalemma (6). Removal of the basal lamina from these epithelia with hyaluronidase is rapidly followed by deposition of a new basal lamina which is ultrastructurally identical with the pre-existing lamina. Labeling studies reveal that both the newly deposited and pre-existing lamina contain similar proportions of newly synthesized GAG; half of this GAG is hyaluronic acid, about a third is chondroitin-4-sulfate and the remainder is chondroitin 6-sulfate with traces of chondroitin (6).

Evidence for the involvement of the basal lamina in the morphogenesis of submandibular epithelia is derived from organ culture studies (Fig. 2) (7,8). Treatment of whole glands with highly purified collagenase and microdissection removes the mesenchyme as well as the fibrillar collagen and amorphous materials, but leaves the epithelia with an intact

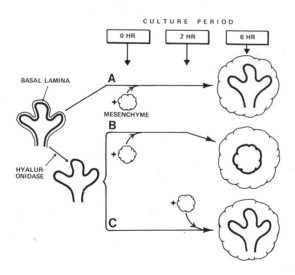

Fig. 2. *Morphogenetic effects of removal and replacement of the basal lamina from 13-day mouse embryo submandibular epithelia. See text for description.*

basal lamina. However, treatment of these isolated epithelia
for 10 min with 50-75 ng per ml of testicular hyaluronidase
completely removes the lamina. In culture combined with
mesenchyme, epithelia retaining the lamina maintain their
lobular morphology (Fig. 2, *see* A), while epithelia stripped
of the lamina lose their lobules and form a spherical mass of
cells (Fig. 2, *see* B). These results indicate that maintenance of lobular morphology is susceptible to removal of the
lamina. However, delaying recombination with mesenchyme for
two hours, during which the epithelium completely replaces
the lamina, prevents the loss of lobular morphology (Fig. 2,
see C). Thus, the mesenchyme may have a property or activity
which is eleterious to epithelial recovery from hyaluronidase
treatment, but which may be required normally for the morphogenetic effect of mesenchyme.

Because the lamina is apparently required for maintenance
of morphology, the mesenchyme may exert its morphogenetic
effect by influencing the lamina. However, mesenchymal cells
are not involved in the synthesis, deposition or organization
of the lamina because epithelia alone produce a lamina with
identical ultrastructural and biochemical characteristics as
the lamina in intact glands (6).

Turnover of Basal Laminar GAG

Following labeling of intact submandibular glands for
2 hours with ^{3}H-glucosamine, radioactivity is seen within the
lobules and at the epithelial surface (Fig. 4a). The label
at the epithelial surface is, at least in part, in basal
laminar GAG because enzyme treatments which remove the lamina
also remove the label and release authentic labeled GAG. The
laminar GAG label accumulates in a unique pattern: much more
GAG label is at the epithelial surfaces of the distal ends
of the lobules than within the interlobular clefts.

To clarify the laminar GAG labeling pattern, the uptake
of ^{3}H-glucosamine into ethanol-soluble precursor pools of the
lobules, of the base of the lobules (which contain the clefts)
and of the stalk was measured under labeling conditions
identical to those used in the autoradiographic experiments
(Fig. 3). Glucosamine was taken up into the lobules and the
stalk at a substantially greater rate than into the base. The
precursor pools in the lobules and stalk were nearly saturated
after 2 hours of labeling. However, the pool in the base did
not saturate until 6-7 hours of labeling, and at saturation,
the steady state size of the labeled precursor pool (cpm per
mg protein) in the base was substantially larger than in the
stalk and lobules (data not shown). These data provide an

explanation for the autoradiographic observations. A larger
steady state precursor pool and less rapid saturation of the
pool in the base would account for the small amount of laminar
GAG label seen in the interlobular clefts following 2 hours
of labeling.

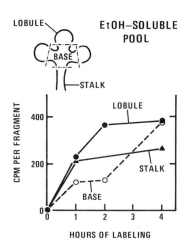

Fig. 3. Uptake of ^3H-
glucosamine into ethanol-
soluble pools. Intact submandibular glands were incubated in complete medium containing 50 μCi per ml ^3H-6-
glucosamine. At the indicated
times, the mesenchyme was removed by collagenase treatment
and microdissection. The
lobules and stalk were then
cut from the base as indicated
in the drawing, and the label
which was soluble in 70%
ethanol was determined for
each fragment.

Glands labeled for 2 hours with ^3H-glucosamine (Fig. 4a)
were chased in non-radioactive medium to assess the metabolism
of laminar GAG. After a 2 hour chase, the label at the distal
ends of the lobules had decreased, while the label at the
lateral aspects had increased. With further periods of chase,
epithelial surfaces which had labeled poorly during pulse became labeled, but those labeling heavily during the pulse lost
label. By 6 hours of chase (Fig. 4b), the great bulk of surface label was deep within the clefts, and after 16 hours of
chase very little surface label was remaining. This change in
the pattern of surface label indicates that soon after its
appearance, there is a rapid decline of laminar GAG at the
distal aspects of the lobules. Labeled GAG increases during
the chase along the lateral aspects of the lobules and within
the clefts, but with prolonged culture, this label also decreases.

Autoradiography after labeling of whole glands for 8 hrs
or longer, periods which result in steady-state labeling,
reveals substantially less label at the distal aspects of the
lobules than at the base of the lobules and within the clefts.
Thus, there is a decline in laminar GAG label at the distal

aspects of lobules even during continuous labeling.

The appearance of a change in distribution of GAG label could result from an increase in the size of lobules or from movement of the epithelial cells. However, ^{3}H-thymidine labeling followed by a chase clearly showed that the change in localization of the GAG label far exceeded the increase in lobule size and was not mimicked by the movement of thymidine-labeled cells. Furthermore, the change in distribution occurred in the presence of DNA synthesis inhibitors which abolished thymidine incorporation. Thus, the change in distribution of laminar GAG label during the chase could not be accounted for by increased lobule size, cell movement, or the effects of differential sites of cell proliferation.

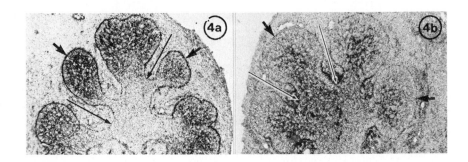

Fig. 4. ^{3}H-glucosamine autoradiography of 13-day mouse embryo submandibular glands. The label at the epithelial surface (arrows) is·susceptible to testicular hyaluronidase; short arrows denote the distal aspects of lobules, the long arrows denote the clefts.
4a. Localization of label after 2 hours of labeling with 50 μCi per ml ^{3}H-glucosamine. See text for description.
4b. Localization of label after labeling as in 4a and chasing for 6 hours in medium containing non-radioactive glucosamine. See text for description.

If the changes in labeled GAG distribution during a chase resulted from rapid turnover of laminar GAG, then preventing GAG deposition into the lamina might reveal a similar change in distribution of the amount of laminar GAG. This possibility was examined by using colchicine, which in other systems, rapidly and reversibly prevents the secretion, and reduces the synthesis of extracellular materials, both collaten and GAG (9,10). Treatment of glands with 2.5 x 10^{-7}M colchicine for

2 hours prevented glucosamine labeling of the epithelial surface, although substantial label accumulated within the epithelium. However, 2 hours following removal of the drug and
isotope, labeled GAG appeared on the epithelial surface in a
pattern identical to that on freshly labeled epithelia, indicating that the drug reversibly inhibited GAG deposition
into the lamina. This effect of colchicine did not result
from inhibition of procollagen secretion since pre-treatment
of glands for as long as 22 hours with inhibitors of procollagen secretion (azetidine-2-carboxylic acid or 3,4-
dihydroproline) did not affect the labeling of the lamina.

The amount of laminar GAG, assessed histochemically with
Alcian Blue, was examined during varying periods of incubation with colchicine. Prior to adding the drug, more histochemically-identified GAG is within clefts than at the distal
ends of the lobules (Fig. 5a). After 6 hrs in the drug, less
stain is on the lobules, but stain remains within the clefts.
After 12 hrs in colchicine (Fig. 5b), stain is nearly gone
from the epithelial surface, and the only stain remaining is
deep within clefts. This result indicates that the amount of
laminar GAG decreases during inhibition of GAG deposition
into the lamina.

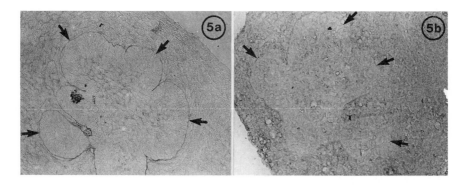

Fig. 5. Alcian Blue staining of 13-day mouse
embryo submandibular glands. Fixation was in Carnoy's
solution and staining was in solutions containing 1%
Alcian Blue, 0.1M MgCl$_2$, pH 2.5. The arrows denote the
epithelial surface at the distal aspects of the lobules.
5a. Gland stained immediately after explantation.
See text for explanation.
5b. Gland stained after 12 hours culture in 2.5 x
10^{-7}M colchicine. See text for description.

These studies strongly suggest that the GAG within the lamina is turning over at a rapid rate. While the relative rates of GAG deposition into the lamina at various sites are unknown, the most rapid turnover appears to be at the distal aspects of the lobules. Compared with the clefts, these sites show less histochemically-identified laminar GAG, a more rapid decline in this GAG during inhibition of new GAG deposition, and the GAG label at these sites disappears more rapidly during a chase. The lobules show a greater rate of precursor uptake and a smaller labeled precursor pool than the base, apparently resulting in the more rapid accumulation of labeled GAG at their surfaces.

Mesenchyme-induced Loss of Basal Laminar GAG

The results diagrammed in Figure 2 suggest that, following removal of the lamina, immediate recombination with mesenchyme inhibits epithelial recovery from the hyaluronidase treatment. Since the lamina is required for maintaining lobular morphology and since GAG in the lamina of intact glands is continually turning over, it is possible that the mesenchyme is degrading laminar GAG. This hypothesis was examined by labeling and chasing epithelia in the presence and absence of mesenchyme, and observing the distribution of surface label autoradiographically and measuring the amount of labeled GAG remaining with the tissue and appearing in the medium.
       Intact glands and collagenase-isolated epithelia were labeled with ^{3}H-glucosamine for 2 hours (Fig. 6). While epithelia labeled as intact glands had label predominantly at the distal aspects of the lobules, epithelia labeled in the absence of mesenchyme showed label over the entire epithelial surface. Epithelia from pre-labeled glands were isolated free of mesenchyme and these, as well as the epithelia labeled in the absence of mesenchyme, were chased for 5 hours either alone, or, recombined with fresh mesenchyme. Chasing in the absence of mesenchyme (Fig. 6, B & D) resulted in a nearly uniform distribution of label over the surface of the epithelia, regardless of the original distribution of surface label. However, when either type of pre-labeled isolated epithelia were recombined with mesenchyme and chased, (Fig. 6, A & C), GAG label decreased at the distal aspects of the lobules, but became prominent within the clefts, precisely duplicating the change in distribution of label previously observed in intact glands. Therefore, the decline of label at the distal aspects of the lobules is dependent on the presence of mesenchyme.

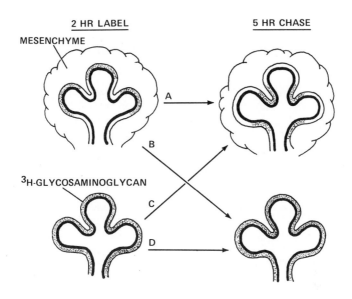

Fig. 6. *Effect of mesenchyme on the autoradio-graphic distribution of laminar GAG label. Labeling of intact and collagenase-isolated epithelia was for 2 hours (left column). Following collagenase treatment, the epithelia were chased in the presence and absence of mesenchyme for 5 hours (right column). See text for description.*

To measure the changes in amount of labeled GAG, epithelia-pre-labeled as intact glands were chased as in experiments A and B, Figure 6. In the presence of mesenchyme, there was a loss of labeled GAG from the tissue during chase, while labeled GAG continued to accumulate in the absence of mesenchyme (Figure 7).

Fig. 7. *Effect of mesen-chyme during a chase on the amount of labeled epithelial and medium GAG. Epithelia were labeled and chased as outlined for experiments A and B, Fig. 6. At the indicated times after labeling, authentic GAG in the tissue and medium were measured as previously described (6).*

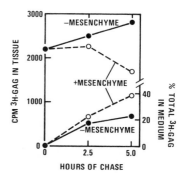

The Basal Lamina and Epithelial Morphogenesis

Collagen of a distinctive type is the major component of the laminae of mature epithelial organs and of certain embryonic epithelia, such as the parietal yolk sac (11). Where examined, these laminae contain no or trace amounts of GAG. On the other hand, the laminae of several embryonic epithelia which undergo branching morphogenesis contain GAG, and the amount of GAG (assessed histochemically) decreases with advancing development (12). Thus, the laminae of embryonic epithelia which undergo repetitive folding and budding appear to change during development from being GAG-rich to being GAG-poor, a sequence which may result in GAG-free laminae after the completion of morphogenesis. The rapidly successive morphological changes shown by branching embryonic epithelia might be more readily accommodated by a flexible and malleable GAG-rich laminar structure than by one composed principally of collagen. This speculation leads to the hypothesis that the laminae of epithelia which undergo branching morphogenesis during post-natal life, such as the mammary gland, may also be rich in GAG.

The basal surfaces of a number of embryonic organs studied as sequential developmental stages show a common pattern of histochemically-identified GAG and of labeled GAG accumulation following 2 hours of ^{3}H-glucosamine (Fig. 8) (12). Unbranched buds do not show site-specific differences

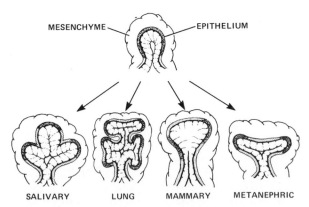

Fig. 8. Patterns of glycosaminoglycan distribution in several mouse epithelial organs. Areas of light stippling are the sites of greatest amount of GAG as revealed by Alcian Blue staining. Areas of dark shading are the sites of greatest accumulation of GAG label observed autoradiographically following 2 hours of ^{3}H-glucosamine labeling.

in either total or labeled laminar GAG. But with tissue-
specific changes in morphology, there is a change in the
distribution of total GAG and in the sites of rapid accumulation of labeled GAG. More total GAG is seen at the surfaces
of stalks, and at the lateral aspects of lobules and branches
than at their distal ends, the regions where subsequent
branching and lobulation will occur. In contrast, the
accumulation of GAG label becomes more rapid at the distal
ends of lobules and projections than at the morphogenetically
quiescent sites. These developmental changes are seen in
submandibular, sublingual, bronchial, metanephric and mammary
epithelia, but are not seen during the development of the
pancreas which develops distinctly (13).

Since the pattern of total GAG and of accumulation of
newly labeled GAG observed in multilobular submandibular
epithelia is due to continuous turnover of laminar GAG,
coupled with an apparently greater rate of GAG loss at morphogenetically active sites, it is possible that the analogous
patterns seen in the other epithelia arise by the same
mechanism. Thus, rapid laminar GAG turnover may be a general
property of branching epithelial morphogenesis.

Because these organs are rapidly growing and changing
shape, it is unlikely that the lamina is ever in the steady-
state condition, i.e., when the amount of lamina is constant
and, by definition, rates of synthesis and degradation are
equal. Therefore, although the mesenchyme is responsible for
the loss of laminar GAG, it is unclear whether the mesenchyme,
the epithelium, or the lamina itself dictates the apparently
preferential loss at the morphogenetically active sites. However, because the epithelium at these sites is expanding most
rapidly, the apparently greater turnover at these sites may
be more illusory than real. On the other hand, ultrastructural
studies of embryonic bronchial (4), metanephric (14) and submandibular (3) epithelia show that the lamina is thin and
interrupted at these sites. The interruptions allow intimate
contact between epithelial and mesenchymal cells, but it is
unknown whether this contact is involved in the morphogenetic
effect of mesenchyme. However, whatever the precise mechanism,
since submandibular epithelia require a basal lamina to maintain morphology, and require mesenchyme for continued branching morphogenesis, it is likely that the effect of the mesenchyme on GAG within the lamina is involved in mesenchymal-
induced changes in epithelial morphology.

REFERENCES

1. Grobstein, C. and Cohen, J., *Science* 150:626-628, 1965.

2. Wessells, N.K., *J. Exp. Zool.* 175:455-466, 1970.
3. Coughlin, M.D., *Develop. Biol.* 43:123-139, 1975.
4. Bluemink, J.G., van Maurik, P. and Lawson, K.A., *J. Ultrastructure Res.* 55:257-270, 1976.
5. Bernfield, M.R., Cohn, R.H. and Banerjee, S.D., *Amer. Zool.* 13:1067-1083, 1973.
6. Cohn, R.H., Banerjee, S.D. and Bernfield, M.R., *J. Cell Biol. (in press)*.
7. Bernfield, M.R., Banerjee, S.D. and Cohn, R.H., *J. Cell Biol.* 52:674-689, 1972.
8. Banerjee, S.D., Cohn, R.H. and Bernfield, M.R., *J. Cell Biol. (in press)*.
9. Diegelmann, R.F. and Peterkofsky, B., *Proc. Nat. Acad. Sci., USA,* 69:892-896, 1972.
10. Jansen, H.W. and Bornstein, P., *Biochim. Biophys. Acta* 362:150-159, 1974.
11. Clark, C.C., Minor, R.R., Koszalka, T.R., Brent, R.L. and Kefalides, N.A., *Develop. Biol.* 46:243-261, 1975.
12. Banerjee, S.D. and Bernfield, M.R., *J. Cell Biol.* 70:111a, 1976.
13. Pictet, R.L., Clark, W.R., Williams, R.H. and Rutter, W.J., *Develop. Biol.* 29:436-467, 1972.
14. Lehtonen, E., *J. Embryol. Exp. Morph.* 34:695-705, 1975.

D. Functions

STRUCTURAL AND FUNCTIONAL EFFECTS

OF GLOMERULAR POLYANION

Manjeri A. Venkatachalam, Helmut G. Rennke,

Marcel W. Seiler and Ramzi S. Cotran

Peter Bent Brigham Hospital,
West Roxbury Veterans Administration Hospital,
and Harvard Medical School
Boston, Massachusetts

It has been recognized for some time that the glomerular capillary wall is a negatively charged structure, and contains glycoproteins rich in sialic acid, also called glomerular polyanion (GPA) (1-5). Polyanionic radicals, particularly sialic acids have been demonstrated cytochemically in epithelial and endothelial cell coats of the glomerulus, through complex formation with electron dense, polycationic reagents such as colloidal iron and alcian blue, but their presence in the glomerular basement membrane (GBM) proper has been controversial (1-5). Biochemically, isolated GBM preparations have been shown to contain complex glycopeptides containing sialic acid as well as cross linked chains of a collagen like protein (6,7). Decreased GPA has been observed in human and experimental nephrotic syndrome, and a regulatory role for GPA in glomerular filtration of proteins, proposed (5,8). Michael et al. also showed a close temporal relationship between the occurrence of proteinuria, decrease of cytochemically demonstrable GPA and flattening of glomerular epithelial foot processes over the GBM in experimental puromycin induced nephrotic syndrome (5). Subsequent biochemical studies on puromycin nephrosis revealed significant decreases in

sialic acid content of whole glomeruli, as well as glomerular
cell membranes and GBM obtained by ultrasonic disruption (9).
Carbohydrate turnover studies indicated an alteration in
sialic acid metabolism (9).   Thus, a single biochemical
abnormality of GPA could be related to glomerular abnormalities of both structure and function.   These interesting
observations have led to a considerable body of new investigation from several laboratories, including our own.   This
new research forms the subject of this review.

EPITHELIAL STRUCTURAL ALTERATIONS INDUCED BY
    NEUTRALIZATION OF GPA

      The intent of the first series of experiments (10,11)
was to determine the effects on the glomerulus of extrinsic
neutralization of GPA.   Rat kidneys were perfused at 120 mm
Hg for 10-30 minutes with solutions of polycations in Krebs-
Ringer's bicarbonate buffer pH 7.4 (KRB), after initial wash-
out of the vascular tree with buffer alone.   The polycations
used included protamine sulfate or free base (20-500 µg/ml),
polylysine (5-50 µg/ml), polyarginine (1-5 µg/ml) and
lysozyme (0.05-20 mg/ml).   As controls, other kidneys were
perfused with KRB alone, or KRB with relatively neutral
macromolecules (poly-DL-alanine 20-50 µg/ml; myoglobin
50 µg/ml) and acidic molecules (heparin 50-100 µg/ml; polyglutamic acid 50 µg/ml; ovalbumin 0.1-20 mg/ml).   Following
exposure to the test substances, kidneys were prepared for
transmission electron microscopy by intravascular perfusion
of fixative.   Perfusion fixed kidneys were also examined by
electron microscopy following freeze-fracture.   In some cases,
protamine perfused kidneys were biopsied and subsequently re-
perfused with heparin to bind tissue bond protamine and
restore GPA.   Both control and experimental kidneys from all
groups were further subjected to colloidal iron staining of
paraffin sections to localize GPA cytochemically.
      Kidneys perfused with polycations exhibited glomerular
epithelial (podocyte) alterations very similar to those
observed in proteinuric disorders, particularly experimental
aminonucleoside nephrosis and human lipoid nephrosis (Fig. 1).
Such changes did not occur after exposure to neutral or
anionic molecules, suggesting that the induced changes were
specific to the electrical charge of the perfused molecule.
Morphogenetic factors in the polycation induced epithelial
lesions included retraction and flattening of foot processes,
narrowing of filtration slits, formation of occluding junctions between foot processes, and cell swelling.   GPA staining

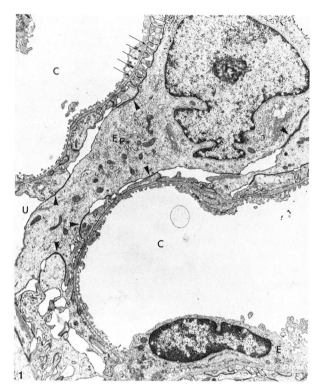

Fig. 1. Electron micrograph of glomerulus exposed to 50 µg/ml of protamine. Most of the glomerular basement membrane in the two capillary loops is covered by flattened masses of epithelial cytoplasm. In some areas, relatively normal foot process configuration is preserved (arrows). Epithelial cell bodies make intimate contact with one another, as well as with underlying foot processes (arrowheads). C - Capillary lumen; E - Endothelium; EP - Epithelium; U - Urinary space.

was simultaneously diminished. On reperfusion with heparin, a polyanion, polycation induced glomerular epithelial alterations were rapidly, and partially reversed. This restoration of foot process integrity was accompanied by a reversal of GPA staining by colloidal iron to normal, presumably by removal of previously perfused polycations from glomerular binding sites. The observations support the notion that GPA is an important determinant of glomerular epithelial cell shape and are consistent with the suggestion (5) that diminution of GPA may be the cause of so-called "fusion of foot processes" observed in nephrotic disorders.

CYTOCHEMICAL LOCALIZATION OF GPA

　　Perfusion of polycationic reagents directly through the
renal vasculature has helped to circumvent some of the problems previously encountered in the ultrastructural localization of GPA, such as the variable penetration of stains
through sections of fixed tissue. These experiments have
been particularly helpful in resolving the controversy surrounding the presence of polyanionic reactive sites in the
glomerular basement membrane (GBM). Perfusion with polycations (protamine, polylysine, polyarginine, lysozyme and
alcian blue), but not polyanions (heparin, polyglutamic acid,
ovalbumin) or neutral molecules (polyalanine, myoglobin)
results in the formation of electron dense precipitates in
various glomerular components including the epithelial and
endothelial cell coats of GBM, indicative of complex formation with polyanionic radicals (10-13). In these experiments,
the lamina rara interna and externa of the GBM (subendothelial and subepithelial layers) showed the presence of
stippled electron dense deposits, in addition to a slight
general enhancement of the staining intensity of the lamina
densa (Fig. 2). Reactive sites were more abundant in the
lamina rara externa than in the lamina rara interna. Similar
results using lysozyme as staining reagent in the perfusate
have been reported by Caulfield and Farquhar (14), and with
ruthenium red by Latta et al. (15).

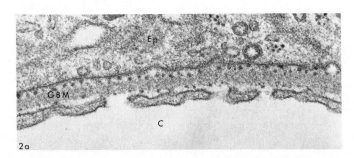

*Fig. 2a. Electron micrograph of glomerulus from
kidney perfused with 50 μg per ml of protamine sulfate
in Krebs-Ringer's bicarbonate buffer for 10 min. In
addition to effacement of foot process architecture,
there are stippled electron dense deposits of protamine
in the subepithelial layer, and to a lesser degree, in
the subendothelial layer of the basement membrane. These
deposits indicate polycation-polyanion complex formation.
EP - Epithelium; C - Capillary lumen; GBM - Glomerular
basement membrane. Fixed by perfusion with buffered
glutaraldehyde. Uranyl Acetate.*

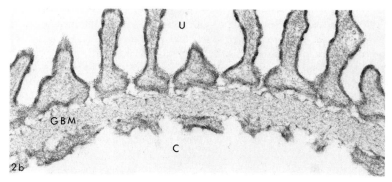

Fig. 2b. Electron micrograph of glomerulus from kidney perfused with 37°C Krebs-Ringer's bicarbonate buffer for 2 min followed by perfusion fixation in buffered glutàraldehyde (pH = 6.5) containing 0.5% alcian blue 8GX. Secondary fixation in 1% osmium tetroxide containing 1% lanthanum nitrate. Electron dense deposits of alcian blue - lanthanum indicate the presence of polyanionic reactive sites in the epithelial and endothelial cell coats and in the glomerular basement membrane. C - Capillary lumen; GBM - Glomerular basement membrane; U - urinary space.

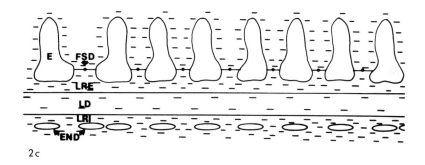

Fig. 2c. Diagrammatic representation of the distribution of polyanionic material (negative signs) in the glomerular extracellular matrix, as inferred from cytochemical studies. The concentration of glomerular polyanion is greatest in the epithelial and endothelial cell coats and in the outer and inner layers of the GBM. END - Endothelium; LRI - Lamina rara interna; LD - Lamina densa; LRE - Lamina rara externa; FSD - Filtration slit diaphragm; E - Epithelial cell.

The localization of polyanions in the GBM has also been
studied using polycationic ferritin as a stain. Thus, mouse
kidneys, when perfused with 3 mg/ml of strongly polycationic
ferritin showed aggregated, fixed stain particles in the
lamina rara interna and externa of the GBM, in addition to
the endothelial cell coat (16). In unpublished experiments,
we have used polycationic ferritin as a stain for GPA in
isolated rat glomeruli fragmented by ultrasonic disruption,
as well as in isolated GBM. The results have confirmed that
abundant polyanionic reactive sites are present in the GBM,
in its subendothelial and subepithelial layers. Prior
digestion of glomerular fragments, or GBM with clostridium
perfringens neuraminidase inhibits this staining pattern
only partially, suggesting that sialic acids form only a
part of total GBM polyanion. The nature of these radicals
was investigated further by staining GBM with polycationic
ferritin at pH values ranging from 1 to 7.4. The results
suggested that the degree of ionization of GBM polyanion is
greater at pH 7.4 than at pH values when sialic acid radicals
(pKa = 2.6) are expected to be maximally ionized. Thus,
GBM polyanion is heterogeneous in its make up and consists,
in addition to sialic acid, of other moieties with relatively
higher pKa values. The identity of the latter remains to be
established. However, the presence, in non-collagen GBM
glycopeptide fractions of large amounts of aspartic acid,
in addition to sialic acid (7,17) suggests that this di-
carboxylic amino acid may represent part of non-sialic acid
GBM polyanion.

The distribution of GPA as inferred from these cytochemical studies is illustrated in Fig. 2c.

MOLECULAR CHARGE AS A DETERMINANT OF GLOMERULAR PERMEABILITY

The glomerular filtration of macromolecules relative to
the filtration rate of water (fractional clearance) decreases
with increasing effective molecular radius (Einstein-Stokes
radius or a_e) (18,19). The relative clearance of proteins
approaches zero at an a_e of 35 Å; for flexible linear molecules such as dextrans and PVP, the corresponding value is
between 45 and 55 Å.

Ultrastructural tracer studies using both dextrans and
proteins have confirmed that glomerular permeability is related to molecular size. Dextrans of increasing molecular
radii encounter progressively greater restriction to transit
across the glomerular filter, the main barrier being localized
at the GBM (20). Proteins smaller than albumin (a_e = 35.5 Å)

such as horseradish peroxidase (a_e = 30 $\overset{o}{A}$) are able to
penetrate the glomerular filter and enter the urine; albumin
and the larger molecules, beef liver catalase (a_e = 52 $\overset{o}{A}$)
and horse spleen ferritin (a_e = 61 $\overset{o}{A}$) are excluded (21-23).
However, tracer studies with two cationic macromolecules
bovine lactoperoxidase (a_e = 36 $\overset{o}{A}$; isoelectric point, pI = 8)
and human leukocyte myeloperoxidase (a_e = 44 $\overset{o}{A}$; pI = 10) have
yielded anomalous results. On the basis of size alone, they
may be expected to be excluded from the urinary space, but
electron microscopic images obtained after intravenous injection have shown that they enter the glomerular filtrate
(21,24). Moreover, they appear to suffer relatively little
restriction to transit across the GBM, but accumulate in the
filtration slits, in between glomerular epithelial foot
processes (21, 24). In contrast, negatively charged proteins
such as catalase and ferritin are retarded by the GBM, as
seen in electron microscopic images (22,23). These variations
in glomerular permeability to macromolecules, predicated on
differences in molecular charge, have been studied recently
under controlled conditions.

Tracer Studies with Ferritin

    This spherical, anionic electron dense protein (a_e =
61 $\overset{o}{A}$; pI = 4.1-4.7) may be cationized to yield similarly sized
molecules with different isoelectric points. We have studied
the ultrastructural localization of differently charged
ferritin molecules within the glomerular filter both in the *in
vitro* perfused mouse kidney, as well as *in vivo*, in mouse and
Munich-Wistar rat kidneys after intravenous injection (16,25).
In both systems, the permeability of the GBM to the ferritin
molecule increases with rising isoelectric point (Figs. 3 and
4). Anionic ferritin is excluded from the filter, very few
molecules being found within its substance; of these few,
most are in the subendothelial layer of the GBM (Figs. 3a and
4a). These results are identical to those obtained in 1961
by Farquhar et al. (23). Cationic molecules penetrate the
filter in greatly increased amounts (Figs. 3b, 3c and 4b).
However, regardless of molecular charge, all ferritins are
excluded from entry into the urinary space at the level of
filtration slits. Images such as that shown in Fig. 3c,
suggested that the limiting barrier in the filtration slits
to these large molecules is constituted by the slit diaphragms. The fate of ferritin particles that traverse the
full thickness of the GBM is shown in Fig. 4b. Molecules of
ferritin, hindered from entry into the urinary space are
taken up by the glomerular epithelial cells through endocytosis.

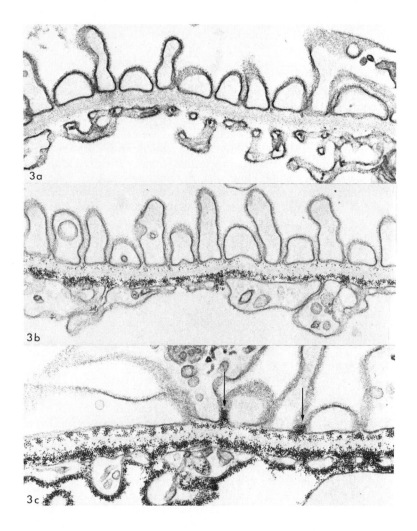

Fig. 3. *Electron micrographs of glomeruli from mouse kidneys perfused with Krebs-Ringer's bicarbonate buffer containing 3 mg/ml of differently charged ferritins and fixed by vascular perfusion with formaldehyde-glutaraldehyde fixative.*

Fig. 3a. *With native anionic ferritin (pI = 4.1-4.7), only occasional molecules are seen in the GBM, mostly in the lamina rara interna. Capillary lumen is empty due to perfusion fixation.*

Fig. 3b. *Cationized ferritin (pI = 8-9) penetrates the filter to a greater extent with numerous particles reaching the lamina rara externa, some of them accumulating proximal to the filtration slits.*

Fig. 3c. Strongly cationized ferritin (pI >8.8) adheres to the endothelial cell surface, accumulates in the lamina rara interna, reaches the lamina rara externa where it forms aggregates. Numerous particles of ferritin are impacted in some filtration slits, proximal to a level corresponding to filtration slit diaphragms (arrows). No ferritin molecules are present in the urinary space.

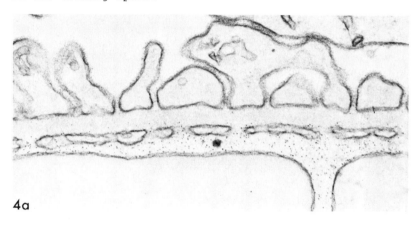

4a

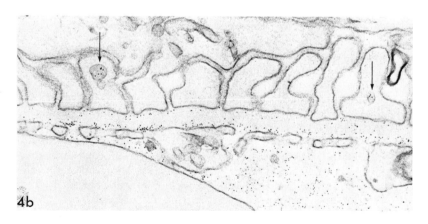

4b

Fig. 4. Electron micrographs of superficial glomeruli from Munich-Wistar rats injected with anionic (4a) or cationized (4b) ferritins and fixed after 15 min by the dripping of formaldehyde-glutaraldehyde fixative on the surface of kidneys.

*Fig. 4a.   Anionic ferritin (pI = 4.1-4.7) is
present in the capillary lumen, and a few molecules
in the subendothelial layer of the GBM.   Only a rare
particle penetrates the total thickness of the GBM.*

*Fig. 4b.   Cationized ferritin (pI = 6.3-8.4)
penetrates the total thickness of the GBM in greatly
increased amounts.   Note that ferritin molecules are
present in endocytic vacuoles in the epithelial cells
(arrows).*

Glomerular epithelial uptake of ferritin particles was greater
at the same time period after intravenous injection of cationized tracer in our experiments (25), compared to that seen
after administration of the native anionic protein, again
suggesting a facilitation of cationized ferritin transport
across the filter.

Clearance Experiments with Differently
    Charged Macromolecules

    In our experiments with ferritin, tracer molecules were
never visualized in the urinary space. Thus, the studies,
while permitting an analysis of GBM permeability, did not
assess charge dependent filtration across the total thickness
of the glomerular capillary wall. In order to study the
latter quantitatively, we have examined the renal clearance
of the filterable protein horseradish peroxidase, a molecule
previously used for ultrastructural tracer studies. The
native protein, obtained commercially, consists of a mixture
of isozymes with different isoelectric points. The main
isozyme, with an isoelectric point of 7.35-7.55 was obtained
in a relatively pure form by gel filtration and ion exchange
chromatography. Negatively charged and positively charged
derivatives of this protein were prepared by succinylation
and cationization, respectively. Molecular size and isoelectric points of the three proteins were determined by gel
filtration chromatography and isoelectric focusing. The renal
clearances of the tracers, relative to inulin clearance
(fractional clearance) were studied by standard methods, using
both the amounts actually excreted in the urine during a 20
minute period, as well as the absolute amounts filtered including the fractions of filtered protein reabsorbed by the
tubules (26). Enzyme reabsorbed by the tubules was estimated
by analyzing the renal parenchymal content of peroxidase in

tissue homogenates. Peroxidase in plasma, urine, and homogenate was assayed by a sensitive enzymatic spectrophotometric method. The calculated fractional clearance of
cationic horseradish peroxidase (a_e = 30 Å; pI = 8.4-9.2) was
5.7 times greater than that of similarly sized neutral
peroxidase (a_e = 29.8 Å; pI = 7.35-7.55) and 49 times that of
negatively charged succinylated peroxidase only 2 Å larger in
effective molecular radius (a_e = 32 Å).

Similar results were previously obtained by Chang et al.
using differently charged dextran molecules in clearance
experiments on Munich-Wistar rats (27). Thus, over a wide
size range, from a_e 20 Å to 44 Å the fractional renal clearances of neutral dextrans greatly exceeded those of similarly
sized anionic dextran sulfate molecules (27).

Clearance of Charged vs. Uncharged Molecules
    in Proteinuria

Reduction of glomerular and GBM polyanion content has
been observed both in experimental puromycin aminonucleoside
nephrosis (5,28) as well as in nephrotoxic serum nephritis
(29,30). It has been postulated that increased filtration of
albumin (a polyanion) in these conditions is related to the
GPA alteration (2,5). This has been experimentally tested
by measuring the fractional renal clearances of uncharged
dextrans and anionic dextran molecules in puromycin aminonucleoside nephrosis (31) and in nephrotoxic serum nephritis
(30). Fractional clearances of anionic, but not neutral
dextrans were found to be increased in both proteinuric
disorders, suggesting that decrease of GPA was causally related to selective increase in polyanion filtration.

MECHANISMS WHEREBY GPA DETERMINES GLOMERULAR PERMEABILITY

Endogenous plasma proteins have been ultrastructurally
localized in rat glomeruli, utilizing an immunoperoxidase
technique. In these studies, Ryan and Karnovsky were able to
show that glomerular restriction of plasma proteins occurs in
the proximal layers of the filter (32,33). Ultrastructural
studies using exogenous tracer molecules have also indicated
that the "primary barrier" to macromolecules is constituted
by the glomerular extracellular matrix in the proximal layers
of the capillary wall, namely the endothelial cell coat and
glomerular basement membrane (16,20,22,23,25). The mode of
action of these filter elements deserves explanation.

Biochemical and cytochemical studies have shown that the
GBM and epithelial and endothelial cell coats are comprised
of polyanionic glycoprotein polymers. In such hydrated gels,
the pathways for water and solute transport are constituted
by the hydrated interior of the polymer matrix, in between
the chains of polymer (34).

Isolated GBM preparations, like porous beads of "sephadex", are able to partition solute molecules between water
compartments in their interior and exterior, according to
size (35-37). Tracer and clearance experiments with differently charged macromolecules clearly indicate that the
glomerular extracellular matrix is in addition able to dis-
criminate between molecules of different net charge (16, 25-
27). Thus, assuming that other factors such as glomerular
hemodynamics and molecular shape remain constant, the following model may be proposed to explain the sieving characteristics of the "primary" glomerular barrier to macromolecules.

The cell coats and GBM may be conceived to be a continuous membrane of polyanionic gel with a hydrated interior of
finite, but as yet unknown spatial configuration and dimensions (Fig. 5). The flow of plasma occurs parallel to this
membrane. The forces that determine filtration operate
across the membrane, with net movement of water, solute, and
macromolecules. In Fig. 5, very large molecules are seen to
be excluded from the matrix, and thus from the filtrate on
the basis of molecular size alone, despite charge based
repulsion or attraction. Filtration of molecules smaller
than the "pores" is governed by charge interactions as well
as the disparity between molecular size and available space
within the membrane. Thus, intermediate sized cationic and
neutral molecules are able to penetrate the filter whereas
polyanionic solutes are excluded due to functional narrowing
of the pores within the matrix by the electrostatic field
generated by GPA. The smallest molecules encounter less
steric hindrance, but the degree of partition between perfusate and gel matrix, and thus entry into the filtrate, is
influenced by molecular charge. Thus, electrostatic forces
act in a direction opposite to that of filtration for anionic
molecules. For neutral molecules charge interactions are
absent or minimal; in the case of cationic molecules the
electrophysical effects act synergistically with the forces
favoring filtration.

Since restriction at the level of the endothelium and
GBM is not absolute, even for large anionic molecules of the
size of catalase (a_e = 52 Å) (22) and ferritin (a_e = 61 Å)
(23,25) a further glomerular barrier, distal to the GBM, has
to be postulated. Ultrastructural studies suggest that the

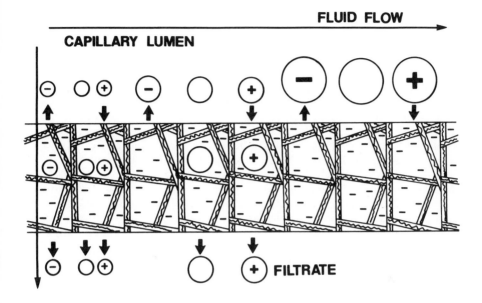

Fig. 5. Model of macromolecule filtration through polyanionic gel of glomerular capillary showing the effects of size and charge. Horizontal arrow depicts blood flow. Filtration occurs across the membrane (vertical arrow). Circles with negative signs, no signs and positive signs represent molecules with negative net charge, zero net charge and positive net charge. The gel is constituted by a framework of glycopeptide chains (rods) with fixed ionized carboxyls. Very large molecules are excluded from the internal environment of the gel due to steric hindrance. Molecules of intermediate size, shown in the middle, suffer restriction due to both charge and size dependent effects. Molecules with net positive charge and zero net charge can penetrate the gel and enter the filtrate. However, the "pores" of the internal environment of the polymer are functionally narrowed by fixed negative charges for anionic molecules which are excluded. Very small molecules can enter the polyanionic gel, but repulsive or attractive forces obviously reduce or increase the rate of movement, and consequently, the quantities filtered.

distal barrier is at the level of the glomerular epithelium
and filtration slits. These protein molecules that penetrate
the GBM appear not to transit via the filtration slits into
the urinary space, but undergo phagocytosis by the epithelial
cells (22,23,25). The barrier in the filtration slits may
have two components. Firstly, the filtration slit diaphragms
may pose a mechanical impediment to entry of molecules into
the urinary space; favorable electron microscopic images show
them to be endowed with rectangular "pores" approximately the
size of serum albumin (38). Electron microscopic images
have indicated that large cationic molecules are able to
penetrate the filter, but are retarded by the slit diaphragms
(16,21). Secondly, the filtration slits are completely filled
by the closely apposed, intensely polyanionic cell coats of
adjacent foot processes. The entry of macromolecules into
this gel may be expected to be governed by the same principles
discussed earlier for the GBM and endothelial cell coats.

An analogy may be drawn between the behavior of anionic
macromolecules in the glomerular extracellular matrix and
synthetic gel systems (34,39). Gelatin membranes exhibit
decreased permeability to albumin when bonded with the
anionic dye Congo Red; conversely, neutralization of Congo
Red treated gelatin membranes by protamine enhances their
permeability (39). The interaction of albumin and beads of
hyaluronate or chondroitin sulfate gels in chromatographic
columns is another example (34). At buffer pH values when
both albumin and the gel beads are ionized, albumin elutes
with the void volume signifying non-entry into the gel; at
lower pH values, elution of albumin from the column is retarded indicating electrophysical interaction between the
molecule and gel (34).

CONCLUSIONS

Electron microscopic, functional and biochemical studies
over the past few years have shown an important role for polyanionic glycoproteins in normal glomerular function and structure. The role of electrophysical interactions between the
negatively charged glomerular capillary wall and macromolecules in glomerular permeability has been clearly demonstrated.
Alterations in glomerular polyanion may cause proteinuria in
certain glomerular diseases and may be related to associated
glomerular epithelial abnormalities. Further chemical characterization of GPA and elucidation of its biosynthetic pathways are indicated.

REFERENCES

1. Rambourg, A. and LeBlond, C.P., *J. Cell Biol.* 32:27-53, 1967.
2. Mohos, S.C. and Skoza, L., *Science* 164:1519-1521, 1969.
3. Jones, D.B., *Lab. Invest.* 21:119-125, 1969.
4. Groniowski, J., Biczyskowa, W. and Walski, M., *J. Cell Biol.* 40:585-601, 1969.
5. Michael, A.F., Blau, E. and Vernier, R.L., *Lab. Invest.* 23:649-657, 1970.
6. Spiro, R.G., *In:* "Glycoproteins: Their Composition, Structure and Function", (edited by A. Gottschalk) Amsterdam, Elsevier, 2nd ed., Part B, pp. 964-999, 1972.
7. Kefalides, N.A., *In:* "International Review of Connective Tissue Research", Academic Press, New York, Vol. 6, pp. 63-104, 1973.
8. Blau, E.B. and Haas, J.E., *Lab. Invest.* 28:477-481, 1973.
9. Blau, E.B. and Michael, A.F., *Proc. Soc. Exp. Biol. Med.* 141:164-172, 1972.
10. Seiler, M.W., Venkatachalam, M.A. and Cotran, R.S., *Science* 189:390-393, 1975.
11. Seiler, M.W., Rennke, H. G., Venkatachalam, M. A. and Cotran, R.S., *Lab. Invest.* 36:48-61, 1977.
12. Behnke, O. and Zelander, T., *J. Ultrastruct. Res.* 31: 424-438, 1970.
13. Seiler, M.W., Rennke, H.G., Cotran, R.S. and Venkatachalam, M.A., *Am. J. Pathol.* 82:54a, 1976.
14. Caulfield, J.P. and Farquhar, M.G., *Proc. Nat. Acad. Sci.,U.S.A.,* 73:1646-1650, 1976.
15. Latta, H., Johnston, W.H. and Stanley, T.M., *J. Ultrastruct. Res.* 51:354-376, 1975.
16. Rennke, H.G., Cotran, R.S. and Venkatachalam, M.A., *J. Cell Biol.* 67:638-646, 1975.
17. Sato, T. and Spiro, R.G., *J. Biol. Chem.* 251:4062-4070, 1976.
18. Renkin, E.M. and Gilmore, J.P., *In:* "Handbook of Physiology", (Orloff, J. and Berliner, R.W., eds.) American Physiological Society, Washington, D.C., pp. 185-248, 1973.
19. Brenner, B.M., Deen, W.M. and Robertson, C.R., *In:* "The Kidney", (Brenner, B.M. and Rector, F.C., eds.), W.B. Saunders, Philadelphia, Vol. 1, pp. 251-271, 1976.
20. Caulfield, J.P. and Farquhar, M.G., *J. Cell Biol.* 63: 883-903, 1974.
21. Graham, R.C. and Karnovsky, M.J., *J. Exp. Med.* 124:1123-1134, 1966.

22. Venkatachalam, M.A., Karnovsky, M.J., Fahimi, H.D. and Cotran, R.S., *J. Exp. Med.* 132:1153-1167, 1970.
23. Farquhar, M.G., Wissig, S.L. and Palade, G.E., *J. Exp. Med.* 113:47-66, 1961.
24. Graham, R.C. and Kellermeyer, R.W., *J. Histochem. Cytochem.* 16:275-278, 1968.
25. Rennke, H.G. and Venkatachalam, M.A., *Kidney Int'l.* 11:44-53, 1977.
26. Rennke, H.G., Patel, Y. and Venkatachalam, M.A., *Clin. Res.* 24:645A, 1976.
27. Chang, R.L.S., Deen, W.M., Robertson, C.R. and Brenner, B.M., *Kidney Int'l.* 8:212-218, 1975.
28. Caulfield, J.P. and Farquhar, M.G., *J. Cell Biol.* 70:92a, 1976.
29. Chiu, J., Drummond, K.N., *Am. J. Pathol.* 68:391-406, 1972.
30. Bennett, C.M., Glassock, R.J., Chang, R.L.S., Deen, W.M., Robertson, C.R. and Brenner, B.M., *J. Clin. Invest.* 57:1287-1294, 1976.
31. Baylis, C., Bohrer, M.P., Troy, J.L., Robertson, C.R. and Brenner, B.M., *Kidney Int'l.* 10:554, 1976.
32. Ryan, G.B. and Karnovsky, M.J., *Kidney Int'l.* 9:36-45, 1976.
33. Ryan, G.B., Hein, S.J. and Karnovsky, M.J., *Lab. Invest.* 34:415-427, 1976.
34. Laurent, T.C., *Fed. Proc.* 25:1128-1134, 1966.
35. Gekle, D., von Bruchhausen, F. and Fuchs, G., *Pflugers Arch. Ges. Physiol.* 289:180-190, 1966.
36. Huang, F., Hutton, L. and Kalant, N., *Nature* 216:87-88, 1967.
37. Igarashi, S., Nagase, M., Oda, T. and Honda, N., *Clinica Chimica Acta* 68:255-258, 1976.
38. Rodewald, R. and Karnovsky, M.J., *J. Cell Biol.* 60:423-433, 1974.
39. Larsen, B., *Nature (Lond)* 215:641-642, 1967.

ANATOMY OF BASAL LAMINA SCAFFOLD

AND

ITS ROLE IN MAINTENANCE OF TISSUE STRUCTURE

Rudolf Vracko

Veterans Administration Hospital
and
University of Washington School of Medicine
Seattle, Washington

ABSTRACT: Basal lamina (BL) is characteristically
located between parenchymal cells and connective
tissue. As an extracellular scaffold it provides
a substrate for attachment of parenchymal cells
with one of its surfaces and for anchorage of
connective tissue with the other. In four anatomic
sites (glomeruli, lung, nervous system and liver)
these relationships are modified. By its presence
BL subdivides the space of the organisms into compartments and defines the spatial relationships among
parenchymal cells and between these and the spaces
occupied by the connective and supportive tissues.
When cell death occurs either from senescence or injuries, BL maintains the spatial plan of the tissue
and provides a substrate for orderly positioning
of parenchymal cells and also defines spaces occupied
by connective tissue. This process appears to be
aided by (a) polarity of BL; (b) an apparent specificity of its surfaces for cell types; (c) harmonious
structural continuity of BL scaffold between injured
and uninjured portions of the organ; and (d) presence

of a surface which devoided of cells could initiate
cell replication and stop it when populated with
cells thus setting the size limit to reconstruction.
Higher vertebrates who have lost the capacity for
regeneration of major organs, i.e., skeletal muscle,
kidney and lung, can, by these mechanisms, reconstitute
histologic structure to what it was prior to loss of
cells. It appears that if BL is destroyed in these
organs the healing results in formation of scar and
associated permanent loss of function. These proper-
ties of BL concerned with maintenance of histologic
order provide new insights into pathogenesis of
several common disorders.

Each organ and tissue of multicellular organisms is
composed of several different types of cells, of extracellular materials and extracellular spaces. These components
are arranged during embryogenesis in precise relationships
within the space of the organism and assume in each tissue and
organ characteristic histologic patterns that remain essentially unchanged throughout the life of the organism. To
maintain this spatial order new cells replace in an orderly
way the old ones and damage caused by injury is repaired by
one of several mechanisms. Some major organs seem to require
a basal lamina (BL) scaffold for proper positioning of cells
as well as for re-establishment of tissue structure.

BL scaffold in normal organisms can be perceived as a
system of extracellular partitions that divide the space of
each tissue and organ into two sets of compartments. With
one of its surfaces BL is a "boundary" (7) of the connective
tissue space and with the other surface it is a substrate or
a "microskeleton" (13) upon which parenchymal cells position
(26) and differentiate (5,16).

The anatomic relationships of BL are diagrammed in
Fig. 1 where BL is depicted as a heavy line. It is typically
located between parenchymal cells on the one side and the
connective and supportive tissues on the other. Parenchymal
cells include epithelial cells of epidermis and epidermal
appendages, of the genito-urinary, respiratory and gastro-
intestinal tracts, exocrine glands, endothelial cells of the
cardiovascular system, mesothelial cells of body cavities,
cells comprising central and peripheral nervous system,
endocrine cells, muscle fibers and fat cells. The space
occupied by connective and supportive tissue (shaded area)
contains bone and associated cells, cartilage and associated
cells, collagen, elastin, microfibrils, "ground substance"
and fibroblasts.

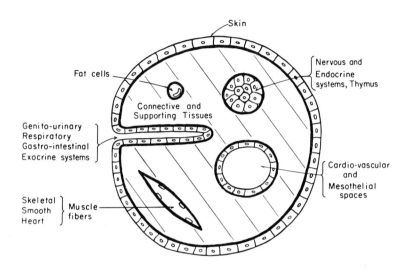

Fig. 1. *Diagram of anatomic distribution of basal lamina (depicted as heavy black line). See text for explanation. Reprinted with permission from Am. J. Path. (22).*

This fundamental pattern of BL distribution is normally modified in several anatomic sites: (a) the BL of endothelial cells in alveoli of lungs, renal glomeruli, central nervous system and rat heart fuses focally with the BL of the adjacent parenchymal cells so that two different types of parenchymal cells, one of them endothelium, occupy opposite sides of a single BL layer (22); (b) in the sinusoids of uninjured liver between liver cells and the cells lining sinusoids BL is almost completely absent (15); (c) during embryogenesis epithelial cell processes transiently extend through gaps in BL forming contacts with cells of the connective tissue (9) and (d) in the lungs of some adult vertebrates, most notably hamsters, type II alveolar pneumocytes are in direct apposition to the alveolar septal collagen through discontinuities in the BL of alveolar epithelium (10).

When injury disturbs these spatial relationships the ensuing repair either reconstructs the tissue design to what it was prior to injury or connective tissue overgrows the injured site to form a scar. In the first instance the function returns while in the second it does not. Both processes can contribute to the healing of the same injury and occasionally

distortion of histologic patterns occurs by disorderly spatial
positioning of newly-formed cells (22).

Orderly tissue structure can grow back in one of at least
three different ways: (a) lower vertebrates, as for example
Amphibia and Reptilia regenerate complex organs and tissues
from blastema that forms at the site of injury; (b) some
complex tissues of higher vertebrates as for example endo-
metrium during menstrual cycle, breast during pregnancy, bone
following fracture or vessels in granulation tissue do not
form blastema but grow by extension from pre-existing tissues
of the same kind and finally (c) most large organs of higher
vertebrates including skeletal muscle, kidneys, skin, lungs
and nerves can neither regenerate as do lower vertebrates nor
can they restructure histologic complexities by outgrowth
from residual tissues. Instead these organs seem to be able
to re-establish histologic order only if a BL scaffold maintains the plan of the tissue in the area of injury.

It has been known for more than 100 years that for
orderly repair of skeletal muscle injuries the sarcolemma
must be intact. Sarcolemma guides and positions newly-formed
muscle cells (18,25) even if the sarcolemmal tubes are positioned by excision and re-implantation at right angles to the
original orientation (2). If sarcolemma is destroyed,
permanent scar forms at the site of injury (1,2,18,25). We
know today that sarcolemma consists of plasma membrane of
muscle cells and of the adjacent BL. That it is BL which
provides the guiding principle was probably first recognized
in renal tubules in cases where re-epithelialization of
acute tubular necrosis occurred along the old BL scaffold (12).

Brief accounts will now be given of the events that take
place during repair of injured skeletal muscle (21) and injured lung (20).

A diagram of BL scaffold of injured muscle is depicted
in Fig. 2. The scaffold can be perceived as consisting of
three parallel sets of pipes made of BL: in the large ones
are located muscle cells and in the smaller ones either endothelial cells and pericytes of capillaries or Schwann cells
and axons of nerves. These pipes are held in spatial relationships by connective tissue which anchors to their outer
surfaces. When cells in skeletal muscle of rat or rabbit
are killed with either freezing, ischemia or *in situ* auto-
grafting (the cells in the excised piece of tissue being
killed by freeze-thaw injury prior to implanting) the BL
scaffold remains intact and preserves the spatial map of
skeletal muscle (21).

In Fig. 3, a portion of skeletal muscle is seen one day
after freezing. There is complete destruction of muscle

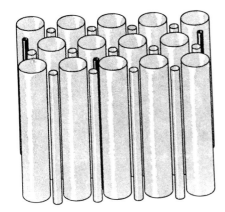

Fig. 2.   Diagram of
basal lamina scaffold of
skeletal muscle.
See text for explanation.

cells (M) and cells of capillaries (C). The cell debris is
contained by intact BL scaffold (arrows). Several hemolyzed
red blood cells are present in the connective tissue space.

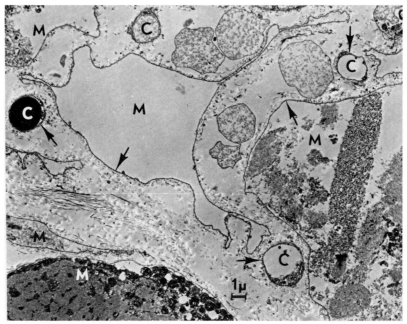

Fig. 3.   Basal lamina scaffold in rat skeletal
muscle four days after freeze-thaw injury. See text
for explanation. Reproduced with permission from
Am. J. Path. (22).

Within seven days, in both animal species and after each type
of injury undifferentiated cell processes presumably myoblasts
grow inside the BL tubes of muscle fibers, as shown in Fig. 4.
The BL of these tubes has partly collapsed forming longitudinal pleats (arrow in inset of Fig. 4). Within a week, the
undifferentiated cell processes fuse to form a single cell
body with multiple centrally placed nuclei. Myofilaments
become apparent in the cytoplasm of these young muscle cells
(M) and, as indicated in Fig. 5, a new layer of BL appears
(open arrow) along those portions of plasma membrane that are
not in close proximity to the old BL layer (solid arrow).
There is also proliferation of large interstitial cells (Ic).

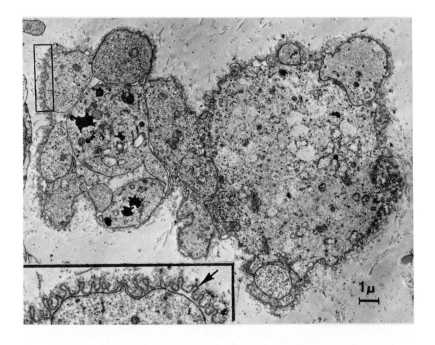

*Fig. 4. Cellular repopulation of basal lamina
scaffold of rat skeletal muscle one week after freeze-
thaw injury. See text for explanation.
Reproduced with permission from Am. J. Path. (22).*

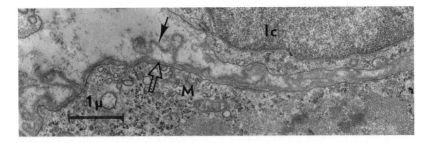

*Fig. 5. Reduplication of basal lamina in rat
skeletal muscle one week after freeze-thaw injury.
See text for explanation.*

As depicted in Fig. 6, interstitial cells (Ic) become aligned
along the surfaces of muscle fibers (M) where BL is redupli-
cated (open arrow). Later, these cells and the redundant BL
disappear, presumably the extra BL having been removed by the
interstitial cells.

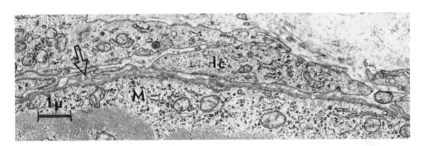

*Fig. 6. Rabbit skeletal muscle one week after
infarction. See text for explanation.*

In capillaries of skeletal muscles similar events take
place. Initially undifferentiated cells grow in spaces defined by capillary BL and later lumens develop within these
cells. As in muscle fibers, a new layer of BL forms where
the cell surface is not in close proximity to the old BL.
The end product of capillary repair in rat muscle five weeks
after a piece of muscle was excised and reimplanted is shown
in Fig. 7. The 20-month-old rat was given 0.15 molar silver
nitrate solution as drinking water from weaning until the
day of operation and only fresh water after operation. The
resulting accumulation of silver granules (open arrows) is
limited to the old, now outer and redundant layer of BL,
indicating that the inner closely fitting silver-free layer

(full arrows) formed after the operation (19). Similar reduplication of BL is also apparent in two of the three muscle
cells (M). Interstitial cells are prominent and presumably
concerned with removal of excessive BL.

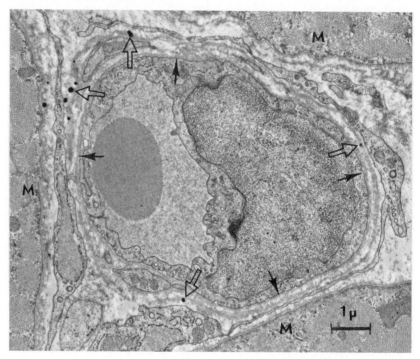

*Fig. 7. Capillary from a piece of rat muscle
five weeks after the piece was excised and reimplanted.
See text for explanation.*

If BL scaffold of skeletal muscle is destroyed or removed
or the defect in the muscle is filled with fibrin or blood
clots, the connective tissue cells overgrow the site of damage
forming a scar and permanently disrupting the orderliness of
muscle structure (1,2,18,24,25).
    Events during healing of injured lung are similar to
those in skeletal muscle. One day after alveolar cells are
killed in the rat by freezing and thawing (24), or by intravenous injection of oleic acid in dogs (12) (Fig. 8) cell
debris of endothelial cells and red blood cells (C) are
contained by an intact capillary BL scaffold (open arrows).
Epithelial BL (solid arrow) is also denuded of cells and forms
a boundary between alveolar spaces (A) and alveolar septae.

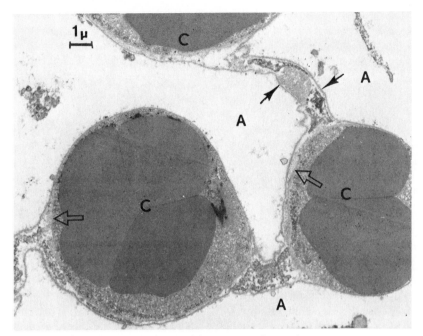

Fig. 8.   Dog lung 24 hours after intravenous
injection of oleic acid.   See text for description.
Reproduced with permission from Am. J. Path. (22).

Within three or four days (Fig. 9) poorly differentiated
cells begin to proliferate from the viable portions of the
tissue inside the capillary (C) BL scaffold (open arrows)
while the alveolar (A) surface of epithelial BL (solid
arrows) is repopulated by flat cells, presumably type I
pneumocytes. Approximately three weeks after injury there
is almost complete cellular reconstitution of the lung injured
by oleic acid (12) while injuries of peripheral lung caused by
freezing leave a flat subpleural scar (24). Although experimental observations documenting formation of scar after
destruction of pulmonary BL scaffold are lacking, the fact
that lung abscesses and infectious granulomas generally heal
with a permanent scar seem to support the notion that once the
scaffold of the lung is destroyed, reconstruction does not
occur.
      Similar relationships between BL scaffold and repopula-
ting cells that occur in muscle and lung have also been noted
in renal glomeruli (17), renal tubules (3,8,12,22), skin (4,
6,22), peripheral nerves (11) and myocardium (24). In most
of these the presence of a BL scaffold appears to be a requirement for orderly positioning of parenchymal cells as well as

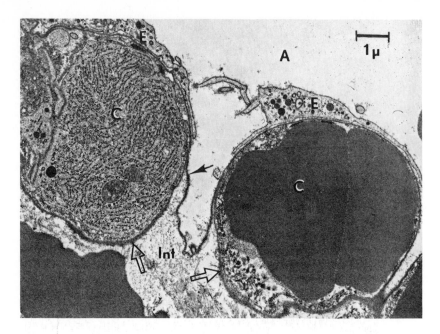

Fig. 9. Dog lung three days after intravenous
injection of oleic acid. See text for explanation.
Reprinted from Virch. Arch. (20).

for containment of connective tissue. If BL is destroyed
during injury in these organs healing is not followed by
orderly repair but rather by formation of a scar. It is of
interest that in some anatomic sites, most notably in muscle
capillaries, muscle fibers, renal tubules, renal glomeruli
and peripheral nerves new layers of BL are deposited by new
cell populations while in others, as for example epidermis,
alveolar capillaries and alveolar epithelium they are not
(23). Of additional interest is the observation that BL
layers are promptly removed in skeletal muscle fibers (21)
but not in skeletal muscle capillaries or renal tubules.

     In summary, there appear to be three general require-
ments for orderly cellular repair of several large organs:
(a) there must be a source of appropriate cell types , (b)
these cells must be able to proliferate, and (c) a BL scaffold
must be present in the damaged portion of the tissue. While
the first two provide the required variety and numbers of
cells, the BL scaffold is concerned with orderliness of repair.
BL is a tough structure (14) that is much less vulnerable to
destruction than are cells. It defines by its presence the
relationship between the spaces occupied by parenchymal cells
and those occupied by connective tissue. When cells are

are killed BL assures a harmonious structural continuity
between injured and uninjured portions of tissue. BL has an
apparent polarity, one surface being devoted to parenchymal
cells and the other to connective tissue. It probably also
has a specificity for cell types (22). It may have something
to do with induction and restriction of cell proliferation
and therefore could represent an important signal to tissue
reconstruction (21,22).

The concept of BL scaffold offers new ways for interpretation of pathogenesis of several different disease
processes (22) and could serve to improve surgical techniques
concerned with repair of damaged tissues.

REFERENCES

1. Allbrook, D., *J. Anat.* 96:137-152, 1962.
2. Clark, W.E. and LeGros,
3. Cuppage, F.E., Neagoy, D.R., Tate, A., *Lab. Invest.*
   17:660-674, 1967.
4. Giacometti, L. and Parakkal, P.F., *Nature* 223:514-515,
   1969.
5. Gustafson, T. and Wolpert, L., *Int. Rev. Cytol.* 15:139-
   214, 1963.
6. Krawczyk, W.S., *J. Cell Biol.* 49:247-263, 1971.
7. Low, F.N., *Anat. Rec.* 159:231-238, 1967.
8. Madrazo, A., Suzuki, Y. and Churg, J., *Am. J. Path.*
   61:37-56, 1970.
9. Mathan, M., Hermos, J.A. and Trier, J.S., *J. Cell Biol.*
   52-577-588, 1972.
10. Huang, T., *Unpublished observations.*
11. O'Daly, J.A. and Imaeda, T., *Lab. Invest.* 17:744-766,
    1967.
12. Oliver, J., *Am. J. Med.* 15:535-557, 1953.
13. Pease, D.C., *In:* "The Basement Membrane: Substratum of
    Histological Order and Complexity", Springer-Verlag,
    Berlin, pp. 139-155, 1960.
14. Ross, M.H and Grant, L., *Exp. Cell Res.* 50:277-285, 1968.
15. Schaffner, F. and Popper, H., *Gastroenterology* 44:239-
    242, 1963.
16. Slavkin, H.C., *IN:* "The Dynamics of Extracellular and
    Cell Surface Protein Interactions, Cellular and Molecu-
    lar Renewal in the Mammalian Body", (Cameron, I.L. and
    Thrasher, J.D., Eds.), Academic Press, New York, pp. 221-
    276, 1971.
17. Thorning, D. and Vracko, R., *Lab. Invest. (in press).*
18. *Volkman, R., Beitr. Pathol. Anat. Allg. Pathol. 12:233-
    332, 1893.*

19. Vracko, R. and Benditt, E.P., *J. Cell Biol.* 47:281-
     285, 1970.
20. Vracko, R., *Virchows Arch. (Pathol. Anat.)* 355:264-
     274, 1972.
21. Vracko, R. and Benditt, E.P., *J. Cell Biol.* 55:406-419,
     1972.
22. Vracko, R., *Am. J. Path.* 77:314-338, 1974.
23. Vracko, R., *IN:* "Proceedings of the First International
     Symposium on the Biology and Chemistry of Basement
     Membranes", (Kefalides, N.A., Ed.) Academic Press,
     New York, p
24. Vracko, R., *Unpublished observations.*
25. Waldeyer, W., *Virchows. Arch. Pathol. Anat. Physiol.*
     Clin. Med. 34:473, 1865.
26. Weiss, P., *Exp. Cell Res. (Suppl. 8)*:260-281, 1961.

Supported by Veterans Administration Research funds and
in part by NIH Grant HL-03174-20.

THE ACELLULAR PERFUSED KIDNEY: A MODEL

FOR BASEMENT MEMBRANE PERMEABILITY

Klaus Brendel, Elias Meezan

and Raymond B. Nagle

Arizona Health Sciences Center
Tucson, Arizona

SUMMARY: Perfusion of rabbit kidneys with a series of solutions containing the detergent Triton X-100 result-ed in the removal of all cells from the organs and yielded acellular preparations in which the basement membranes preserved the anatomical architecture of the kidneys, thus demonstrating the major role of these boundary structures in tissue organization. These acellular perfused kidneys were used as a model for examining the permeability of basement membranes with-out the cellular contribution. Perfusion of the acel-lular organs with colloidal carbon and gold showed that these particulate tracers were excluded from the urinary space indicating the absence of breaks in the continuous basement membrane "vasculature". Ferritin penetrated the glomerular basement membrane to a variable, but appreciable extent with particles appearing in the urin-ary space and tubular lumens, and in the urinary outflow of perfused kidneys. Both colloidal gold and ferritin freely penetrated peritubular capillary basement mem-branes, but not tubular basement membranes. Dextran of 500,000 weight average molecular weight rapidly appeared in the urinary outflow of perfused kidneys at concentra-tions equal to those in the venous outflow, with the size

*distribution of the filtered molecules being identical
to that in the perfusate. These results are consistent
with the view of the basement membrane as a significant,
but not absolute barrier for proteins with the charge
and size characteristics of ferritin, but as a poor
barrier for neutral molecules the size of large dextrans.*

   *The relative contributions of the structural ele-
ments comprising the glomerular capillary wall to the
selective permeability of the glomerulus to macromolecules has been the subject of intensive investigation
and some controversy (1-3). Central to this issue has
been the role of the glomerular basement membrane,
which forms a continuous barrier separating the capillary space from the urinary space. Since this basement
membrane is lined on the capillary side by fenestrated
endothelial cells and on the urinary side by the foot
processes of epithelial cells which are connected by
slit diaphragms, any investigation of basement membrane
permeability in vivo or in the isolated perfused kidney
will be influenced by the contributions of these cellular layers. Thus, it has been difficult if not impossible to examine the intrinsic permeability of the basement membrane layer divorced from its accompanying
cellular components. Our ability to obtain ultrastructurally and chemically intact basement membrane preparations from kidney glomeruli and tubules by selective
extraction of cellular components with detergents (4,5)
suggested that this might be a feasible procedure to
obtain a whole organ preparation in which all intact
cellular elements had been removed, leaving a basement
membrane skeleton of the kidney which could serve as a
model in which to study basement membrane permeability.
Using this "acellular" perfused kidney preparation, we
have examined the intrinsic permeability of glomerular,
peritubular and tubular basement membranes to particulate
tracers, ferritin and dextrans.*

METHODS

   All kidneys were obtained from 1-1.5 Kg New Zealand White
rabbits. The animals were sacrificed by shooting in the base
of the head with a pellet gun. The abdominal cavity was
immediately opened and both kidneys were removed and cannula-
ted at the renal arteries with a 19 gauge blunt needle. The
cannulas were connected to a perfusion system (Fig. 1) and
both kidneys were perfused in a non-recirculatory fashion with

0.9% NaCl containing 0.2 mg isoproterenol and 5 mg isoxsuprine per liter.

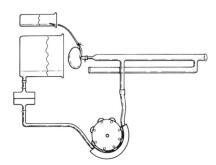

Fig. 1. Perfusion apparatus showing attachment of a single kidney. See Methods for details.

        The perfusion system consisted of a Harvard peristaltic pump which circulated the perfusion medium (50 ml) from a reservoir into two 150 cm tall open-ended glass columns to which the kidneys were attached. After circulating through the kidneys, the perfusate flowing from the renal vein was either returned to the reservoir to be recycled (recircula- tory perfusion) or directed into the drain and replaced with new perfusion medium in the reservoir (non-recirculatory perfusion). By allowing any perfusate rising higher than 150 cm in the glass columns to overflow back into the reservoir, the perfusion pressure could be kept within phyio- logical limits at all times. Thus at constant pressure, a variable flow rate of perfusion medium was circulated through the kidneys. The flow rate of the perfusate increased as the cellularity of the kidneys and thus the vascular resistance decreased. At the conclusion of the procedure the flow rate through the kidneys was 40-50 ml per minute. About 3-4 ml of "urine" per hour could be obtained by cannulating the ureter.
        Perfusion with saline was continued for 10-15 minutes until all blood had been washed out of the organs and the vasculature had been maximally dilated. The outside of the kidneys were cleaned of adhering fatty tissue during the saline perfusion and the ureters were cannulated with PE-60 tubing. The kidneys were then perfused with a series of detergent solutions to disrupt and remove all possible cellu- lar material. All solutions contained 0.025% sodium azide. The order of perfusion was: 1) 4 liters of 0.5% Triton X-100

for 1-2 hrs non-recirculatory; 2) 4 liters of 0.5% Triton
X-100, 5 mM $CaCl_2$, 5 mM $MgSO_4$ for 1-2 hrs non-recirculatory;
3) 3% Triton X-100, 5 mM $CaCl_2$, 5 mM $MgSO_4$ for 4 hrs recirculatory; 4) 3% Triton X-100, 1 M NaCl, 5 mM $CaCl_2$, 5 mM $MgSO_4$
for 16 hrs recirculatory; 5) distilled H_2O for 2-6 hrs non-
recirculatory. The color of the kidneys became progressively
lighter during the detergent perfusions turning a translucent
off-white color at the completion of the procedure (Fig. 2).
This treatment disrupted all intact cells in the kidneys, but
did not solubilize or wash out all cellular debris. This

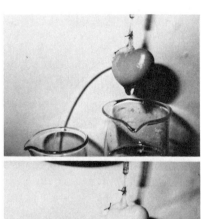

*Fig. 2. Photograph of
isolated perfused kidney before (upper) and after (lower)
perfusion with detergent
series. See Methods for details.*

could be achieved by a final perfusion with pancreatic deoxyribonuclease (.025%) in pH 5 acetate buffer in a recirculatory
fashion for 4 hours followed by recirculatory perfusion with
4% sodium deoxycholate for 1-2 hours, and an extensive water
wash. However, these final cleanup steps were omitted in
these experiments since the objective was to obtain an acellular kidney with ultrastructurally intact basement membranes
under minimal conditions of detergent treatment.
    Red blood cells were perfused as a 5% suspension in
physiological saline. India ink (colloidal carbon) was perfused as a suspension of a few drops of ink in distilled water.

Colloidal gold was a clear non-turbid solution (.093 g/liter) obtained from Harleco and used directly for perfusion. The appearance of colloidal gold in the urinary and venous outflows was minotored by optical absorption at 520 nm. Ferritin isolated from horse spleen was obtained from Calbiochem or Miles and was perfused at concentrations of 0.5-2.0 mg/ml in water. The appearance of ferritin in the urine and the perfusate was followed by optical absorption at 410 nm. Dextran T-500 having a weight average molecular weight of 496,000 and consisting of a weight distribution from 50,000-2,000,000 molecular weight was obtained from Pharmacia and was perfused in water solution at a concentration of 0.1 mg/-ml. The concentration of Dextran T-500 in the urinary and venous outflows was measured at 620 nm by the anthrone method (6). FITC-Dextran T-500 was synthesized according to the method of De Belder and.Granath (7) and size distributed on a 2.5 x 90 cm Sepharose 6B column eluted with water at a flow rate of 15 ml/hr. The effluent fractions were monitored for fluorescence with a Turner 110 fluorometer. The FITC-Dextran T-500 was perfused at a concentration of 1 mg/ml in water.

At the end of perfusion, the kidneys were fixed by perfusion and/or immediate sectioning and immersion in Karnovsky's fixative (4% paraformaldehyde, 5% glutaraldehyde in 0.05 M cacodylate buffer). Tissues were subsequently washed and stored in 0.15 M cacodylate buffer pH 7.4. Samples of tissue were embedded for routine light microscopy and sections were alternately stained with hematoxylin eosin, periodic-acid-Schiff (PAS) or Gomori's iron stain. Small 1 mm cubes were dehydrated in graded alcohols and embedded in Spurr's low viscosity epoxy resin. Sections were cut at 1 micron with glass knives and stained with toluidine blue. Ultra thin sections were cut 80 nm with diamond knives and sequentially stained with uranyl acetate and lead citrate. Sections were examined on a Hitachi HU-12 transmission electron microscope. Material for scanning electron microscopy was critical point dried in CO_2, coated with a thin covering of gold-palladium and viewed in a ETEC scanning electron microscope.

RESULTS AND DISCUSSION

The kidneys perfused with the series of detergents described were similar in appearance, size and shape to the isolated kidneys except for the almost white color of the organs (Fig. 2). The use of a graded series of solutions of

Triton X-100 made possible the stepwise removal of cellular material from the perfused kidneys to yield acellular organ preparations which could be perfused at good flow rates and in which the urinary outflow could be collected from a cannula inserted in the ureter. The inclusion of first Ca^{++} and Mg^{++} and then 1 M NaCl in the later perfusion solutions is believed to aid first in the digestion of DNA by an endogenous nuclear endonuclease and then in the disruption of nuclear membranes. This sequence of treatments minimized problems resulting from the sudden release of high molecular weight viscous DNA into the perfusion solution. The resulting acellular kidneys could be perfused at normal physiological pressures of about 110 mm Hg (8) and retained the anatomical appearance of normal kidneys grossly (Fig. 2), histologically (Figs. 7 and 8), and ultrastructurally (Figs. 3, 4 and 6), except that they lacked intact cells. Scanning electron microscopy of the detergent treated perfused fixed kidneys indicated that the histoarchitecture of the organs had been maintained and that they were free of cells (Fig. 3). Glomeruli enclosed by Bowman's capsule and aggregates of tubules were readily identifiable and were present in their normal spatial arrangement to one another. Light microscopy of stained sections revealed that the spatial relationships of the glomerular, peritubular capillary, Bowman's capsule and tubular basement membranes had been preserved (Figs. 7 and 8). Transmission electron microscopy confirmed the retention of ultrastructurally intact basement membranes (4,5) in the acellular kidneys (Figs. 4, 5, 6, 9 and 10). With the exception of some unbroken nuclear envelopes, tissue debris and interstitial collagen, the kidney was a skeleton of its basement membranes. This is a striking demonstration of the important structural role which the basement membranes play in the form and structure of the kidney and its subfractions (9-11). Treatment with Triton X-100 and sodium deoxycholate has been shown not to alter the ultrastructure, physical properties or chemical characteristics of isolated basement membranes (4,5,9).

Perfusion of the detergent treated acellular kidneys with a suspension of red blood cells in physiological saline resulted in the progressive entrapment of all of the blood cells in the kidneys with none appearing in the clear "urine". Perfusion of the bright red kidneys at this point with distilled water resulted in hemolysis of the red blood cells and the appearance of hemoglobin in the venous outflow and the urine. This indicated that a protein of molecular weight of 68,000 was freely accessible to the urinary space of the acellular kidney. Hemoglobin has been shown to traverse the intact glomerular capillary wall in rats (12).

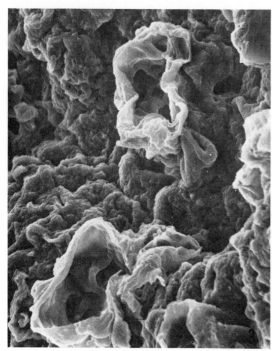

Fig. 3. Scanning electronmicrograph showing the appearance of the cut surface of the acellular renal cortex. The two large central structures represent partially opened Bowman's capsules. A portion of the denuded glomerular capillary tuft can be seen within the lower one. Note also the appearance of the convoluted tubular profiles. X 600.

Fig. 4. Transmission electronmicrograph showing portions of three glomerular capillaries from a kidney perfused with colloidal carbon particles. Note that the acellular capillary basement membrane is completely impermeable to the carbon although the particles enter freely into the mesangial regions. Also note the absence of carbon particles in the interstitial space (IS) and tubular lumens (T). X 6,600.

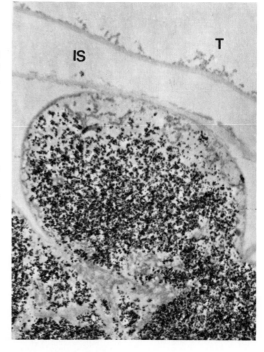

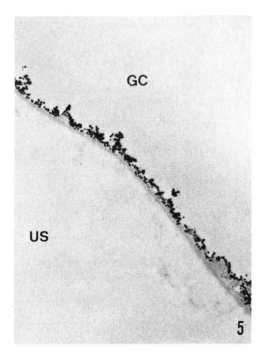

Fig. 5. Transmission electronmicrograph showing a portion of an acellular glomerular capillary wall taken from a kidney perfused with colliidal gold. The electron dense gold particles are completely restricted by the membrane to the capillary side (GC). X 19,900.

Fig. 6. Transmission electronmicrograph taken from acellular kidney perfused with colloidal gold. The gold particles are seen throughout the entire thickness of the peritubular capillary basement membranes (C), free within the interstitial space (IS) and adherent to the outside of the tubular basement membranes (%). X 9,450.

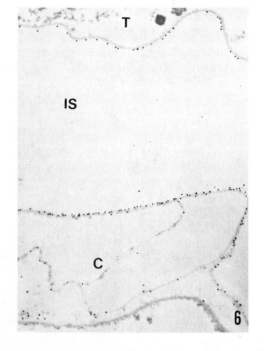

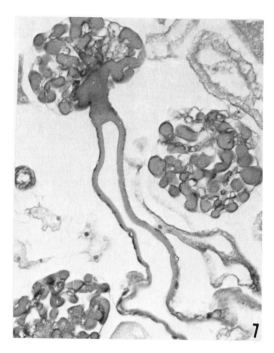

Fig. 7. Light micrograph of acellular renal cortex perfused for 10 min with ferritin. Note the lightly stained ferritin filling the glomerular capillary lumens of the three glomeruli present as well as lining the wall of the afferent arteriole supplying the central glomerulus. PAS X 1,200.

Fig. 8. Light micrograph of the acellular renal cortex of a kidney perfused for 90 min with ferritin. The section was stained with Gomori's iron method to accentuate the ferritin localization. Note the presence of ferritin within the Bowman's space of the two glomeruli present as well as within the convoluted tubular lumens (T) and peritubular capillaries. X 1,200.

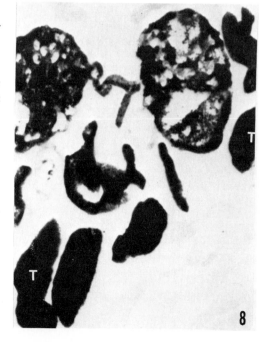

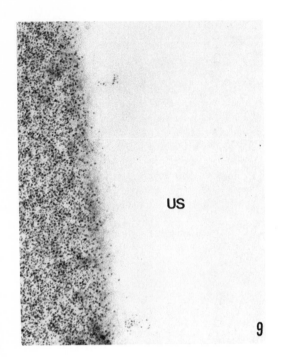

US

Fig. 9. Transmission electronmicrograph of a portion of glomerular capillary wall from kidney perfused 90 min with ferritin. Note partial penetration of the electron dense ferritin particles which are seen penetrating the wall with some particles free within the urinary space (US). X 79,000.

9

Fig. 10. Transmission electronmicrograph showing post glomerular distribution of ferritin after 90 min of perfusion. The basement membrane of the peritubular capillary (C) at the top is completely permeated by ferritin particles. At the bottom is a portion of a proximal tubule (T). Note the adherence of the ferritin to the outside of the tubular basement membrane as well as to probable collagen fibrils present within the interstitial space. X 81,000.

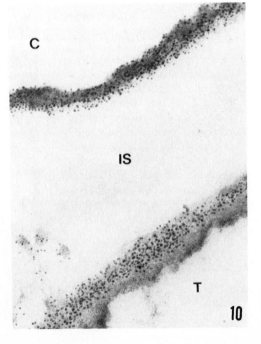

C

IS

T

10

Perfusion of the acellular kidneys with India-ink solutions caused the kidneys to turn dark black as particles of
colloidal carbon were trapped in the vasculature. No colloidal carbon was observed in the urinary outflow. Transmission
electron microscopy of the colloidal carbon perfused kidneys
confirmed that all of the carbon particles were restricted to
the vasculature, predominantly the glomerular capillaries,
with none gaining access to the interstitial spaces or the
tubular lumens (Fig. 4), indicating that there were no breaks
in the continuous basement membranes of the vasculature which
would permit the passage of particles of about 200-500 Å in
diameter.

Colloidal gold perfusion of the acellular kidneys
resulted in a similar trapping of gold particles in the
kidneys turning them a maroon color. As with the carbon, no
colloidal gold particles were detected in the urinary outflow
by optical density measurements (Table I). Electron micrographs of glomerular capillary basement membrane showed that

TABLE I

Urinary and Venous Concentrations of Tracers
in the Acellular Perfused Kidney

Tracer	Time of Perfusion	Urinary Outflow	Venous Outflow
	(min.)	*(O.D.)*	*(O.D.)*
Colloidal gold	30	None	.25
.093 mg/ml	60	None	.10
Ferritin	35	.16	.35
0.5 mg/ml	70	.18	.36
Ferritin	1	--	.565
1.0 mg/ml	19	.023	.562
	51	.032	.348
	91	.031	.252
Dextran 500,000 MW	3	.25	.77
0.1 mg/ml	6	.43	.76
	9	.59	.74
	13	.60	.72
	17	.66	.71

all of the gold particles were restricted to the capillary
space with almost no penetration into the basement membrane
(Fig. 5). This observation is similar to the behavior of
these particles of 40-200 A in diameter *in vivo* (13). The
gold particles appeared to line up along the capillary side
of the glomerular basement membrane (Fig. 5). Electron micrographs of peritubular capillary basement membrane, however,
showed penetration of the gold particles into and through the
basement membrane with the appearance of particles in the
interstitial space (Fig. 6). No gold particles were observed
within tubular lumens. However, particles were seen adhering
to the interstitial side of the tubular basement membrane.

When a solution of ferritin was perfused through the
acellular kidneys, they gradually assumed a light orange
color as ferritin was trapped within the kidneys. Light
microscopy clearly indicated that much of the ferritin filled
afferent arterioles and became localized in the glomerular
capillaries (Figs. 7 and 8). Little or no ferritin was seen
within tubular lumens or in the collecting ducts after
short (10 min) periods of perfusion (Fig. 7). In kidneys perfused for longer periods of time (1-2 hrs), ferritin could
be seen within proximal tubular lumens in sections stained
for iron (Fig. 8). Electron micrographs of the acellular
kidneys showed variable but appreciable penetration of the
ferritin into and through the glomerular capillary basement
membranes, although most of the tracer was retained in the
capillary lumens (Fig. 9). In some areas ferritin penetrated
through to the urinary space whereas in others the particles
were either restricted to the capillary side of the basement
membrane or retained by the lamina densa. This behavior of
the basement membrane towards the passage of ferritin is
qualitatively similar to that observed *in vivo* where ferritin
shows variable penetration into the glomerular capillary
basement membrane with most particles retained on the capillary side (14,15). However, whereas in the intact glomerular
capillary wall ferritin seldom penetrates into the lamina
densa and beyond, in the glomerular basement membrane of the
acellular kidneys many ferritin particles were seen penetrating through the entire width of the basement membrane with
some appearing free in the urinary space (Fig. 9). The
endothelial cells which normally line the glomerular basement
membrane limit access to (14) and modulate hydrodynamic
flow (15-20) through the membrane. The absence of these
cellular elements probably accounts for the observed greater
permeability to ferritin in our experimental model.

Electron micrographs of peritubular capillary and
tubular basement membranes showed that ferritin freely

penetrated the peritubular capillary basement membrane appearing in significant amounts in the interstitial space (Fig. 10). Although ferritin penetrated to the central portion of the tubular basement membranes, none was observed to pass through its entire thickness (10). Ferritin particles were frequently seen in association with interstitial collagen. The greater permeability of peritubular vs. glomerular capillaries to ferritin and catalase (21) and neutral high molecular weight dextrans (22) has been reported. In the acellular perfused kidneys the basement membranes of the peritubular capillaries were distinctly more permeable to both colloidal gold and ferritin than those of the glomerular capillaries indicating an intrinsic difference between the ability of these two barriers to retain macromolecules. In some areas, however, penetration of particulate markers from the peritubular capillaries into the interstitial space may be due to the known thin and discontinuous nature of the peritubular capillary basement membrane in rabbit kidneys (23). Monitoring of the urinary and venous outflows of kidneys perfused with ferritin by means of optical density measurements, indicated that ferritin appeared in the urine in variable, but significant amounts (Table I). The concentration of ferritin in the venous outflow sometimes decreased with time indicating sequestration of ferritin within the kidneys. This extraction of ferritin from the perfusate is in agreement with the electron microscopic observations of extensive penetration and probable trapping of ferritin particles in the basement membranes of glomeruli, tubules, peritubular capillaries and Bowman's capsule (Figs. 9 and 10). The extraction of red blood cells and colloidal gold (Table I) and carbon from the perfusate by the acellular kidneys was also observed and appears to be a common feature of this preparation when perfused with particulate tracers. This trapping along with possible occlusion of narrow portions of the collecting ducts with debris not removed by the detergent procedure may sometimes result in only minimal appearance by tracer macromolecules such as ferritin in the urinary outflow even though light (Fig. 8) and electron (Fig. 9) microscopy indicates that these particles are gaining access to the urinary space and the tubules.

Perfusion of acellular kidneys with dextran of 500,000 weight average molecular weight and monitoring of the urinary and venous outflows for glucose by the anthrone method revealed that dextran appeared in the urine within a few minutes with the concentration rising until it equaled that in the venous outflow of the perfusate (Table I). When FITC-Dextran was perfused through an acellular kidney and the

material appearing in the urine was fractionated on a Sepharose 6B column, the size distribution obtained was identical
to that of the material originally present in the perfusate
(Fig. 11).

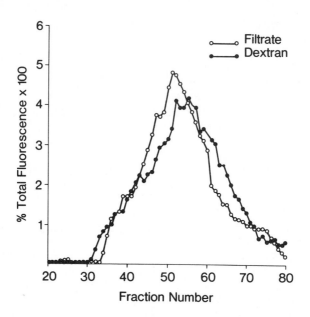

*Fig. 11. Size distribution of FITC-Dextran
of 500,000 MW on Sepharose 6B. Material before
perfusion (dextran) is compared with that found in
urinary outflow after perfusion (filtrate).
See Methods for details.*

These results indicate that the basement membranes of the
kidney are freely permeable to dextrans of high molecular
weight in the size range of 100-300 Å in diameter, compared
with ferritin which has an average molecular weight of 600,000
including iron, and a diameter of 122 Å. Very little dextran
appeared to be extracted from the perfusate by the kidneys,
indicating that little trapping of these macromolecules in
the basement membranes occurs. Although confirmation by
electron microscopy is essential, and leakage of dextran
through large vessels in the renal pelvis into the urine has
not been ruled out, it is probable that the main route of
passage of dextran into the urinary space in the acellular
kidney is through the glomerular basement membrane. The
considerably greater permeability of the glomerular basement

membrane to dextrans in this system as compared to ferritin
is consistent with the known behavior of the glomerular
basement membrane as a polyanion (24-26) which restricts the
passage of anionically charged molecules such as ferritin (27)
and dextran sulfate (28) more than it does cationically
charged and neutral molecules (27-29). The intrinsic barrier
characteristics of the glomerular basement membrane without
any cellular contribution as seen in this model system,
therefore, appears to be one of a poor size filter and a more
selective charge filter. This is in agreement with the
findings of Venkatachalam and co-workers (27,29) that the
permeability of the glomerular basement membrane to cationized
derivatives of ferritin and horseradish peroxidase is much
greater than to the native molecules, even though there are
no appreciable size differences between the differently
charged forms of these tracers. Since dextrans of the
molecular size observed to pass freely through the basement
membranes in the acellular kidney are virtually excluded from
the urinary space in the intact kidney (3), it would appear
that the cellular linings of the glomerular capillary wall
must contribute substantially to the barrier to this type
of macromolecule. The difficulty of staining dextran molecules in the glomerular basement membrane has made the
contribution of this component of the capillary wall difficult
to assess in *in vivo* perfusion experiments monitored by
electron microscopy (22).

Our results, along with those obtained by others in
intact kidneys, suggest that no single component of the
glomerular capillary wall is an effective barrier to macromolecules by itself. Rather, the glomerular barrier function
is probably a composite of limitation of access to the
glomerular basement membrane by the fenestrated endothelial
layer (1,14) and by concentration-polarization effects (18,
19,30), of passage through the basement membrane by charge
(27-29) and molecular sieving effects (16), and of limitation
to the urinary space by the epithelial cell lining which
with its foot processes and slit pore complex (17) modulates
the permeability of the filter by controlling hydrodynamic
flux (16-20) and recovers macromolecules traversing the
basement membrane by cellular ingestion (1,14). The basement
membrane plays an essential but not autonomous role in this
barrier. The acellular perfused kidney appears to be an
interesting and useful model system to define this role, independent of the cellular contributions present *in vivo*.

ACKNOWLEDGEMENTS

    We would like to thank Ms. Ann Griffen for valuable
technical assistance in the preparation of the acellular
perfused kidneys, and Dr. Claire Payne and Ms. Virginia
Satterfield for help in the preparation of samples for light
and electron microscopy. This work was supported by grants
AM-15394 and AM HL 14977 from the National Institutes of
Health. Dr. Meezan is a Research Career Development Awardee
of the National Institute of Arthritis, Metabolism and
Digestive Diseases.

REFERENCES

1. Farquhar, M.G., *Kidney Int'l*. 8:197-211, 1975.
2. Karnovsky, M.J. and Ainsworth, S.K., *Adv. Nephrol*.
    2:35-60, 1973.
3. Brenner, B.M., Baylis, C. and Deen, W.M., *Physiol. Rev*.
    56:502-534, 1976.
4. Meezan, E., Hjelle, J.T., Brendel, K. and Carlson, E.C.,
    Life Sci. 17:1721-1732, 1975.
5. Meezan, E., Brendel, K., Hjelle, J.T. and Carlson, E.C.,
    (this volume)
6. Seifter, S., Dayton, S., Novic, B. and Muntwyler, E.,
    Arch. Biochem. 25:191-200, 1950.
7. DeBelder, A.N. and Granath, K., *Carbohydrate Res*. 30:
    375-378, 1973.
8. Kozma, C., Macklin, W., Cummins, L.M. and Mauer, R.,
    In: "The Biology of the Laboratory Rabbit" (Weisbroth,
    S.J., Flatt, R.E. and Kraus, A.L., eds.) Academic Press,
    New York, p. 57, 1974.
9. Welling, L.W. and Grantham, J.H., *J. Clin. Invest*. 51:
    1063-1075, 1972.
10. Murphy, M.E. and Johnson, P.C., *Microvasc. Res*. 9:242-
    245, 1975.
11. Vracko, R., *IN:* "Proceedings of the First International
    Symposium on the Biology and Chemistry of Basement
    Membranes" (Kefalides, N.A., ed.) Academic Press, New
    York
12. Ericsson, J.L.E., *Nephron* 5:7-23, 1968.
13. Farquhar, M.G. and Palade, G.E., *Anat. Rec*. 133:378-
    379, 1959.
14. Farquhar, M.G., Wissig, S.L. and Palade, G.E., *J. Exp.
    Med*. 113:47-66, 1961.
15. Ryan, G.B. and Karnovsky, M.J., *Kidney Int'l*. 8:219-232,
    1975.

16. Renkin, E.M. and Gilmore, J.P., *Handbook Physiol.* 8:185-248, 1974.
17. Ryan, G.B., Rodewald, R. and Karnovsky, M.J., *Lab. Invest.* 33:461-468, 1975.
18. Ryan, G.B. and Karnovsky, M.J., *Kidney Int'l.* 9:36-45, 1975.
19. Ryan, G.B., Hein, S.J. and Karnovsky, M.J., *Lab. Invest. 34:415-427, 1976.*
20. *Shea, S.M. and Morrison, A.B.,* J. Cell Biol. 67:436-443, 1975.
21. Venkatachalam, M.A. and Karnovsky, M.J., *Lab. Invest.* 27:435-444, 1972.
22. Caulfield, J.P. and Farquhar, M.G., *J. Cell Biol.* 63:883-903, 1974.
23. Bulger, R.E. and Nagle; R.B., *Am. J. Anat.* 136:183-204, 1973.
24. Seiler, M.W., Venkatachalam, M.A. and Cotran, R.S., *Science* 189:390-393, 1975.
25. Caulfield, J.P. and Farquhar, M.G., *Proc. Natl. Acad. Sci.* 73:1646-1650, 1976.
26. Caulfield, J.P. and Farquhar, M.G., *IN:* "Proceedings of the First International Symposium on the Biology and Chemistry of Basement Membranes", (Kefalides, N.A., ed.), Academic Press, New York, p.
27. Rennke, H.G., Cotran, R.S. and Venkatachalam, M.A., *J. Cell Biol.* 67:638-646, 1975.
28. Chang, R.L.S., Deen, W.M., Robertson, C.R. and Brenner, B.M., *Kidney Int'l.* 8:212-218, 1975.
29. Venkatachalam, M.A., Rennke, H.G., Seiler, M.W. and Cotran, R.S., *IN:* "Proceedings of the First International Symposium on the Biology and Chemistry of Basement Membranes", (Kefalides, N.A., ed.), Academic Press, New York
30. Deen, W.M., Robertson, C.R. and Brenner, B.M., *Biophys. J.* 14:412-431, 1974.

ROLE OF THE BASEMENT MEMBRANE COLLAGEN

IN THE

BLOOD-BRAIN BARRIER PERMEABILITY

A. M. Robert, M. Miskulin,

F. Moati and G. Godeau

*Institut de Recherche
sur les Maladies Vasculaires
C. H. U. Henri Mondor
Creteil, France*

*SUMMARY: Several proteases are able to increase the
permeability of the blood-brain barrier (BBB) to
several traces, such as trypan blue or horseradish
peroxidase. We found that collagenase administered
in the lateral brain ventricles exerted the strongest
and the longest lasting action on BBB permeability.
We found after intraventricular injection of collagenase an increased OH-Pro level in the CSF. We also
found that after treatment of the animals with inhibitors of protein synthesis, puromycin and cycloheximide,
twice as much time is necessary (140 hours) for the
recovery of normal BBB permeability after collagenase
action, than in control animals.*

*Electronmicroscopic studies showed that the tight
junctions between endothelial cells remain closed after
collagenase action, but more horseradish peroxidase
enters the endothelial cells by pinocytosis than in
rats which did not receive collagenase.*

INTRODUCTION

The study of intercellular matrix has received these last few years a great deal of attention and has been carefully studied in many fields. However very little is known of the contribution of intercellular matrix macromolecules to physiological junctions and pathological alterations of the central nervous system. The purpose of our work is to obtain information on the functional role of the intercellular matrix macromolecules in the central nervous system.

We began our work with the study of the permeability of the blood-brain barrier. The reason for this choice was the fact, that it was predictable, that in this field some connective tissue elements may interfere with the exchange of metabolites between the capillary blood and the brain tissue. As a matter of fact, a molecule on its way from the capillary lumen to the brain tissue has to cross several anatomical structures where the intercellular matrix macromolecules are present: the endothelial cell wall, the intercellular cleft with its tight junctions, and the basement membrane.

Our first approach was based on the fact that proteolytic enzymes are able to increase the permeability of the blood-brain barrier to tracers such as trypan blue, which in normal conditions does not enter the brain tissue. We worked out a quantitative colorimetric method (1) based on the measurement of the optical density of i.v. injected trypan blue extracted from the brain after protease action. We found that several proteases were able to increase the permeability of the BBB for trypan blue, but collagenase had the most intense action of all of them, when it was injected in the CSF, in the lateral ventricle of the brain (1). When injected that way, the enzyme does not have to cross the BBB itself, before it acts on its substrate. We also established that increase of the BBB permeability after enzyme action was reversible. When we studied the time necessary for the recovery of normal BBB function, it turned out that collagenase also exerted the most long-lasting action: normal BBB function was recovered 72 hours after intraventricular injection of 100 μg of collagenase, against only 18 hours for orally administered 20 mg of α-chymotrypsin (2).

The very intense and long-lasting action of collagenase on BBB permeability suggested the hypothesis that its action is due to its specific action on basement membrane collagen: the increase of the permeability would be due to the partial hydrolysis and the recovery of normal permeability to the resynthesis of this collagen. In the present work, we intend

to present some experimental results which are in favor of
the above hypothesis.

MATERIALS AND METHODS

White Wistar male or female rats of 250 to 300 mg were
used.

Intraventricular injections were carried out under light
ethylcarbamate anesthesia with a stereotaxic instrument.

CSF was collected also under anesthesia by puncturing
the cysterna (3).

Hydroxyproline (OH-Pro) was determined according to
Bergman and Loxley (4).

The functional value of the BBB was determined with a
quantitative colorimetric method using trypan blue as a
tracer, as described before (1).

All substances used in this work were dissolved in
NaCl 0.9% for injection. All the reagents used were of
analytical grade.

In electronmicroscopical studies horseradish peroxidase
was used as a tracer; the rat brains were perfused with the
glutaraldehyde fixative and processed as described by
Karnovsky and co-workers (5).

The following chemicals and enzymes were used:

Collagenase, Calbiochem, Bacterial B grade and A
grade.

Collagenase, Worthington CLSPA.

Collagenase, Achromobacter iophagus, highly purified
(a gift of Professor Keil, Pasteur Institut,
Paris.)

Pronase, Calbiochem B grade.

Alpha-chymotrypsin, Worthington 3 X cryst.

Trypsin, Nutritional Biochemicals 3 X cryst.

Pepsin, Nutritional Biochemicals 3 X cryst.

Puromycin, Sigma.

Cycloheximide, Sigma.

Trypan Blue, Gurr.

RESULTS

Hydroxyproline in the Cerebrospinal Fluid

Many byproducts of the cerebral metabolism are carried by
the cerebrospinal fluid (CSF); a way to check on the degree of
collagen degradation in the brain consists of the measurement
of the amount of hydroxyproline (OH-Pro) released into the

cerebrospinal fluid.

We punctured the cysterna of rats and obtained 0.1 to
0.15 ml of CSF from one animal. When we determined the
OH-Pro content of CSF of untreated control rats, we found
levels from 1 to 2 µg/ml (Table I). When proteases were
introduced in the lateral brain ventricles and CSF is withdrawn 1 hour later, the level of OH-Pro was increased with
two of them, namely after injection of collagenase (6.75
µg/ml on the average) and of pronase (3.8 µg/ml). Injection
of pepsin and trypsin in the ventricles or orally administered alpha-chymptrypsin did not increase the level of
Oh-Pro in the CSF (Table I).

Collagenase had the strongest action on the BBB permeability, and it also provoked the highest increase of the
OH-Pro level in the CSF. Next comes pronase as well for its
capacity to increase the BBB permeability as well as for the
increase of the OH-Pro level in the CSF. Both of these
enzymes were found by Kefalides (6) to be able to hydrolyse
basement membrane collagen. Alpha-chymotrypsin, trypsin and
pepsin were much less effective on BBB permeability, they
also act much less on basement membrane and they do not increase the level of OH-Pro in the CSF (Table I).

These results seem to support the idea that the intense
action of collagenase on BBB permeability is in connection
with collagen degradation in brain capillaries (7).

Influence of Inhibition of Protein Synthesis on the Time
   Necessary for the Recovery of Normal BBB Function after
   Protease Induced Increase of the BBB Permeability

We have shown in another study that intraventricular
injection of 100 µg of collagenase is followed by a significant increase of BBB permeability which decreases progressively from the fourth hour following the injection, and normal
BBB permeability can be found again at about 70 to 72 hours
after collagenase injection. We present above arguments in
favor of the first part of our hypothesis concerning the
mechanism of action of collagenase degradation. The second
part of the hypothesis postulates that the recovery of a
normal permeability requires the re-synthesis of the degraded
substrate.

In order to check on this part of the hypothesis we
submitted groups of rats to injection of inhibitors of protein
synthesis, namely puromycin and cycloheximide. Collagenase
was injected in the lateral brain ventricles after 2 days of
treatment with the inhibitors. The permeability of the BBB
was then followed with the trypan blue test. We could show

TABLE I

The Relationship Between the Effect of Proteases
on the Level of OH-Pro in the CSF, on BBB Permeability
and on Their Ability to Hydrolyse Basement Membrane

	Action on BBB Permeability		OH-Pro in CSF µg/ml	Percentage of Peptides Hydrolysed* from Basement Membrane	
	Dose µg	Trypan Blue O.D. x 10^3		Renal Glomeruli	Descemet's Membrane
NaCl 0.9%	900	0	1.5**	--	--
Collagenase	100	90	6.75	90	96
Pronase	100	70	3.8	96	81
α-Chymotrypsin	510^3	45	2.0	--	--
Trypsin	10^3	35	1.0	71	71
Pepsin	10^3	12	1.0	70	84

* from Kefalides (6).

** normal values range from 1 to 2 µg/ml.

that the time necessary for the recovery of normal BBB
function was twice as much as in the absence of treatment
with protein synthesis inhibitors (7).

These results are also in agreement with the above
mentioned hypothesis as far as the connection is concerned
between the protein synthesis and the recovery of the normal
BBB permeability.

Role of the Tight Junctions in the Action of Collagenase
    on the BBB Permeability

Karnovsky and Reese, as well as others, showed in an
elegant series of electronmicroscopic work the existence of
the tight junctions between neighboring endothelial cells in
the "true" brain capillaries (8). Using horseradish peroxidase as a tracer, they demonstrated, that peroxidase given
intravenously as well as through the lateral brain ventricles,
diffuses through the intercellular spaces and the basement
membrane, and is stopped only at the level of the tight
junctions. This finding suggested to these authors the
conclusion that the true anatomical site of the BBB would be
the tight junctions.

Rapoport and Thompson reported the possibility of the
tight junctions opening after injection by osmotic shock,
and according to these authors, hypertension may also open
the tight junctions by causing distention. In the osmotic
shock the opening would be caused by shrinkage (9).

It was important for us to know what happens to the
tight junctions after injection of collagenase. The opening
of these junctions by the action of the enzyme was a possibility. We used the methodology of Karnovsky and co-workers
and injected horseradish peroxidase intravenously 1 hour
after the intraventricular injection of collagenase. We
found that the intravenously injected tracer penetrates
through the BBB (this is not the case in untreated control
animals), but the tight junctions remain closed (10).

This result indicates that the increase of BBB permeability following collagenase injection does not involve the
opening of tight junctions and some other explanation has to
be found.

Influence of Intraventricularly Injected Collagenase on the
    Rate of Pinocytosis

It is known that intravenous injection of horseradish
peroxidase is followed by the appearance of pinocytotic

vesicles in the endothelial cells. As a matter of fact, the peroxidase enters the endothelial cell, but when BBB function is normal, it does not migrate through the cells and does not reach the basement membrane.

Two groups of 20 rats were injected with horseradish peroxidase as above. One of the groups received, 1 hour before the tracer, 100 µg of collagenase by intraventricular injection. Half an hour after the injection of peroxidase the brain of the rats was perfused with glutaraldehyde, then the brains were excised and treated according to the above mentioned method in order to observe the brain capillaries on thin sections.

We counted the dark spots corresponding to horseradish peroxidase which entered the endothelial cells by pinocytosis in the control group and the group injected with collagenase. In the control group we found 1 to 5 spots for a capillary against 8 to 14 in the collagenase treated group. The difference between (Fig. 1) both averages, 3.3 for the controls and 10.4 for the collagenase groups is statistically highly significant (P <0.01).

DISCUSSION

The exchange of metabolites between blood and brain tissue is of great importance for the maintenance of normal brain function. The most important exchanges take place between capillary blood and the adjacent tissue through the capillary wall. This latter consists of a layer of endothelial cells and the basement membrane.

We studied the permeability of the BBB with two tracers, trypan blue and horseradish peroxidase. We found that several proteases were able to increase the permeability of the BBB to trypan blue, but collagenase had the strongest and the longest lasting effect. We thought that this intense action on the BBB permeability may be due to its action on the basement membrane collagen. In that case the increase of the permeability would be due to the partial hydrolysis of basement membrane collagen, and the recovery of normal permeability to the re-synthesis of the degraded substrate.

We collected some experimental evidence which seems to support this hypothesis:

Intraventricular injection of several proteases is followed by an increase of the level of OH-Pro in the CSF. The highest increase follows collagenase administration.

Treatment of the experimental animals by inhibitors of protein synthesis (puromicin and cycloheximide) doubled the

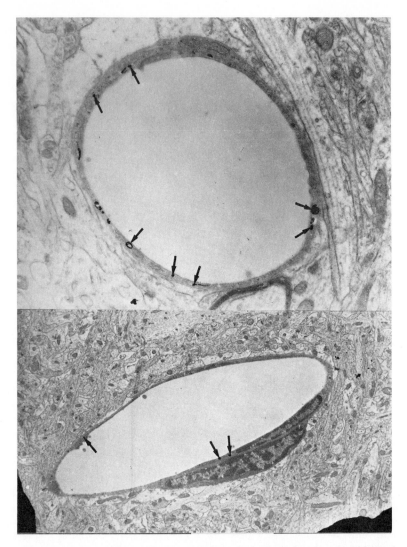

*Fig. 1. The picture shows two rat brain capillaries.
Top: control animal. Bottom: rat injected with collagenase. Both animals received intravenous peroxidase.
Dark spots, indicated by the arrows, represent the tracer
entered by pinocytosis, in the endothelial cells. After
intraventricular injection of collagenase, an increased
pinocytotic activity is seen in this capillary (not all
of the peroxidase spots are indicated by the arrows).*

time necessary for the recovery of normal BBB function. We
checked in control experiments that the inhibitors used had
no action on the BBB permeability by themselves.

We could show with horseradish peroxidase as a tracer
that intraventricular injection of collagenase does not open
the tight junctions between neighboring endothelial cells.

We also demonstrated with the same tracer that collagenase induces an increase in the number of peroxidase spots
visible in the endothelial cells. This finding indicates an
increase of the pinocytotic activity of the endothelial cells.

According to these results, at the present state of
our work, we can consider that collagenase increases the
permeability of the BBB to different tracers; this increase
of permeability is accompanied by an increase of the collagen
degradation products in the CSF; and the recovery of normal
BBB permeability requires protein synthesis. Opening of
tight junctions does not appear to be necessary for increased
BBB permeability. It appears however possible that the
pinocytotic activity of endothelial cells is controlled by
the structure of the basement membrane.

REFERENCES

1. Robert, A.M. and Godeau, G., *Biomedecine Express* 21:36-
    39, 1974.
2. Robert, A.M., Godeau, G. and Miskulin, M., *IN:* "Protides
    of Biological Fluids", 22nd Colloquium, (Peeters, H.,
    Ed.), Pergamon Press, Oxford and New York, pp. 343-347,
    1975.
3. "The Rat in Laboratory Investigation", 2nd Edition,
    Farris, E.J. and Griffith, J.Q., Eds., Lippincott,
    Philadelphia, London, Montreal, pp. 196-199, 1970.
4. Bergman, J. and Loxley, R., *Anal. Chem.* 35:1961, 1961.
5. Karnovsky, M.F., *J. Cell Biol.* 35:213, 1967.
6. Kefalides, N.A., *IN:* "International Review of Experiment-
    al Pathology" (Richter, G.W. and Epstein, M.A., Eds.),
    Vol. 10, p. 1, 1971, Academic Press, New York.
7. Robert, A.M., Godeau, G., Miskulin, M. and Moati, F.,
    (in press) *Neurochemical Research*, 1977.
8. Reese, T.S. and Karnovsky, M.J., *J. Cell Biol.* 34:207-
    217, 1967.
9. Rapoport, S.J. and Thompson, H.K., *Science* 180:971, 1973.
10. Robert, A.M., Godeau, G. and Miskulin, M., *Ann. Med.
    Reims.* 13:89-92, 1976.

Supported by CNRS (ER No. 53), D.G.R.S.T. (72-7-0823),
I.N.S.E.R.M. (20.75.43) of France.

MOLECULAR SIEVING PROPERTIES OF ISOLATED

GLOMERULAR BASEMENT MEMBRANE

N. Gunnar Westberg

Sahlgren's Hospital
University of Göteborg
Göteborg, Sweden

SUMMARY: The permeability to macromolecules of the glomerular capillary filter (GCF) can be studied in man or in experimental animals using conventional clearance techniques for macromolecules of varying sizes. This approach does not give any information regarding the role of the different layers of the GCF for the permeability. Electron microscopy after infusion of tracer molecules into experimental animals has been used in a large number of studies. As has recently been emphasized by Ryan and Karnovsky (1), hemodynamic factors influence the results to a major degree. The methods for fixation may introduce arti- facts and are open to objections. These problems are discussed by Venkatachalam at this symposium.

   To reduce the complexity of the problem, an attempt to study the permeability of isolated GBM in vitro was made using columns with packed GBM. The importance of solute concentration, sialic acid and chaotropic ions was studied in this system.

METHODS

Fresh bovine kidneys were obtained from the slaughterhouse and transported on ice. The preparation was started
the same day and the tissue was never frozen. GBM was
prepared using the same method as previously described from
human kidneys (2,3), but using 150 mesh sieves to collect the
glomeruli. After ultrasonication, low speed centrifugation
in 1 M sodium chloride and passage through a fine mesh sieve,
the GBM was twice suspended in the phosphate buffer used for
the first run on the chromatography column. The GBM suspension was poured into a glass column,   8 x 300 mm with a
small amount of glass wool at the tapered lower end. After
the GBM had settled, the column was washed with at least two
column volumes of buffer. The pressure was increased by the
connection of the column to a Mariotte flask. All runs were
performed at a hydrostatic pressure of 100 to 200 cm water
and a flow of one or two drops per hour.

Buffers

The following buffers were used:
Tris-HCl with 0.02% sodium azide, pH 7.4, osmolarity
300 mOsm/kg water.
Tris-HCl-Ca^{2+}, same as above with $CaCl_2$ 0.005 M added.
Tris-HCl-EDTA, as Tris-HCl with Na_2EDTA 0.005 M.
Tris-HCl-Ca^{2+}, higher concentration: NaCl added to
Tris-HCl-Ca^{2+} to get an osmolarity of 400 mOsm/kg water.
Tris-HCl, lower concentration: Water was added to
decrease the osmolarity to 180 mOsm/kg water at pH 7.4.
Acidic phosphate buffer, Na_2PO_4 0.05 M with NaCl added
to osmolarity 280 mOsm/kg water, pH 5.6.
Bovine serum albumin was added to all buffers at a con-
centration of 5 grams per·liter, except in two runs using
Tris-HCl.

Sample

To the column was applied 100 µl of the following
mixture: Frozen human serum from one donor, 35 µl; Inulin-
(methoxy-^{3}H) 35 µl; ^{51}Cr-EDTA, 25 µl; 1 M KCl x 10 µl; β_2-
microglobulin, 10 µl, or about 15 µg (a gift from Prof.
Göran Lindstedt).

Column

At least 50 fractions of ten drops each were collected. The test tubes were covered to prevent evaporation. The run was performed in the cold room. Between each run, the GBM was washed in the column with at least two volumes of buffer.

Quantification of Fraction Constituents

Immunoglobulin G, α_2-macroglobulin, and β_2-microglobulin were quantified using the Mancini radial diffusion technique. Albumin was measured with the Laurell rocket immunoelectrophoresis. ^{3}H-Inulin radioactivity was measured in a Packard Tri-Carb liquid scintillation counter and ^{51}Cr-EDTA was counted in a Selectronic gamma counter. Potassium was quantified in a flame photometer. In preliminary experiments, when tritiated water but not ^{3}H-inulin was used, using buffer Tris-HCl, osmolarity 300 mOsm/kg water, potassium was found to elute in the same place as THO.

Preparation of glomerular basement membrane. Most fragments are long and folded. A large piece of Bowman's capsule is seen.

RESULTS

The majority of the GBM fragments obtained with this method of sonication and low speed centrifugation are relatively large, as can be seen in the figure. Bowman's capsule, aggregated endothelial and epithelial cells and some mesangial material are present in the preparations, but to a very small extent.

In all experiments, there was a separation according to molecular weight, with the heavier molecules coming first. The degree of separation was relatively poor, and somewhat variable from run to run, even under apparently identical conditions on the same column. There was generally a separation into three peaks. The first peak contained α_2-macro-

globulin, immunoglobulin G and albumin.   There was a tendency
for α_2-macroglobulin to precede IgG and albumin, both with
the peak and the inflexion point of the up-slope, and α_2-
macroglobulin was never delayed after IgG.   The second peak
contained ^{3}H-inulin, ^{51}Cr-EDTA and potassium and there was a
tendency for these three tracers to separate out also according to molecular weight.

β$_2$-microglobulin was delayed and eluted in a third peak
in all experiments.

A total number of 21 experiments on four columns are
reported.

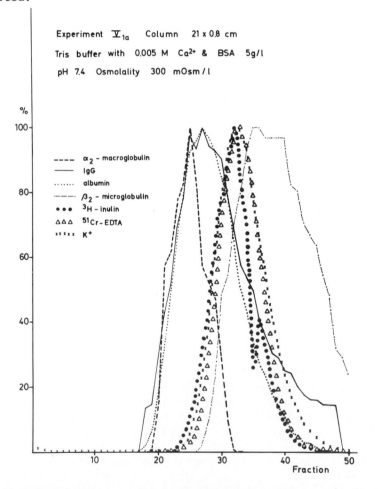

Experiment V_{1a}, with calcium ions and bovine serum
albumin in the buffer.   To facilitate comparison between
the different curves, the maximum value for each tracer
is set to 100%.   Three groups of peaks are seen.

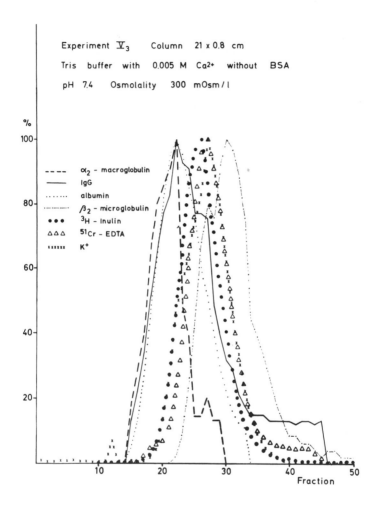

Experiment V_3, with calcium ions but not bovine serum albumin in the buffer. The pattern is similar to that seen in Experiment V_{1a}.

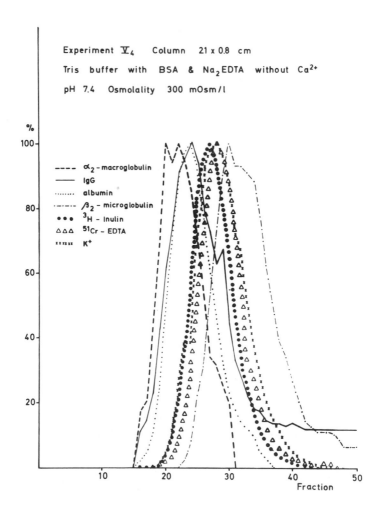

Experiment V_4. Bovine serum albumin, but not calcium is present. Same pattern as in Experiment V_{1a} and V_3 but the α_2-macroglobulin elutes clearly before IgG and albumin.

In 14 experiments at pH 7.4 the solute concentration
was close to that of blood plasma, in three it was lower, in
two it was higher. Two runs were made at pH 5.6 (acidic
phosphate buffer), but all other experiments at pH 7.4. Of
19 experiments at pH 7.4, twelve were done in the presence
of calcium ions, five without calcium but with EDTA and two
experiments without either. Two runs made without BSA in the
buffer (Tris-HCl) showed the same pattern as runs with BSA.

The solute concentration, pH, presence or absence of
calcium or of BSA had no observable influence on the type or
degree of separation.

DISCUSSION

When the GBM fragments are packed in a glass column
as in the experiments described here, there are four
readily apparent possibilities for the passage of macromolecules through the column:

1. *Peribetes* (Περι, around + βαινειν, to go, cf.
diabetes). In this situation the tracers will not significantly pass through the GBM fragments, but only around them.
No significant separation will occur. If the fragments are
sufficiently tightly packed, the water channels between the
fragments might be narrow enough to impede the passage of
larger molecules at certain points, and frictional drag
could conceivably be of importance. These mechanisms would
tend to retard the larger molecules preferentially.

2. *Filtration* Here channels through the GBM fragments
would provide "short-cuts" for molecules small enough to pass
through these holes. If the pores are not much bigger than
α_2-macroglobulin molecules, this macromolecule would be retarded and eluted after the smaller molecules.

3. *Diffusion* This mechanism has been eloquently
championed by Chinard (4) as an alternative to the filtration
hypothesis. Here too, small molecules would travel faster
than larger molecules.

4. *Gel filtration* The GBM matrix is here visualized as
a network of molecules with water spaces of variable sizes
between. The distribution volume for smaller molecules will
be larger than for larger molecules. In a system of random
movements of molecules, the tracers will be dispersed through
all the available volume. Thus, the smaller molecules will
have a longer way to travel during their passage through the
column, and will be retarded.

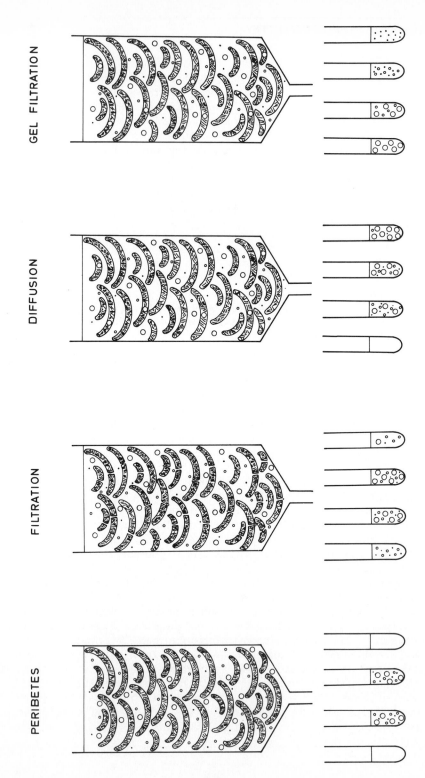

GEL FILTRATION

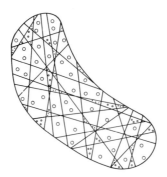

In addition there are of course obvious possibilities
for the delay of molecules through binding phenomena of
various kinds.

The two explanations of filtration and diffusion would
both give a separation with the smaller molecules eluting
before the larger molecules. Only the gel filtration mechanism is thus compatible with the observations obtained in the
experiments presented here. However, the retardation of
β_2-microglobulin must have a different explanation. A binding
to some group in the GBM seems to take place. This binding
could not be overcome by the lowering of pH to 5.6 or the
increase of osmolarity to 400. The nature of this binding
awaits further study. For the remaining molecules, there is
no evidence of binding, or at least, of differences in binding
to the GBM matrix.

The solute concentration or the presence or absence of
calcium ions had no influence on the elution pattern, as far
as could be ascertained with technique. This is in contrast
to the effect obtained on the domain of hyaluronic acid (5).
One interpretation would then be that the GBM network is not
maintained by the repulsion of negative charges.

Studies of the separation of macromolecules on columns
packed with GBM were first reported by Huang, Hutton and
Kalant (6). They prepared GBM from rat kidneys and filled
capillary tubes, only five millimeters in length. They
obtained a minute degree of separation between ^{131}I-albumin,
eluting ahead of ^{14}C-mannose. Separation was not obtained
with GBM prepared from the kidneys of rats made nephrotic with
pupuromycin aminonucleoside. This single experimental observation remains to be confirmed.

Igarashi and collaborators (7) recently reported on similar experiments using rabbit kidney. The degree of separation between ^{131}I-albumin and ^{125}I-globulin from ^{125}I-insulin and tritiated water obtained by these authors were far superior to the results obtained by me. One experiment only was reported.

Gekle and collaborators (8) approached the problem of permeability of GBM *in vitro* in a study of the equilibrium partition co-efficient of macromolecules in a mixture of GBM and buffer. I have tried to use this direct and fast method, but have been unable to get reproducible results.

To study the physical chemistry of GBM in isolated form in chromatography column is certainly not a very physiologic approach. There is no blood flow past the capillary fragments; the GBM may well be severely damaged during the preparation procedure; the pressure fall per unit path length is much lower than *in vivo*. Venkatachalam has pointed out in the discussion during this conference that isolated GBM is permeable to ferritin, when exposed *in vitro* to this tracer. The main value of the type of experiment described here would be to compare the separation of macromolecules by isolated GBM under different physical conditions, and to compare GBM obtained from healthy kidneys and kidneys with e.g., diabetic nephropathy. This may provide some insight into the internal structure of GBM, in health and disease (9).

This work was partly reported in abstract form at the 6th International Congress of Nephrology, Firenze, 1975 (10).

REFERENCES

1. Ryan, G.B. and Karnovsky, M.J., *Kidney Int*. 9:36, 1976.
2. Westberg, N.G. and Michael, A.F., *Biochemistry* 9:3837, 1970.
3. Spiro, R.G., *J. Biol. Chem*. 242:1915, 1967.
4. Chinard, F.P., *In:* "Renal Function" (Bradley, S.E., ed.) Macy, New York, p. 40, 1961.
5. Laurent, T.C., *In:* "Chemistry and molecular biology of the intercellular matrix", (E.A. Balasz, ed.) Academic Press, New York, Vol. 2, p. 703, 1970.
6. Huang, F., Hutton, F. and Kalant, N., *Nature* 216:87, 1967.
7. Igarashi, S., Nagase, M., Oda, T. and Honda, N., *Clin. Chim. Acta* 68:255, 1968.
8. Gekle, D., v. Bruchhausen, F. and Fuchs, G., *Pflügers Archiv*. 289:180, 1966.
9. Westberg, N.G. and Michael, A.F., *Acta Med. Scand*. 194: 39, 1973.
10. Westberg, N.G., *6th Int. Cong. Nephr.*, Firenze, Abst. 320, 1975.

II. Chemistry and Metabolism of Basement Membranes
A. Composition and Structure

CURRENT STATUS OF CHEMISTRY AND STRUCTURE

OF BASEMENT MEMBRANES

Nicholas A. Kefalides

University of Pennsylvania
Philadelphia, Pennsylvania

SUMMARY: Basement membranes are extracellular matrices synthesized by a variety of cells which line up all epithelial and endothelial surfaces. Basement membranes in the mature animal are free of lipids, DNA and proteoglycans and are composed of dissimilar protein subunits. One of these is a procollagen-like molecule associated with non-collagenous matrix glycoprotein(s). Initially, the collagenous component of basement membranes, isolated after limited pepsin digestion alone, resulted in a molecule having three identical α-chains, whose molecular weight was about 115,000, rich in hydroxylysine, 3- and 4-hydroxyproline, low content of alanine and arginine and from 4 to 8 residues of half-cystine. However, reduction and alkylation of the initially isolated collagen under non-denaturing conditions, followed by a second pepsin treatment of the reduced product resulted in a collagen molecule having three identical α-chains with a molecular weight of 95,000 but without half-cystine. It contains 38 residues of glucosyl-galactosyl-hydroxylysine per chain. Newly synthesized basement membrane collagen is secreted in the extracellular space as the precursor molecule "procollagen". This molecule does not undergo conversion to collagen, as is the case with interstitial collagens, but interacts with the matrix glycoprotein(s) to give rise to the appropriate structure.

*Small variations in the amino acid composition of
the basement membrane collagens from different tissues
were noted. There are insufficient data available,
however, to state with certainty whether there are
primary structural differences among them, although
this may turn out to be the case.*

*The proportion of the non-collagen glycoprotein(s)
varies among basement membranes. The collagen and non-
collagen subunits are stabilized by hydrogen bonds,
disulfide bonds and aldehyde-derived cross-links which
are so extensive that they render these structures
highly insoluble.*

*Immunochemical studies show that antibodies to
basement membrane collagen (Type IV) are type specific
and do not cross-react with interstitial collagen
Types I, II or III.*

INTRODUCTION

Studies on the chemistry and structure of basement membranes have been reported by this and other laboratories and
form the basis of two reviews (1,2). The purpose of this
article is to summarize some earlier findings on the chemistry and structure of basement membranes and to present new
data on the structure of the protein components.

Since basement membranes are practically ubiquitous and
the structure and chemistry of a basement membrane in one
tissue may vary greatly from that found in another, I must
state that when I speak of the chemistry and structure of
basement membranes, I refer to those which we and others (1,
2) have isolated from mammalian species free of other tissue
components and proceeded to characterize them subsequently.
Data on invertebrate basement membrane from *Ascaris suum* are
presented elsewhere in this volume by Hudson et al. Table I
lists the basement membranes which were isolated from various
tissues, were freed of the cells with which they are associated and then analyzed.

In our hands, the only basement membranes which have
been isolated in a relatively pure state are those of the
glomerulus, the lens capsule, Descemet's membrane and
Reichert's membrane (1,2,5). Relatively "clean" tubular
basement membrane has been isolated from bovine kidneys by
Ferwerda et al. (3) and from rabbit kidneys by Meezan et al.
(25, and this volume). Preparations of basement membrane
from the choroid plexus and other brain vessels as well as
from the lung alveolus usually contain variable amounts of

TABLE I

Types of Mammalian Basement Membranes
Isolated and Analyzed

Organ	Tissue	Cell(s)	Basement Membrane
Kidney	Glomerulus	Epithelium & Endothelium	Glomerular (1,2)
Kidney	Tubule	Epithelium	Tubular (3,25)
Eye	Lens	Epithelium	Lens Capsule (1,2)
Eye	Cornea	Endothelium	Descemet's Membrane (1,2)
Brain	Choroid Plexus	Epithelium & Endothelium	Choroid Plexus (4)
Lung	Alveolus	Epithelium & Endothelium	Alveolar (4)
Embryo	Parietal Yolk Sac	Endoderm	Reichert's Membrane (5)

contaminating interstitial collagen making interpretation of structural studies difficult (4,25).

CHEMICAL STUDIES

Composition

Basement membranes are generally thought to be composed of peptide and carbohydrate moieties. The presence of phospholipid and cholesterol as well as nucleic acids and sulfated mucopolysaccharides has been attributed mainly to contamination by cellular debris. During embryonic development, however, sulfated mucopolysaccharides may be localized within a basement membrane (Bernfield and Banerjee, this volume), but their association with intrinsic basement membrane components may be transient.

The amino acid and carbohydrate composition of basement membranes from a variety of tissues and species has been examined by several investigators (see reviews 1 and 2). Tables II and III summarize the amino acid and carbohydrate composition of 4 representative mammalian basement membranes.

TABLE II

Amino Acid Composition of Mammalian Basement Membranes[a]

| Amino Acid | Human[b] | | Canine[b] | Rat[c] |
	Glomerulus	Lens Capsule	Descemet's Membrane	Reichert's Membrane
Hydroxylysine	24.5	34.5	20.0	18.9
Lysine	26.4	19.4	24.0	32.3
Histidine	18.7	15.2	11.6	18.3
Arginine	48.3	39.5	39.5	46.0
3-Hydroxyproline	12.0	21.3	6.8	6.6
4-Hydroxyproline	53.0	85.0	77.0	45.3
Aspartic	70.0	57.0	58.0	83.0
Threonine	40.3	31.0	36.4	46.0
Serine	54.2	43.4	42.0	64.0
Glutamic	101.3	94.4	94.0	10.0
Proline	64.1	67.3	95.0	62.2
Glycine	225.2	260.0	230.0	180.0
Alanine	58.6	40.6	53.0	64.0
Half-Cystine	22.0	21.0	11.0	19.2
Valine	36.0	33.2	45.0	44.0
Methionine	7.0	5.0	8.2	11.8
Isoleucine	28.6	32.0	28.5	32.0
Leucine	60.3	57.7	75.2	73.0
Tyrosine	20.5	13.0	21.0	17.5
Phenylalanine	28.3	29.8	25.0	31.0

[a]*Residues/1,000 residues.*

[b]*Kefalides and Denduchis (4).*

[c]*Clark et al. (5).*

Reprinted with permission from Biochemistry 8:4613-4621, 1969. Copyright by the American Chemical Society.

TABLE III

Carbohydrate Composition of Mammalian
Basement Membranes[a]

| Carbohydrate | Human[b] | | Canine[b] | Rat[c] |
	Glomerulus	Lens Capsule	Descemet's Membrane	Reichert's Membrane
Hexose	6.8	11.8	8.2	7.0
Glucose	2.5	5.5	3.5	2.2
Galactose	2.6	5.6	3.7	3.0
Mannose	1.7	0.7	2.0	1.8
Glucosamine	1.7	0.8	1.2	4.6
Galactosamine	0.3	0.2	0.3	0.4
Fucose	0.7	0.6	0.6	0.4
Sialic Acid	1.5	0.5	0.7	1.7
Hexuronic Acid	–	–	0.05	0.2

[a] Gm/100 gm.

[b] Kefalides and Denduchis (4).

[c] Clark et al. (5).

Reprinted with permission from Biochemistry 8:4613-4621, 1969. Copyright by the American Chemical Society.

Characteristic of their amino acid composition is the high content of the imino acids, proline and 3- and 4-hydroxy-proline, of the nonpolar amino acid glycine, and the basic amino acid hydroxylysine. Although in most mammalian colla-gens the sum of proline plus hydroxyproline accounts for al-most 22% of the amino acid residues, the sum of these two amino acids accounts for 14%, 17.4% and 17.7% for glomerular, lens capsule, and Descemet's membranes, respectively. Similar-ly, whereas glycine accounts for one-third of the amino acid residues in interstitial collagens, it accounts for about one-fourth in whole basement membranes. Since hydroxyproline and hydroxylysine are found almost exclusively in interstitial collagen, one of the protein components in basement membranes must be collagenous. The lower hydroxyproline and glycine content indicates that proteins other than collagen are also present. The heterogeneity of the molecular composition of basement membranes was suggested by significant amounts of cysteine and tyrosine, by a low total imino acid content, and

by the fact that in addition to glucose and galactose, hexosamine, mannose, fucose, and sialic acid, sugars not present
in soluble collagens, are found in basement membranes (1,2).
Solubility studies and analyses of the solubilized fractions
indicate that basement membranes are composed of dissimilar
protein subunits. Table IV shows that the ratio of hydroxylysine to hexosamine varies according to the conditions of
solubilization. Extraction of glomerular basement membrane

TABLE IV

Hydroxylysine/Hexosamine Ratio in
Soluble Fractions of Canine Basement Membranes[a]

	Glomerulus	Descemet's Membrane	Anterior Lens Capsule
Intact	1.8	1.7	10
Reduction and alkyla- tion in 8 M urea	2.0	1.6	9
Reduction and alkyla- tion without urea	0.4	0.1	9
8 M urea alone	0.4	0.1	9

[a]Kefalides (6).

and Descemet's membrane with 8 M urea alone or by reduction
and alkylation alone resulted in a lower hydroxylysine:hexos-
amine ratio than in the intact membrane, whereas with anterior
lens capsule the ratio in the soluble fractions was the same
as in the intact membrane. Reduction and alkylation, in the
presence of 8 M urea, of all three types of membranes, solubilized practically the whole membrane. The ratio of hydroxy-
lysine:hexosamine in this fraction was the same as in the
intact membrane. These data support the hypothesis that a
non-collagenous protein component or components are associated
with the collagen component via hydrogen and disulfide bonds.
Lysyl-derived and hydroxylysyl-derived cross-links are also
present and contribute to the total structural organization
and stability of basement membranes (7).

Nature of the Structural Components

Further evidence of collagenous and non-collagenous protein components came from isolation of a collagen after basement membranes from various tissues had been treated with pronase or pepsin at low temperatures (1,8). This collagen met the chemical and physical criteria for collagens, i.e., proline and hydroxyproline accounted for 20 to 22% and glycine for 33% of all the amino acid residues (Table V); the

TABLE V

Amino Acid Composition of α-Chains
Isolated from Basement Membrane Collagens[a]

Amino Acid	Human[b] Glomerulus	Sheep[b] Anterior Lens Capsule	Sheep[b] Descemet's Membrane	Canine[c] Tendon
Hydroxylysine	44.6	57.0	43.0	6.0
Lysine	10.0	10.0	15.2	23.0
Histidine	10.4	8.0	7.8	4.0
Arginine	33.0	27.0	30.0	45.0
3-Hydroxyproline	11.0	12.0	8.0	1.0
4-Hydroxyproline	130.0	120.0	157.0	93.0
Aspartic	51.0	50.0	30.0	46.0
Threonine	23.0	20.0	18.0	18.0
Serine	37.0	38.0	25.0	33.0
Glutamic	84.0	92.0	78.0	73.0
Proline	61.0	67.0	90.0	130.0
Glycine	310.0	330.0	320.0	331.0
Alanine	33.0	32.0	32.0	122.0
Half-Cystine	8.0	8.0	8.0	0.0
Valine	29.0	26.0	25.0	20.0
Methionine	10.0	10.0	9.5	7.0
Isoleucine	30.0	20.0	24.0	110.0
Leucine	54.0	43.0	52.0	21.0
Tyrosine	6.0	2.0	3.0	4.0
Phenylalanine	27.0	30.0	22.0	13.0

[a]*Residues/1000 residues.*

[b]*Kefalides (9).*

[c]*Kefalides, unpublished data.*

isolated collagen molecule had a triple-helical configuration
suggested by the high intrinsic viscosity and the high negative specific optical rotation (8).

When the collagen prepared after the basement membranes
had been treated with pepsin was denatured and chromatographed
on carboxymethyl-cellulose, only a single major component
emerged, eluting in the region of α-chains. We, therefore,
concluded that the basement membrane collagen molecule is
composed of three identical α-chains (9). The amino acid
composition of the α-chains isolated by carboxymethyl-
cellulose chromatography from various basement membrane
collagens shows that they are all characterized by unusually
high amounts of hydroxyproline and hydroxylysine. The sum
of lysine and hydroxylysine is higher than that of interstitial collagens, but the sum of proline and hydroxyproline is
closer to the one observed for interstitial collagens. A
significant percentage of the total hydroxyproline is 3-hy-
droxyproline. In early amino acid analyses, the color value
for 4-hydroxyproline was used to calculate 3-hydroxyproline.
This resulted in values for 3-hydroxyproline that were 2-3
times higher than actual ones. Later, the color value for
proline was used but more recently a 3-hydroxyproline
standard is being used routinely. Using the latter standard,
the calculated value for this amino acid averages 9 ± 1.5
residues per 1000 residues. Glycine accounts for one-third
of all the amino acid residues. Alanine and arginine, however, are low and correspond to about 30% and 60% of the
values found in interstitial collagen. Another unusual
feature is the presence of 4-8 residues of half-cystine. The
chains from all three types of basement membranes contain from
10.0 to 12.5% hexose composed of equimolar amounts of glucose
and galactose (Table VI). About 95% of all the hexose is in
the form of the disaccharide unit, glucosyl-galactosyl-hydroxylysine; the rest in the form of galactosyl-hydroxylysine.
Less than 0.2% of mannose and glucosamine was also found.

The presence of 4-8 residues of half-cystine in the basement membrane collagen isolated after limited pepsin digestion
suggested the possibility that persistence of disulfide bonds
in the non-helical carboxyl and amino termini of the procollagen molecule of basement membrane may have prevented pepsin from digesting the non-helical peptide extensions completely. This possibility was strengthened by the fact that
the molecular weight of the isolated α-chains was about
115,000 (8). Dehm and Kefalides (10 and this volume) succeed-
ed in isolating basement membrane collagen α-chains with a
molecular weight of 95,000 by a three-step procedure. The
procedure involves an initial treatment with pepsin followed

TABLE VI

Carbohydrate Composition of α-Chains
Isolated from Basement Membrane Collagens[a]

| | Human | Sheep | |
| | | Anterior Lens | Descemet's |
	Glomerulus	Capsule	Membrane
		gm/100 gm	
Hexose	12.0	12.5	10.0
Glucose	5.5	6.0	5.0
Galactose	6.0	6.3	5.2
Mannose	>0.2	>0.2	>0.2
Fucose	0	0	0
Hexosamine	>0.1	>0.1	>0.1
		μmoles/μmole α-chain	
Glc-Gal-Holy	34	34	26
Gal-Holy	2	2	1.8

[a]Kefalides (9).

by reduction and alkylation of the isolated collagen, under
non-denaturing conditions, and then another pepsin treatment
of the reduced and alkylated product. The amino acid and
carbohydrate composition of this α-chain is typical of basement membrane collagen but contains half-cystine.

Soluble glycoprotein fractions which lack hydroxyproline,
hydroxylysine, and glucose can be obtained from glomerular
basement membrane and Descemet's membrane by extraction with
8 M urea at 40°C for 48 hr. and then by gel filtration in
Sephadex G-200 with 8 M urea (11,12). In another method the
membranes are solubilized by reduction and alkylation without
8 M urea and isolated by gel filtration on Sephadex G-200.
These fractions have an amino acid composition unlike that
of the collagen component. Consistent with the absence of
hydroxylysine is the absence of glucose; galactose, mannose,
hexosamines, fucose, and sialic acid are present in significant amounts. The molecular weight of these fractions is
greater than 200,000. The sequence and linkage of the sugars
in the polysaccharide chains of the high molecular weight

protein are still unknown.

Non-collagen glycoprotein fractions of low molecular
weight were obtained after prolonged digestion of the basement
membranes with bacterial collagenase (11,12). This procedure
solubilized 72 to 90% of the membranes and rendered over 90%
of the hydroxyproline dialyzable. Two glycoprotein fractions,
free of hydroxyproline and hydroxylysine, were obtained from
glomerular basement membrane (12); lens capsule (14), and
Descemet's membrane (1). A low-molecular-weight glycoprotein was isolated from the soluble, undialyzable collagenase digest by gel filtration on Sephadex G-200. Amino acid
analysis of this fraction showed no hydroxyproline or hydroxylysine but significant amounts of half-cystine. Carbohydrate
analysis showed significant amounts of hexose composed mainly
of galactose, mannose, hexosamine, fucose, and sialic acid.

The fraction which remains undigested after the basement
membrane has been treated with collagenase can be partially
solubilized by reduction and alkylation in 8 M urea and further purified on Sephadex G-200 (12). On the basis of its
amino acid and carbohydrate composition, this fraction contains a non-collagen-type glycoprotein or glycoproteins;
since it is excluded on Sephadex G-200, its molecular weight
must be 200,000 or greater and may form a highly cross-linked
polymer (15).

Structural Studies

For the eventual determination of amino acid sequences
in collagen, several investigators resorted to cleavage of
the α-chains with cyanogen bromide (13). Cyanogen bromide,
which cleaves at peptide bonds involving methionyl residues,
provides limited cleavage with a high degree of selectivity.
The reaction of cyanogen bromide with denatured collagen
yields from 7 to 12 peptides, depending on the number of
methionyl residues.

Treatment of basement membrane collagen from either
sheep or bovine anterior lens capsules with cyanogen bromide
yields 12 peptides whose molecular size varies from about
500 to 20,000 (14). Homologies in terms of size and amino
acid composition were observed between peptides from the two
species. It was not surprising to observe an almost uniform
distribution of hydroxylysine and disaccharide units along
the entire length of the α-chain. Hydroxylysine was present
in 9 of 12 peptides, the same peptides containing the disaccharide glucosyl-galactose. Only one peptide contained the
monosaccharide galactose. The proportion of hydroxylysine
which was glycosylated varied between 54 and 100%.

An exciting finding during these studies was the isolation of
a peptide which contained almost all the cysteic acid, was
rich in the hydroxylysine-linked disaccharide unit, and
contained, in addition, all the mannose and hexosamine
present in the intact α-chain. These observations indicate
that a non-collagen polypeptide chain which contains mannose
and hexosamine is linked to the collagen polypeptide chain
either via disulfide bonds or via a peptide bond. Recent
studies support the latter view (10). Since the collagen
component is obtained after pepsin treatment of the basement
membrane, we must assume that pepsin cleaves a major portion
of the non-collagen polypeptide chains, leaving only a small
portion still linked to the collagen α-chain (10,14).

An extension of our cyanogen bromide studies was the
use of 5,5'-dithiobis-2-nitrobenzoic acid, followed by KCN at
alkaline pH, which results in the cleavage of basement membrane proteins at cysteine peptide linkages (15). This
procedure has enabled us to isolate and characterize a unique
region in the anterior lens capsules consisting of a number
of polypeptide chains derived from collagen and non-collagen
components joined to one another through covalent, non-
disulfide cross-linkages. Cysteine residues of basement
membrane collagen are located near the amino terminal region
of the α-chain (14).

The presence of a cross-linkage region in basement
membranes has several implications. Because of its amino acid
composition, it appears to be composed of both collagen-
derived and non-collagen-derived peptides. If these peptides
are derived from the amino terminal region of the basement
membrane collagen, the collagen in basement membranes may be
a procollagen. Alternatively, this may indicate cross-
linking between collagen and non-collagen molecules which
would explain why the non-collagenous proteins cannot be
isolated from anterior lens capsules without resorting to
proteolysis.

In support of the former view is the electron microscopic
observation of Olsen, Alper and Kefalides (16), that a citrate-
soluble fraction of lens capsule basement membrane has two
morphologically distinct components: one represented by thin
strands measuring 2 to 3 nm in diameter, the other by globular
units measuring 5 to 20 nm in diameter. Pepsin treatment
destroyed the globular component and left the filaments intact. The amino acid composition of the pepsin-treated extract was similar to that of basement membrane collagen.
Treatment with bacterial collagenase destroyed the filamentous components and left the globular components intact.
These data indicate that the filamentous structure was

collagenous whereas the globular component was not.   Reduction
of disulfide bonds resulted in an apparent dissociation of
the globular unit from the filamentous component.

More recently, Minor et al. (23) demonstrated that the
initially synthesized procollagen by parietal yolk sac endoderm, in organ culture, did not undergo a time dependent
conversion to a smaller molecular weight species.   Using the
rat lens system, Heathcote et al. (24) showed that the basement membrane procollagen synthesized by the lens epithelium
was not converted to a smaller molecular weight species.
Thus, in lens capsule, Reichert's membrane and most probably
in all basement membranes the collagen molecules may exist
in the procollagen state, i.e., with the non-helical
extensions still present.

The recent studies by Chung and Miller (17) and Trelstad
and Lawley (18) are interesting.   Both these groups of
investigators treat whole tissues, which contain a mixture of
interstitial collagens and some basement membrane, with
pepsin at low temperature and through a series of salt
fractionations obtain collagenous peptides of varying composition which, they claim, derive from basement membranes.   One
of these peptides has a molecular weight of 55,000 and
another of 100,000 (17).   None of these studies presents
evidence that the alleged collagen chains arise from basement
membranes.   The variations in molecular size and composition
may indicate that their peptide chains arise from non-specific
cleavage of the collagen molecules.   Approaches where whole
tissues, containing several types of collagens of varying
proportions, are digested with pepsin and the resultant
products fractionated to isolate the alleged basement membrane
collagen chains are likely to yield confusing results.   To
study more precisely the nature of the basement membranes and
their constituent protein subunits we must still rely on
first isolating the membrane from a tissue and then proceeding
with its characterization and fractionation.

Evidence that our conclusions are valid comes from a
series of biosynthetic studies of basement membrane collagen
using chick lenses (19), rat glomeruli (20), rabbit corneal
endothelium (21) and rat parietal yolk sac (22).   These
studies confirm that epithelial, endothelial and endodermal
cells synthesize and secrete both collagenous and non-collagenous components of basement membranes.   The collagenous
component is secreted in the form of procollagen and is
deposited in the extracellular matrix without further reduction in molecular size.   The molecular weight of the newly
synthesized basement membrane procollagen ranges between
140,000 and 160,000 depending on the tissue under examination.

The procollagen interacts with the non-collagen components to form the structure we call basement membrane.

At present, we cannot state with certainty whether the procollagens found in basement membranes of various tissues within a given species are identical or distinct genetic products. We have immunochemical evidence, however, which shows that the collagen of bovine anterior lens capsule and that isolated from the human glomerulus are distinct genetic products (1,2).

ACKNOWLEDGEMENT

This work was supported by NIH Grants AM-14526, AM-20553 and HL-16058.

REFERENCES

1. Kefalides, N.A., *Int. Rev. Exp. Pathol.* 10:1-39, 1971.
2. Kefalides, N.A., *Int. Rev. Connect. Tissue Res.* 6:63-104, 1973.
3. Ferwerda, W., Feltcamp-Vroom, T.M., Smit, J.W., *Biochem. Society Trans.* 2:640-642, 1974.
4. Kefalides, N.A. and Denduchis, B., *Biochemistry* 8:4613-4621, 1969.
5. Clark, C.C., Minor, R.R., Koszalka, T.R., Brent, R.L. and Kefalides, N.A., *Dev. Biol.* 46:243-261, 1975.
6. Kefalides, N.A., *J. Invest. Derm.* 65:85-92, 1975.
7. Tanzer, M.L. and Kefalides, N.A., *Biochem. Biophys. Res. Comm.* 51:775-780, 1973.
8. Kefalides, N.A., *Biochemistry* 7:3103-3112, 1968.
9. Kefalides, N.A., *Biochem. Biophys. Res. Comm.* 45:226-234, 1971.
10. Dehm, P. and Kefalides, N.A., *Fed. Proc.* 36:680, 1977.
11. Denduchis, B. and Kefalides, N.A., *Biochim. Biophys. Acta* 221:357-366, 1970.
12. Kefalides, N.A., *Connect. Tissue Res.* 1:3-13, 1972.
13. Piez, K.A., *IN:* "Treatise on Collagen" (G. N. Ramachandran, Ed.) pp. 207-252, Academic Press, New York, 1967.
14. Kefalides, N.A., *Biochem. Biophys. Res. Comm.* 47:1151-1158, 1972.
15. Alper, R. and Kefalides, N.A., *Biochem. Biophys. Res. Comm.* 61:1247-1253, 1974.
16. Olsen, B.R., Alper, R. and Kefalides, N.A., *Eur. J. Biochem.* 38:220-228, 1973.

17. Chung, E., Rhodes, R.K. and Miller, E.J., *Biochem. Biophys. Res. Comm.* 71:1167-1174, 1976.
18. Trelstad, R. and Lawley, K.R., *Biochem. Biophys. Res. Comm.* 76:376-384, 1977.
19. Grant, M.E., Kefalides, N.A. and Prockop, D.J., *J. Biol. Chem.* 247:3539-3544, 1972.
20. Grant, M.E., Harwood, R. and Williams, I.F., *Eur. J. Biochem.* 54:531-540, 1975.
21. Kefalides, N.A., Cameron, J.D., Tomichek, E.A. and Yanoff, M., *J. Biol. Chem.* 251:730-733, 1976.
22. Clark, C.C., Tomichek, E.A., Koszalka, T.R., Minor, R.R. and Kefalides, N.A., *J. Biol. Chem.* 250:5259-5267, 1975.
23. Minor, R.R., Clark, C.C., Strause, E.L., Koszalka, T.R., Brent, R.L. and Kefalides, N.A., *J. Biol. Chem.* 251:1789-1794, 1976.
24. Heathcote, J.G., Sear, C.H.J. and Grant, M.E., *IN:* "Proceedings of the First International Symposium on the Biology and Chemistry of Basement Membranes" (N.A. Kefalides, Ed.) Academic Press, New York, pp.   , 1977.
25. Meezan, E., Hjelle, J.T., Brendel, K. and Carlson, E.C., *Life Sci.* 17:1721-1732, 1975.

ISOLATION OF BASEMENT MEMBRANE COLLAGENS

USING A

NEW HEAT GEL FRACTIONATION METHOD

Robert L. Trelstad,

Karen R. Lawley and Kimiko Hayashi

Shriners Burns Institute
Massachusetts General Hospital
Harvard Medical School
Boston, Massachusetts

Although basement membranes are ubiquitous cell surface structures present on a wide number of different cell types including adipose, smooth and skeletal muscle, cardiac muscle and most epithelia, their isolation to date has been principally accomplished from two basement membrane rich structures, the renal glomerulus and the lens. We have developed a simple method for isolating basement membrane like collagens from all of the above mentioned tissues using a fractionation method which involves heat gelation of pepsin solubilized collagens.

Supported by the Shriners Burns Institute and grants from the NIH (HL 18714 and AM 18729).

Basement membrane collagens were isolated according to the following scheme:

Heat Gel Fractionation

Homogenize tissue in 0.5M acetic acid
Adjust pH to 2.0-2.5 with HCl
Add Pepsin at 20-40 mg/gm wet weight
Digest at 4°C for 3 days

Centrifuge

Residue, discard   Supernatant

Neutralize to pH 7.0-8.0
with NaOH
Add NaCl to 20%
Sit overnight
Centrifuge

Pellet   Supernatant,
Resolubilize in 0.4 ionic   discard
strength pO_4 buffer, pH 7.6
Dialyze versus buffer to remove
salt
Heat gel at 37°C for 16 hours
Centrifuge

Pellet   Supernatant, Type IV collagen
Process for collagen   Precipitate with 20-30% NaCl
Types I-III   Resolubilize in 0.1M acetic
                                acid, desalt by dialysis
                                and store in acid solution

The materials isolated by this procedure showed anomolous behavior on SDS polyacrylamide gels and molecular sieve columns. In addition, they were not able to form SLS (segment-long-spacing) crystallites for visualization in the electron microscope. Reduction under native conditions followed by re-pepsinization significantly improved their electrophoretic, chromatographic and ultrastructural properties.

The reduction and repepsinization scheme is as follows:

Reduction - Repepsinization

*Dialyze Type IV collagen against
0.4 ionic strength PO_4 buffer
pH 7.6 containing 0.1% mercaptoethanol at 4°C overnight.*

*Dialyze against 0.5M acetic acid, pH
2.0-2.5 (adjusted with HCl) containing 0.1% mercaptoethanol.*

*Add pepsin (20-40 mg/ml) and digest
at 4°C overnight.*

*Neutralize with NaOH, and add NaCl
to 20-30%. Precipitate contains
the Type IV collagen.*

The above method has been successfully applied to a
variety of human tissues and has recently been successfully
employed on embryonic chick tissues. The basement membrane
collagen derived from these various tissues have been evaluated by amino acid analysis, electrophoresis, chromatography
on molecular sieve columns and ion exchange columns and by
SLS crystallite formation. The differences between heat gel
supernatant and the heat gel precipitate is readily discerned
by the hydroxylysine/4-hydroxyproline ratio derived from amino
acid analysis as noted in the following table.

	KIDNEY	SPLEEN	AORTA	BOWEL	LUNG	HEART	SKIN	LIVER
			OHLYS/4-OHPRO					
SUP	.40	.39	.39	.35	.25	.38	.25	.33
PPT	.10	.15	.06	.07	.11	.13	.08	.08

The amino acid composition of materials derived from human kidney, spleen and lung and embryonic chick tissues are noted in the following table.

Amino Acid Composition
Residues/1000

| | Human - Native | | | Chick Embryo |
	Kidney	Spleen	Lung	α-chain
3-OHPro	9	7	2	5
4-OHPro	97	69	107	103
Asp	54	55	55	48
Thr	25	31	26	23
Ser	36	43	41	29
Glu	85	83	90	90
Pro	62	54	84	114
Gly	326	356	302	356
Ala	41	41	38	51
1/2 Cys	8	11	5	0
Val	29	32	27	19
Met	13	12	14	11
Ileu	24	23	24	21
Leu	55	54	49	42
Tyr	13	21	15	3
Phe	27	28	25	7
OHLy	39	27	41	34
Lys	16	17	15	18
His	10	10	7	5
Arg	31	26	33	39

Comparison of the amino acid composition of human materials with those recently isolated by Chung et al. (1) and Burgeson et al. (2) show significant differences in respect to the degree of hydroxylation of the proline residues and the arginine content. The composition resembles that described previously from renal glomeruli and lens capsule by Kefalides (3) and from a mouse tumor as reported by Orkin et al. (4). The material from the chick embryo resembles an α-chain initially described in the scleral cartilage of the chick eye by Trelstad and Kang (5).

Segment-long-spacing crystallites viewed in the electron microscope of the collagen obtained from the chick embryo reveals a distinctive band pattern different from that previously described for collagen Types I, II and III as illustrated in the following figure.

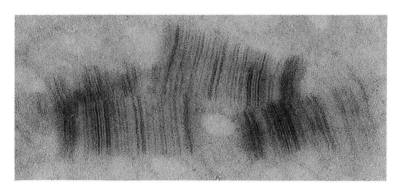

The complete characterization of collagen types isolated by this new heat gelation method will require further study including cyanogen bromide peptide analysis and sequence determinations. This method combined with salt fractionation (6) should be effective tools in further study of this cell surface-extracellular matrix material.

BIBLIOGRAPHY

1. Chung, E., Rhodes, R.K. and Miller, E.J., *Biochem. Biophys. Res. Commun.* 71:1167-1174, 1976.
2. Burgeson, R.E., El Adli, F.A., Kaitila, I.I. and Hollister, D.W., *Proc. Natl. Acad. Sci. USA* 73:2579-2583, 1976.
3. Kefalides, N.A., *Biochem. Biophys. Res. Commun.* 45:226-234, 1971.
4. Orkin, R.W., Gehron, P., McGoodwin, E.B., Martin, G.R., Valentine, T. and Swarm, R., *J. Exp. Med.* 145:204-220, 1977.
5. Trelstad, R.L. and Kang, A.H., *Exp. Eye Res.* 18:395-406, 1974.
6. Trelstad, R.L., Catanese, V.M. and Rubin, D.F., *Analyt. Biochem.* 71:114-118, 1976.

ISOLATION OF COLLAGEN α-CHAINS

FROM LENS CAPSULE BASEMENT MEMBRANE APPLYING

A THREE-STEP PROCEDURE OF PEPSINIZATION,

REDUCTION-ALKYLATION AND RE-PEPSINIZATION

Peter Dehm and Nicholas A. Kefalides

University of Pennsylvania
Philadelphia, Pennsylvania

SUMMARY: A method has been devised for the excision of a collagen triple-helix in high yield from lens capsule basement membrane. This method consists of three steps: limited proteolysis by pepsin is followed by reduction and alkylation of disulfide bonds under non-denaturing conditions of the solubilized product, and subsequently by a second treatment with pepsin. Upon denaturation and gel filtration, a collagenous component is obtained whose size was determined as 95,000 daltons by polyacrylamide gel electrophoresis in the presence of sodium dodecylsulfate. The amino acid composition of this α-chain, αl(IV), shows the high 3-hydroxyproline, 4-hydroxyproline and hydroxylysine contents and the low alanine and arginine content, characteristic for basement membrane collagen. The low content or absence of cysteine is noteworthy.

METHODS

Twelve bovine anterior lens capsules were homogenized in
15 ml of .5 M acetic acid containing protease inhibitors.
The homogenate was dialyzed against .5 M acetic acid and then
100 μg/ml of pepsin were added and the digestion carried out
for 24 hours at 16°C. After neutralization, the digest was
dialyzed against .2 M Tris·HCl, pH 8.5 (22°C), containing
.4 M NaCl and 2 mM EDTA. The retentate was then reduced with
10 mM dithiothreitol for 16 hours at 16°C and subsequently
alkylated in the dark with 40 mM iodacetamide also for 16
hours at 16°C. The reduced and alkylated sample was dialyzed
against .5 M acetic acid. A second pepsin treatment was then
carried out under conditions identical to those of the first
limited proteolysis with pepsin. The collagen was subsequently precipitated with 7% (w/v) NaCl overnight and collected by
centrifugation for 50 min. at 18,000 g. The pellet was resuspended in 1 M CaCl$_2$, containing .5 M Tris. HCl, pH 8.5
(22°C), and denatured at 45°C for 15 minutes. The sample was
clarified by centrifugation and was then placed on a 6%
agarose column in 1 M CaCl$_2$, .05 M Tris·HCl, pH 7.5.

RESULTS AND DISCUSSION

The aim of this investigation was the isolation and
characterization of the collagenous component of lens capsule
basement membrane. It is well known that basement membranes
are not easily solubilized, and breaking of covalent bonds is
necessary to bring a major portion of the basement membrane
into solution (for review see ref. 1). Limited proteolysis
and reduction of disulfide bonds were therefore chosen as a
means to release and isolate a major structural element of
lens capsule basement membrane.

Bovine lens capsules were subjected to a three-step-
procedure, which included limited proteolysis by pepsin,
reduction and alkylation of disulfide bonds under non-denaturing conditions, and finally a second pepsin treatment.

To make certain that the collagenous component was not
destroyed during these steps, the recovery of hydroxyproline
was determined. In the final salt precipitate after the three-
step-procedure (see METHODS) about 70% of the originally
present hydroxyproline was recovered. From this it was concluded that most of the collagenous portion of the lens capsule was recovered in this fraction.

Gel filtration on 6% agarose in 1 M CaCl$_2$ of the salt-precipitable material in the final digest revealed one main peak at an elution position identical to or very close to that of α-chains of Type I collagen. The peak fractions electrophoresed as a single band on SDS-polyacrylamide gels with an electrophoretic mobility identical to that of αl(I). The same molecular weight, namely 95,000 was therefore tentatively assigned to it.

The amino acid composition of this peptide is shown in Table I. It demonstrates the collagenous nature of the molecule. The high content of 3- and 4-hydroxyproline and hydroxylysine, and the low content of alanine, and arginine are indicative for a basement membrane collagen (1). Cysteine is either absent or present at less than 1 residue/-1000.

TABLE I

Amino Acid Composition of the αl(IV)chain from Bovine Lens Capsule

	Residues/1000		Residues/1000
Hydroxylysine	49	Proline	65
Lysine	8.6	Glycine	328
Histidine	6.3	Alanine	37
Arginine	26	1/2 Cystine	0 to <1*
3-Hydroxyproline	6.9	Valine	28
4-Hydroxyproline	133	Methionine	13
Aspartic Acid	51	Isoleucine	29
Threonine	20	Leucine	52
Serine	37	Tyrosine	2.3
Glutamic Acid	79	Phenylalanine	29

(Mean of 5 analyses)

*Sum of cysteic acid, cysteine and carboxymethyl cysteine.

Previous investigations have established that basement membranes contain a collagenous portion (for review see ref. 1), and that limited proteolysis by pepsin can be used for its release from the basement membrane (2). The basement membrane collagen α-chains isolated after a single pepsin digestion were, however, larger than the α-chains of interstitial collagens. The larger size might be explained by the presence of a pepsin-resistant, non-collagenous region at the end(s) of the molecule. As reported here, successive treatment of lens capsule with pepsin, reduction and alkylation of the solubilized product, and a second treatment with pepsin released most of the collagenous protein as α-chains identical in size to that of interstitial collagen α-chains. These results support the earlier notion (2) that one biosynthetic unit of basement membranes is a procollagen-like molecule. Future work will have to establish whether the proposed procollagen-like unit is indeed a true procollagen, even though it is not converted (3), or whether it is a molecule of an analogous structure but of a different genetic type.

This work was supported by USPHS grants AM-14526, AM-20553, and HL-15061.

REFERENCES

1. Kefalides, N.A., *Int. Rev. Conn. Tissue Res.* 6:63-104, 1973.
2. Kefalides, N.A., *Biochem. Biophys. Res. Comm.* 45:226-234, 1971.
3. Minor, R.R., Clark, C.C., Strause, E.L., Koszalka, T.R., Brent, R.L. and Kefalides, N.A., *J. Biol. Chem.* 251:1789-1794, 1976.

THE USE OF HYDROXYALMINE CLEAVAGE AS A PROBE

OF THE SUPRAMOLECULAR ORGANIZATION OF

BOVINE ANTERIOR LENS CAPSULE BASEMENT MEMBRANE

Robert Alper and Nicholas A. Kefalides

University of Pennsylvania
Philadelphia, Pennsylvania

*ABSTRACT: Bovine anterior lens capsule basement membrane (BALC) was solubilized by treatment with 2 N Hydroxy-almine for 16 hours at 35°C. After passage through a Bio-Gel A 15m column, three fractions were obtained. Fraction A which appeared in the void volume of the column had an amino acid composition suggesting that it was rich in non-collagen derived peptides. Upon reduction and alkylation, and re-passage through the A 15m column, a major fraction reappeared in the void volume and a second peak appeared which was well retarded on the column. The major fraction had the amino acid composition of a non-collagen protein. The retarded fraction contained most of the collagen derived amino acids found in Fraction A. Fraction B had an amino acid composition comparable to basement membrane collagen. Upon reduction and alkylation, it broke down to yield a series of lower molecular weight collagen fractions. Fraction C contained a mixture of low molecular weight peptides. Its amino acid composition was similar to that of the original lens capsule.
These data suggest that: 1. BALC contains a very large molecular weight non-collagen region which may be associated with the collagen region through disul-fide bonds. It is likely that this fraction is composed of a series of peptides associated through some type of*

239

non-disulfide crosslinkage. 2. Inter-α-chain disulfide crosslinkages are present in BALC suggesting the existence of a procollagen-like structure.

Studies on the supramolecular organization of basement membranes have revealed the presence of two types of protein components, collagen and non-collagen (1). These components can be prepared free of one another using certain proteolytic techniques, such as pepsinization as described in the preceding paper by Dr. Dehm, for the isolation of the collagen components or treatment with bacterial collagenase for the isolation of non-collagen proteins (2).

The nature of the association of these two components is not yet clear. It appears that disulfide bonds are of great importance in this interaction, yet the reduction of disulfide bonds has not been sufficient to allow the separation of the non-collagen from the collagen components of basement membranes (3). This can be explained, in part, by the observations of Tanzer and Kefalides (4) that collagen-type crosslinkages can be identified in various basement membranes and by the more recent studies of Alper and Kefalides (5) demonstrating the presence of a crosslinkage region in a non-collagen region of the anterior lens capsule. Thus it appears that non-disulfide crosslinkages may also play an important role in the stabilization of the basement membrane matrix.

Recently, Minor et al. (6) reported that there does not appear to be a conversion of basement membrane procollagen to collagen during the synthesis and secretion of the rat embryo parietal yolk sac basement membrane as has been reported for Types I and II collagens from other tissues and this may be true for other basement membrane synthesizing systems as well. If this is the case, it would follow that basement membranes contain intact procollagen molecules. If one examines the amino acid composition of the anterior lens capsule, it would appear that unlike most other basement membranes, the complement of amino acids could almost be accounted for by the presence of basement membrane procollagen without having to invoke the presence of any non-procollagen molecules. Thus the lens capsule conceivably could be a specialized type of basement membrane consisting of a crosslinked polymer of basement membrane procollagen. The present study was conceived in an attempt to obtain evidence for or against this hypothesis.

Hydroxylamine is capable of cleaving proteins at aspara-
ginyl residues (7,8). The sequence which appears to be most
sensitive to cleavage is asparaginylglycine. It was reasoned
that on a probability basis, this sequence would be found
predominantly in the collagenous regions of the anterior lens
capsule and that the non-collagen regions, although still
susceptible to cleavage whereever asparaginylglycine is found,
would be less likely to contain this sequence. Thus, it was
anticipated that hydroxylamine cleavage might result in the
formation of many peptides derived from the collagen regions
of the lens capsule and that the non-collagen components
would be comparatively unaffected. Since this type of cleavage should preserve the integrity of disulfide and other types
of covalent crosslinkages, it was further anticipated that
some peptides could be isolated in which the association between different molecular species was preserved and that from
compositional analysis and the study of the behavior of these
peptides after reduction of disulfide bonds one might be able
to draw some inferences concerning the supramolecular organization of the lens capsule.

METHODS

One hundred bovine anterior lens capsules (ALC) were
isolated by dissection from beef eyes obtained at a local
slaughterhouse. Care was taken to remove the ciliary fibers
and the posterior lens capsule. The capsules were then soni-
fied for a total of 2 minutes in distilled water at 0°C using
15 second bursts from a Branson sonifier cell disruptor equip-
ed with a microtip. The supernate was decanted and the capsules were washed three times with cold distilled water. The
supernates were discarded.

Hydroxylamine Cleavage

The lens capsules were suspended in 50 ml of 2 M hydroxylamine in $0.2M$ K_2CO_3 at pH 9.0 and the mixture was incubated
at 37°C with stirring for 16 hours. At the end of this period
the pH had risen to 9.5. The reaction mixture was centrifuged
at 10000 x g at 4°C for 15 minutes. The supernate was then
adjusted to pH 6.0 by the drop-wise addition of 6 N HCl,
dialyzed at 4°C against two changes (20 liters each) of distilled water and the retentate was lyophilized.

Gel Filtration

Fifty mg portions of the hydroxylamine treated ALC were dissolved in 1 M $CaCl_2$ in 8 M urea. The solution was applied to a column (1.5 x 130 cm) of Bio-Gel A15m (200-400 mesh) equilibrated with 1 M $CaCl_2$ in 0.05 M Tris·HCl pH 7.5. The peptides were eluted using the same buffer. Two ml fractions were collected and the absorbance of the effluent at 230 nm was monitored.

Reduction and Alkylation of Disulfide Bonds

Reduction of peptide fractions was performed in 0.5 M Tris·HCl pH 8.6 in 8 M urea containing 5% (w/v) 2-mercapto-ethanol under N_2 at 37°C for 16 hours. A 1.5 molar excess of iodacetamide in the urea-tris buffer was then added, the pH was readjusted to pH 8.6 and alkylation was allowed to proceed in the dark at room temperature for 30 minutes. The reduced and alkylated material was then exhaustively dialyzed in distilled water and lyophilized.

Collagenase Digestion

Bacterial collagenase digestion of peptide fractions was performed in 1-2 ml of 0.2 M Tris·HCl pH 7.5 in 0.4 M NaCl containing 10 mM $CaCl_2$ and 25 mM N-ethylmaleimide at 37° for 24 hours using a peptide to collagenase ratio of 50:1 (w/w). The collagenase digest was directly applied to the Bio-Gel A 15m column and chromatographed as described in the previous section.

Amino Acid Analysis

Amino acid analyses of peptide fractions were performed on a Jeol 6 AH amino acid analyzer after hydrolysis of the peptides in constant boiling HCl for 22 hours at 110°C in vacuo followed by evaporation to dryness under reduced pressure.

RESULTS AND DISCUSSION

After the treatment with hydroxylamine, more than 98% of the material had entered solution at the end of 16 hours. Gel filtration of the hydroxylamine digest on Bio-Gel A 15m resulted in the isolation of three peptide fractions A, B and C (Figure 1). Fraction A, comprising about 12-13%

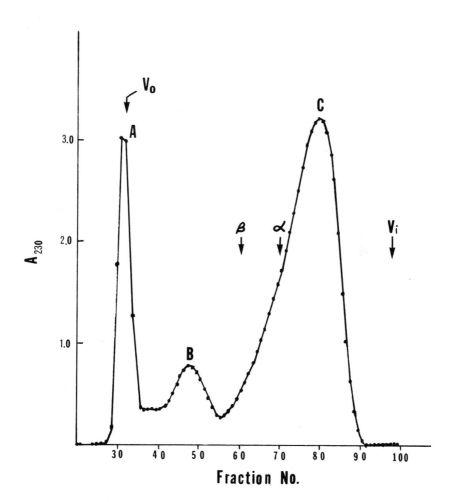

Fig. 1. Gel filtration on Bio Gel A15m of hydroxylamine-treated bovine ALC.

of the material was eluted at the void volume of the column.
This suggests a molecular weight for this fraction of in
excess of 10^6 daltons. Fraction B eluted with an apparent
molecular weight of about 400,000 daltons and comprised about
15% of the total material. Fraction C eluted as a broad peak
with an apparent molecular weight at the peak of about 45,000
daltons.

The amino acid composition of these peptides is given
in Table I. Fraction A contained low levels of the collagen-
derived amino acids, hydroxyproline and hydroxylysine, and
appeared to be predominantly non-collagen. Fraction B had an
amino acid composition typical of basement membrane collagen
(about 1/3 glycine and high levels of hydroxylysine and
hydroxyproline). Fraction C had an amino acid composition
similar to that of the original anterior lens capsule. All
three fractions contained significant amounts of cysteine.

For the purpose of this presentation, we will focus on
the properties of Fraction A. Fraction A was very insoluble
at acid pH and had a pale yellow color. Upon dialysis vs
distilled water, it tended to precipitate out. Fraction A
was reduced and alkylated and passed back through the Bio-Gel
A 15m column. The results are shown in Figure 2. A major
peak appeared in the void volume of the column and two, very
retarded, minor fractions of low molecular size were observed.
The amino acid composition of the void volume fraction is
presented in Table I.

Two major features were observed. First, reduction and
alkylation caused the disappearance of all of the hydroxylysine and hydroxyproline present prior to reduction and
alkylation. These amino acids were found in the two retarded
fractions (data not shown). This suggested that collagen
components are associated with non-collagen components through
disulfide linkages. Second, the carboxymethylcysteine formed
accounts for only about 20% of the cysteine found in unreduced
Fraction A. This suggested that about 60% of the disulfide
linked peptides are not linked directly to the non-collagen
protein through disulfide bonds (i.e., only 40% of the total
cysteine could be accounted for by the sum of the cysteine
contributed by reduced Fraction A and that in the peptide(s)
attached to it through the disulfide linkages). In addition,
reduced Fraction contained no methionine.

Unreduced Fraction A was treated next with bacterial
collagenase in order to establish whether the collagenous
positions of Fraction A were actually associated through covalent linkages to the non-collagen fraction or if the collagen portion consisted of a series of peptides joined by
disulfide linkages which resulted in a fraction sufficiently

TABLE I

Amino Acid Component of
Hydroxylamine Cleavage Products
(Residues/1000)

			Fraction		
	A	B	C	A (Reduced and Alkylated)	A (Collagenase)
Hydroxylysine	17.8	34.2	36.5	0	0
Lysine	24.7	7.7	11.0	39.1	24.9
Histidine	27.1	4.5	9.9	19.2	21.0
Arginine	55.3	24.8	29.4	35.5	36.4
3-Hydroxyproline	6.1	10.7	5.9	0	0
4-Hydroxyproline	29.6	106.0	99.1	0	0
Aspartic	77.4	55.2	58.9	74.1	96.3
Threonine	48.5	29.7	35.2	54.7	57.3
Serine	62.5	29.6	51.2	136.0	90.8
Glutamic	111.8	100.3	89.5	135.5	131.4
Proline	77.2	76.5	68.6	70.5	67.5
Glycine	147.5	323.0	269.1	151.2	126.1
Alanine	69.4	31.2	42.6	85.8	86.7
Half-Cystine	29.9	15.2	12.4	6.0*	40.6
Valine	43.3	33.0	31.0	41.3	59.5
Methionine	16.0	4.2	17.8	0	0
Isoleucine	24.2	24.1	29.3	51.2	50.0
Leucine	74.5	54.2	58.0	68.1	68.3
Tyrosine	21.9	10.6	11.0	9.7	8.7
Phenylalanine	35.1	25.2	33.3	21.9	34.3

*Carboxymethylcysteine

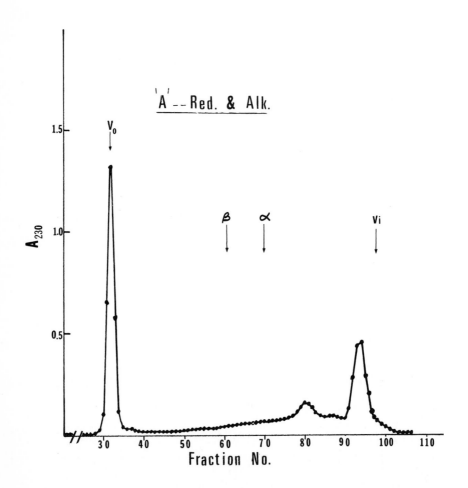

Fig. 2.   Gel filtration on Bio Gel A15m of reduced and alkylated Fraction A.

large to co-elute with the non-collagen fraction. The collagenase treated material was re-chromatographed on A 15m (Figure 3).

Again, a void volume fraction and a very retarded fraction was obtained. Examination of the amino acid composition of the void volume fraction (Table I) revealed that the collagenase removed all of the hydroxylyse and hydroxyproline and that methionine is absent. Significantly, however, the content of cysteine was increased by about 25% over that found in the original Fraction A. This indicated that collagenase removed all of the collagen components from Fraction A but that little, if any, cysteine was removed.

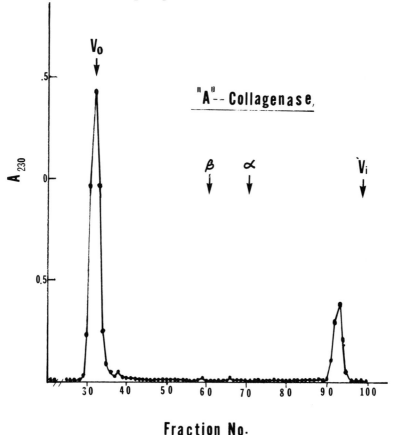

Fig. 3. Gel filtration on Bio Gel A15m of bacterial collagenase digested Fraction A.

On the basis of the very high apparent molecular size of Fraction A and its derivatives, it is unlikely that this is a single polypeptide chain. We feel that it is more probably a polymer of non-collagen peptides joined by non-disulfide crosslinkages to form a macromolecular complex. This is consistent with the observations of Alper and Kefalides (5) in a study on the effects of selective cleavage of ALC at cysteine residues.

In Figure 4, a model is presented for one possibility for the structure of Fraction A based upon the present data. Small non-collagen peptides are represented as being cross-linked to one another through non-disulfide crosslinkages. Attached to these through disulfide linkages are peptides similar to what has been observed for the carboxyl terminal portion of Type I procollagen. Upon reduction, all of the disulfide linked peptides are removed leaving behind the crosslinked polymer. On the other hand, treatment with collagenase removes only the collagen portions of the peptide leaving the disulfide-linked extensions intact. If the collagenase treated fraction is reduced, a peptide identical to that formed from reduction of the original Fraction A should be formed and this has proven to be the case (data not shown).

On the basis of these interpretations, and that based on data from Fractions B and C (not presented) se suggest a model for the supramolecular organization of the bovine anterior lens capsule (Figure 5). This consists of procolla-gen-like molecules which are bound to the non-collagen com-ponents via the linkages outlined in Figure 4 and to each other via intermolecular disulfide crosslinkages located within one or more collagenous portions of the α chains. At the amino terminal regions of the α chains are disulfide crosslinkages (which may be intrachain, interchain, or both) and some collagen-type crosslinkages. The highly-crosslinked non-collagen peptide described for Figure 4 conceivably could be derived from procollagen extension peptides but there is no direct evidence available as yet for this possibility.

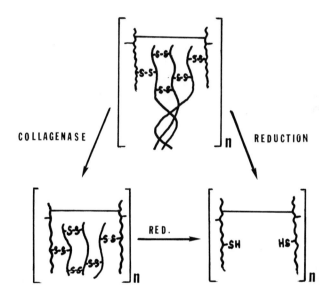

*Fig. 4. Explanation for the effects of
reduced and alkylated or collagenase digestion
of Fraction A based upon a hypothetical model for
the molecular origin of Fraction A.*

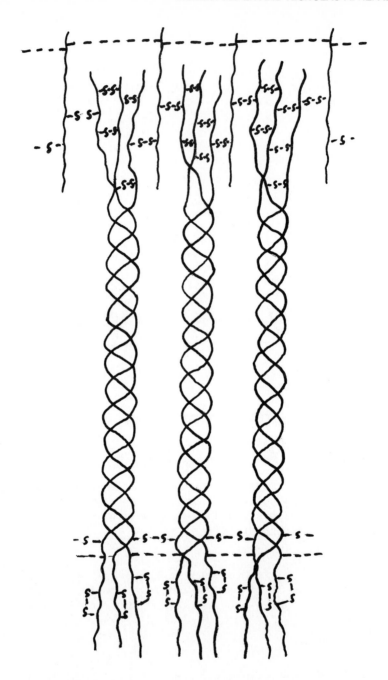

Fig. 5. Hypothetical model for the molecular organization of anterior lens capsule basement membrane.

ACKNOWLEDGEMENTS

    This work was supported by NIH Grants AM-14526 and AM-20553.

REFERENCES

1. Kefalides, N.A., *Int. Rev. Exptl. Pathol.* 10:1, 1971.
2. Kefalides, N.A., *Connect. Tiss. Res.* 1:3, 1972.
3. Hudson, B.G and Spiro, R.G., *J. Biol. Chem.* 247:4239, 1972.
4. Tanzer, M. and Kefalides, N.A., *Biochem. Biophys. Res. Comm.* 51:775, 1973.
5. Alper, R. and Kefalides, N.A., *Biochem. Biophys. Res. Comm.* 61:1297, 1974.
6. Minor, R.R., Clark, C.C., Strause, E.L., Koszalka, T.R., Brent, R.L. and Kefalides, N.A., *J. Biol. Chem.* 251: 1789, 1976.
7. Bornstein, P., *Biochemistry* 9:2408, 1970.
8. Bornstein, P. and Balian, G., *J. Biol. Chem.* 245:4854, 1970.

CHEMISTRY OF *ASCARIS SUUM*

INTESTINAL BASEMENT MEMBRANE

Billy G. Hudson

University of Kansas Medical Center
Kansas City, Kansas

Our present knowledge of basement membrane structure is based primarily on studies of two vertebrate membranes, renal glomerular basement membrane and lens capsule, both of which have complex biochemical structures. We have recently extended our structural studies to include an invertebrate basement membrane in an effort to find a membrane of less complex structure that would serve as a model for obtaining basic structural information. Moreover, a comparison of the detailed structure of a phylogenetically distant membrane to that of the vertebrate membranes would define those properties which have been conserved during evolutionary development, thus, providing information about the structural features required for function.

The intestinal basement membrane of the helminth, *Ascaris suum,* was chosen for study because the morphological simplicity of the intestinal wall indicated that the membrane could easily be isolated and without possible contamination with other basement membrane types or classical collagen components. The wall consists of a single layer of high columnar epithelial cells positioned onto a basement membrane, 3-4 μ in width (Fig. 1). This membrane can be obtained in pure form by ultrasonic disruption of small segments of the

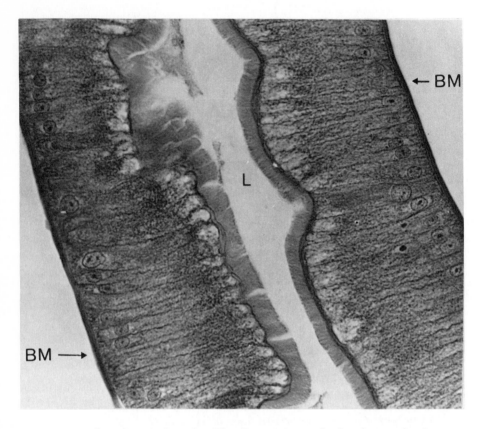

Fig. 1. A longitudinal section of the intestinal wall of Ascaris suum viewed by light microscopy. The lumen (L) of the intestine appears at the center and the pseudocoelomic cavity appears at the right and left. The intestinal wall consists of high columnar cells attached to a thick basement membrane (BM). X 500.

of the intestine. Its transparent sheet-like character (Fig. 2) closely resembles that of lens capsule (2). A comparison of its amino acid composition with that of a glomerular basement membrane, a representative vertebrate membrane, reveals close similarities and certain notable differences (Table I). These differences mainly reflect a decrease in the amount of hydroxyproline, hydroxylysine, and glycine and an increase in the acidic amino acids of the intestinal membrane as compared to glomerular membrane.

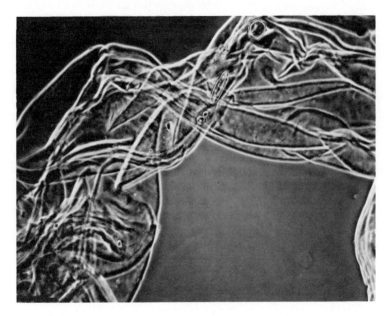

Fig. 2.   A dark-field light micrograph of isolated Ascaris suum intestinal basement membrane.
X 100.

The carbohydrate composition of the intestinal membrane is
qualitatively similar to the glomerular membrane, with the
exception of sialic acid, but differs in the relative proportion of the individual monosaccharides as well as the total
amount of carbohydrate (Table I).

The monosaccharide constituents of the intestinal membrane are distributed between two different carbohydrate units
(4).   One unit is identical to the glucosyl-galactosyl-hydroxylysine unit found in glomerular and lens capsule basement
membranes (5,6).   The identity is based on stability to
alkaline hydrolysis, retention time on the amino acid analyzer, chemical composition, graded acid hydrolysis, methylation
analysis and periodate oxidation.   The other unit is a heteropolysaccharide consisting of fucose, mannose, galactose,
glucosamine, and galactosamine.   The structural information

TABLE I

Composition of Intestinal and Glomerular
Basement Membranes

Component[a]	Ascaris suum Intestinal basement membrane[b]	Bovine glomerular basement membrane[c]
	Residues per 1000 amino acid residues	
Amino acid		
4-Hydroxyproline	24	69
Aspartic acid	82	68
Glutamic acid	124	96
Proline	87	69
Glycine	139	208
Alanine	61	61
Half-cystine	38	31
Leucine	61	59
Hydroxylysine	11	22
Lysine	22	27
Carbohydrate		
Mannose	2.6	5.1
Galactose	13.3	20.2
Glucose	10.6	16.4
Glucosamine	3.8	10
Galactosamine	2.2	1.4
Fucose	3.3	1.6
Sialic acid	0	4.4
	Percentage of dry weight	
Amino acids	91	86
Carbohydrate	4.9	8.8

[a] Key components.

[b] Data taken from Pczon et al. (1).

[c] Data taken from Spiro (3).

about these three units is summarized in Table II and compared to that of glomerular membrane. These membranes share the feature of two distinctly different units but differ in the nature of the heteropolysaccharide unit and the number of dissacharide units.

TABLE II

Carbohydrate Units of Intestinal and Glomerular Basement Membranes[a]

	Ascaris suum Intestinal basement membrane	Bovine glomerular basement membrane
	Residues per 1000 amino acid residues	
Disaccharide unit[b]	9.7	17.3
Heteropolysaccharide unit	2.6	1.7
Disaccharide/heteropolysaccharide unit	3.7	10
Composition of heteropolysaccharide unit		
	Residues/mole	
Mannose	1	3
Galactose	1	4
Fucose	1	1
Sialic acid	0	3
Glucosamine	2	5
Galactosamine	1	–

[a]*Data taken from Peczon et al. (4).*

[b]*2-0-D-glucopyranosyl-0-D-galactopyranosylhydroxylysine.*

[c]*Data taken from Spiro (5).*

The intestinal basement membrane consists of at least 17 reduced polypeptides (Fig. 3 and 4) which range in molecular weight from 22,500 to >400,000 (Table III), as determined by SDS polyacrylamide gel electrophoresis (7). At least 5 of these contain carbohydrate as determined by PAS stain (Fig. 4).

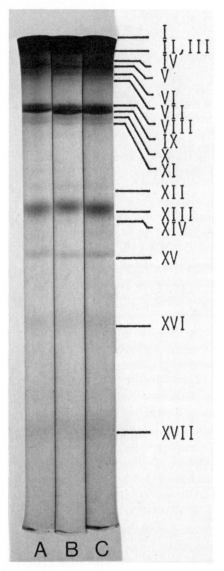

Fig. 3. Analysis of poly-peptides of Ascaris suum intestinal basement membrane by SDS polyacrylamide gel electrophoresis. Membrane was isolated at 0-3° (gel A) and 27° (gel C) without proteolytic inhibitors and at 0-3° with 10 mM ethylemediaminetetraacetic acid 25 mM N-ethylmaleimide-10 mM diisopropylfluorophosphate (gel B). Numeral designation of the polypeptides in gels A, B, and C is based on both visual inspection and spectrophotometric scan of gels (Fig. 4).

This multiplicity of polypeptides is apparently a true characteristic of the membrane, as opposed to being generated artifactually during membrane isolation by the action of tissue-associated proteases, because the polypeptide composition of membrane prepared at 0-3° or 27° without specific enzyme inhibitors is identical to that of membrane prepared at 0-3° in the presence of ethylenediamininetetraacetic acid, diisopropylfluorophosphate, and N-ethylmaleimide (Fig. 3). Their

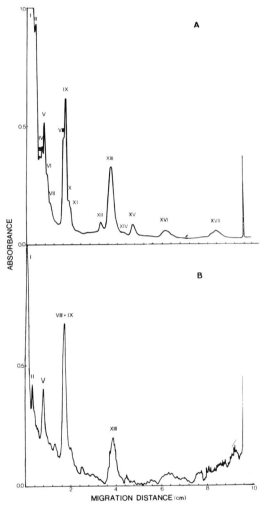

Fig. 4. Spectro-
photometric scans of
polyacrylamide gels of
Ascaris suum intestinal
basement membrane ana-
lyzed in SDS. A (upper)
gel was stained for pro-
tein with Coomassie blue
and B (lower) gel was
stained for carbohydrate
with periodic acid Schiff
reagents. Numerals give
migration position of
polypeptides designated
in Fig. 3.

molecular weights were also determined by gel filtration chromatography using 6 M urea as eluent (Table III). The molecular weight values for certain polypeptides, such as VIII, IX
and XIII, as determined by gel filtration chromatography
differ greatly with the values determined by SDS electrophoresis (Table III). This indicates anomalous behavior of these
polypeptides in SDS analogous to that observed for collagen
chains (8). The relative amounts of these various polypeptides were determined by combining the electrophoresis and
gel filtration results (Table III).

TABLE III

*Summary of Molecular Weight, Carbohydrate Staining
and Relative Amount of Reduced Polypeptides of
Ascaris suum Intestinal Basement Membrane*[a]

| Polypeptide Number[b] | SDS Electrophoresis | | Gel Filtration Chromatography | |
	PAS Stain	Molecular Weight	Molecular Weight	Relative Amount (%)
I	+	>400,000	I	
II	+	400,000	c	>450,000 ... 40.3
III		365,000	III	
IV		350,000	IV ⎫	
V	+	330,000	c ⎬	260,000 ... 19.1
VI		320,000		
VII		300,000	VII ⎭	
VIII		240,000	VIII ⎫	
IX	+	235,000	c ⎬	150,000 ... 14.6
X		220,000		
XI		215,000	XI ⎭	
XII		135,000	XII	100,000 ... 1.9
XIII	+	115,000	XIII	70,000 ... 14.8
XIV		96,000	XIV ⎫	
XV		77,000	c ⎬	35,000 ... 9.2
XVI		48,500		
XVII		22,500	XVII ⎭	

[a] *Data taken from Hung et al. (7).*

[b] *Polypeptide designated in Figs. 3 and 4.*

[c] *Unresolved polypeptides.*

These polypeptides differ greatly with regard to the
amino acid caracteristic of collagen and to the distribution
of the two types of carbohydrate units (Table IV) (7). Polypeptides VIII or IX or both (molecular weight 150,000) are the
most collagen-like, being characterized by the presence of the
largest amount of hydroxyproline, hydroxylysine, and glycine.
In contrast, polypeptide XIII (molecular weight 70,000) and
possibly XIV-XVII are least collagen-like, being characterized
by the absence of hydroxyproline and the presence of the
least amount of glutamic acid and proline. The disaccharide

TABLE IV

Amino Acid and Carbohydrate Composition of
Partially Purified Polypeptides[a] of Ascaris suum
Intestinal Basement Membrane

Component[b]	Polypeptide Number[c]		
	I (99%)[d]	VIII + IX (75%)	XIII (93%)
	Residues/1000 amino acid residues		
Amino acid			
4-Hydroxyproline	24	44	0
Aspartic acid	88	64	62
Glutamic acid	115	111	175
Proline	64	99	132
Glycine	166	257	134
Alanine	67	73	59
Half-cystine	5	6	35
Hydroxylysine	18	18	5
Lysine	34	18	16
Monosaccharide			
Fucose	3.4	1.8	2.0
Mannose	4.6	2.6	2.0
Galactose	18.2	16.7	5.0
Glucose	13.3	15.6	3.1
Disaccharide/hetero-			
polysaccharide	2.9	6	1.6

[a] Data taken from Hung et al. (7).

[b] Key components.

[c] Polypeptides designated in Figs. 3 and 4.

[d] Values in parentheses are percent purity.

and heteropolysaccharide units are constituents of both the
collagen-like (VIII and IX) and less collagen-like polypeptide
(XIII) but the proportion of the disaccharide unit relative to
the heteropolysaccharide unit is greater in the more collagen-
like polypeptides.

The more collagen-like polypeptides (VIII and IX) appear
to be linked with the less collagen-like polypeptide (XIII)
and probably to the high molecular weight components (I-VIII)

by disulfide bonds in the native membrane, whereas polypeptides XII, XIV-XVII interact in the membrane exclusively
through noncovalent forces (7). A substantial amount of the
reduced membrane (40%) exists as very high molecular weight
(>400,000) components (I-III) in both SDS and urea. These
may represent aggregates of the more collagen-like and less
collagen-like peptides that are covalently crosslinked by
bonds other than disulfide, such as the lysyl-derived cross-
link present in vertebrate collagen and basement membranes
(9).

These findings extend the observation from vertebrates
to invertebrates that basement membranes consist of multiple
polypeptides which have composition of varying degrees of
relatedness to collagen (less collagen-like to more collagen-
like) (10-12). This suggests a conservation of this structural characteristic during evolutionary development. Recent
studies (11-12) are consistent with the proposal (10,13) that
the relatedness reflects the proportion of collagen-like to
noncollagen segments in each polypeptide. The identification
of multiple collagen components of varying sizes from pepsin
digest of vertebrate basement membrane (14,15) further
supports this proposal.

ACKNOWLEDGEMENT

This work was supported by grant AM 18381 from the
National Institutes of Health.

REFERENCES

1.   Peczon, B.D., Venable, J.H., Beams, C.G.,Jr., and
     Hudson, B.G., *Biochemistry* 14:4069-4975, 1975.
2.   Fukushi, S. and Spiro, R.G., *J. Biol. Chem.* 244:2041-
     2048, 1969.
3.   Spiro, R.G., *J. Biol. Chem.* 242:1915-1922, 1967.
4.   Peczon, B.D., Wegener, L.J., Hung, C.H. and Hudson, B.G.,
     J. Biol. Chem. (in press), 1977.
5.   Spiro, R.G., *J. Biol. Chem.* 242:1923-1932, 1967.
6.   Spiro, R.G. and Fukushi, S., *J. Biol. Chem.* 244:2049-
     2058, 1969.
7.   Hung, C.H., Ohno, M., Freytag, J.W. and Hudson, B.G.,
     J. Biol. Chem. (in press) 1977.
8.   Furthmayr, H. and Timpl, R., *Anal. Biochem.* 41:510-516,
     1971.
9.   Tanzer, M.L. and Kefalides, N.A., *Biochem. Biophys. Res.
     Commun.* 51:775-780, 1973.

10. Hudson, B.G. and Spiro, R.G., *J. Biol. Chem.* 247:4239-4247, 1972.
11. Ohno, M., Riquett, P. and Hudson, B.G., *J. Biol. Chem.* 250:7780-7787, 1975.
12. Sato, T. and Spiro, R.G., *J. Biol. Chem.* 251:4062-4070, 1976.
13. Spiro, R.G., *N. Eng. J. Med.* 288:1337-1342, 1973.
14. Kefalides, N.A., *Biochem. Biophys. Res. Commun.* 45:226-234, 1971.
15. Daniels, J.R. and Chu, C.H., *J. Biol. Chem.* 250:3531-3537, 1975.

COMPARISON OF BASEMENT MEMBRANE COLLAGENS

WITH INTERSTITIAL COLLAGENS

Edward J. Miller

University of Alabama Medical Center
Birmingham, Alabama

SUMMARY: The known interstitial collagens collectively
contain four genetically distinct, but clearly homologous, α chains each of which has a molecular weight of
about 95,000 daltons. Current information on basement
membrane collagen lends itself to the hypothesis that
the latter structures contain two additional classes
of collagen molecules: one in which the polypeptide
chains of the molecules contain collagenous sequences
that are 15-20% longer than those observed for the
α chains of interstitial collagens; and another class
of molecules which contain relatively short collagenous
sequences. The data leading to this hypothesis is discussed and evaluated and note is made of several crucial
areas in which there is a singular lack of information
concerning the molecular organization of basement membranes.

This presentation will be devoted to a concise summary
of pertinent information on the biochemical properties of the
genetically-distinct collagens in vertebrate organisms with
special emphasis on the relationship of the interstitial collagens to the basement membrane collagens. Although there
exists considerable controversy with respect to the nature

and number of the collagenous constituents in basement membranes, there are sufficient data to allow at least preliminary comparisons with the relatively well-characterized interstitial collagens. In addition, it is possible to use the current data to construct a general working hypothesis with respect to the nature of the collagens in basement membranes. This hypothesis which has been generated in part from recent work in the author's laboratory is also presented here as it may prove useful in the design and interpretations of future experiments on the nature of basement membrane collagens.

The Interstitial Collagens.

     To date, four distinct types of collagen molecules have been identified in studies on a variety of tissues and organs (1). They are designated as the Type I molecule, the Type I-trimer molecule, the Type II molecule, and the Type III molecule. Since fibers derived from these molecules are apparently alwayd deposited between the cellular components of a given tissue or organ, they may be referred to collectively as the interstitial collagens.
     The three polypeptide chains of the Type I collagen molecule include two identical $\alpha 1(I)$ chains and one $\alpha 2$ chain. Fibers derived from Type I molecules are prevalent in virtually all the major connective tissues (most notably dermis, tendon, bone, and fibrocartilages) as well as in the connective tissue stroma of a variety of organs. Type I collagen is, however, most conveniently prepared in high yield and in native monomeric form by extracting the dermis or tendons of relatively young organisms with neutral salt or dilute acid solvents (2). The relative ease with which Type I collagen can be extracted from such tissues is apparently due to at least two factors: the accumulation of a substantial pool of newly-synthesized collagen in which intermolecular cross-linking is delayed; and the prevalence of aldimine intermolecular cross-links which are readily cleaved on exposure to low pH.
     The Type I-trimer collagen molecule is comprised of three $\alpha 1(I)$ chains. It therefore resembles the Type I molecule with respect to the genetic locus or loci involved in its synthesis, but differs from the latter molecule in that it does not contain an $\alpha 2$ chain. At the moment, the possibility that Type I-trimer molecules are synthesized and actually accumulated in connective tissues under normal circumstances has not been adequately investigated. These molecules are, however,

synthesized by cultured chick embryo chondrocytes when the
cells are maintained in culture under conditions leading to
the assumption of a fibroblast-like morphology (3), and
studies on the collagens synthesized in these systems led to
the definitive isolation and characterization of the Type I-
trimer molecule. In addition, preliminary evidence has
recently been offered for the occurrence of Type I-trimer
molecules among the collagens synthesized by cultured fibroblasts from inflamed human gingivae (4). Since Type I
collagen is also synthesized along with Type I-trimer in all
of these culture systems, the occurrence of Type I-trimer
molecules might be attributed either to a selective decrease
in the synthesis of α2 chains or an alteration in the chain
assembly mechanisms normally operative in the production of
Type I collagen molecules. On the other hand, Type I-trimer
molecules may represent normal biosynthetic products of
several cell types and it is apparent that further work will
be required to distinguish among these various possibilities.

The Type II collagen molecule is comprised of three
α1(II) chains, and fibers derived from Type II molecules have
a much more restricted tissue distribution than Type I fibers
as they are confined largely to hyaline cartilages. The
latter tissues are, then, the most suitable sources for the
preparation of Type II collagen (1). Due to the prevalence
of the more stable keto-amine intermolecular cross-links in
cartilage collagen fibers, a sizeable pool of readily ex-
tractable collagen does not accumulate in these tissues. The
collagen may, however, be readily extracted as native monomeric molecules from the cartilagenous tissues of lathyritic
animals in which cross-linking has been inhibited by the
administration of a lathyrogen. Alternatively, cartilage
collagen may be rendered soluble in relatively young specimens
by employing limited proteolysis with enzymes such as pepsin
which function at low temperature to selectively remove the
non-helical extremities of the molecules where cross-links
originate, thereby allowing solubilization of the fibers as
somewhat truncated molecules (5). Depending on the cartilage
used in the preparation of the collagen, some Type I collagen
molecules may also be present in extracts of cartilage
collagen. The Type I and II collagens may, however, be
readily separated in native form by differential salt precipitation (6), an approach which subsequently proved useful in
separating mixtures of Type I and III collagens (7) as well
as Type I and Type I-trimer collagens (3).

The Type III collagen molecule is likewise comprised of
three identical chains, designated α1(III). Fibers containing Type III molecules are prevalent in the more distensible

connective tissues such as dermis, vessel walls, the uterine
wall, and the intestinal wall. They also occur prominently
in the connective tissue stroma of several organs. In all
locations, the fibers derived from Type III collagen
generally appear as a fine reticular network along with larger
fibers comprised of Type I molecules (8). It is not as yet
clear, however, whether the two types of molecules always
form separate fibers or are, in many instances, combined in
a random fashion to form what might be termed as "mixed"
fibers.

Although Type III collagen is similar to Type I collagen
with respect to its abundance in a variety of tissues, it
closely resembles Type II collagen with respect to extracta-
bility and is quite resistant to extraction in native form.
This accounts for the observations that preparations of
Type I collagen obtained by extracting tissues with neutral
salt or dilute acid solvents commonly contain little, if
any, Type III collagen. It would appear that the relative
resistance to extraction on the part of Type III collagen
may be ascribed to a cross-linking pattern similar to that
noted above for Type II collagen, although definitive data
on this point are not as yet available. Due to its resistance to extraction, Type III collagen molecules were
isolated initially from several human tissues following
solubilization by means of limited proteolysis with pepsin,
and separation from Type I molecules in the extracts by
differential salt precipitation (7). Although this procedure
remains the method of choice in the preparation of relatively
large quantities of Type III collagen from a variety of
sources, it was subsequently shown that some Type III collagen
as well as a form of Type III procollagen could be extracted
in neutral salt solvents from the skin of young animals (9).
On denaturation, Type III collagen is found to be comprised
largely of γ-components (molecular weight, 285,000 daltons)
due to interchain disulfide bonding between the $\alpha 1$(III)
chains, and the latter are readily recovered following
reduction and alkylation of the higher molecular weight components.

The information outlined above with respect to the interstitial collagens, their chain composition, and general tissue
distribution is summarized in Table I.

Each of the constituent α chains of these collagens exhibits a molecular weight of about 95,000 daltons and contains
approximately 1050 amino acid residues. In the native molecules, the chains are parallel, traverse the full length of
the molecule, and exist throughout most of their length in
the conformation of a polyproline-like helix with the

TABLE I

The Interstitial Collagens

	Chain Composition	Distribution
Type I	$[\alpha 1(I)]_2 \alpha 2$	All Major Connective Tissues
Type I-trimer	$[\alpha 1(I)]_3$	Unknown
Type II	$[\alpha 1(II)]_3$	Hyaline Cartilages
Type III	$[\alpha 1(III)]_3$	Selected Connective Tissues

individual helices further coiled around a central axis. This produces a rigid, highly asymmetric molecule with dimensions approaching 3000Å in length and 15Å in diameter. On denaturation, the constituent chains of a given collagen may be readily isolated and purified by ion-exchange chromatography. The primary structure of each chain is comprised largely of repetitive triplets of the form, Gly-X-Y, where X and Y represent any amino acid. This type of sequence in which glycyl residues occur at every third position, is an absolute requirement for participation of the sequence in the triple-chain collagen fold. Exceptions to the repetitive triplet structure of the collagen α chains occur only at relatively short sequences located at the NH_2- and COOH-terminal extremities of the chains. These regions, then, cannot assume the collagen fold in native molecules and represent remnants of much larger non-helical peptide extensions occurring at both the NH_2- and COOH-terminal ends of a biosynthetic precursor, the procollagen molecule (1). As far as is known, all the interstitial collagens are synthesized initially as relatively large procollagen molecules comprised of pro α chains, each of which exhibits a molecular weight in the range of 160,000 daltons. Subsequent to secretion the procollagen peptide extentions are removed by apparently specific proteases, although scission of the extension peptides of Type III procollagen may in certain instances be delayed or incomplete as indicated in recent studies showing that specific antibodies to Type III procollagen and Type III collagen exhibit no large differences in their staining patterns when employed to localize collagen in skin sections by indirect immunofluorescence (10). Following scission of the procollagen extension peptides, interstitial collagen molecules

apparently precipitate under physiological conditions to form
more or less well-defined fibrous elements. The eventual
size and disposition of the fibers that are formed, however,
probably depends on the environment in which fiber formation
occurs as well as on the proportion of procollagen extension
peptides which remain associated with the precipitating
molecules.

Of particular interest and relevance to the present
discussion are the chemical characteristics of the interstitial collagen α chains. These are summarized in Table II
which lists a partial amino acid composition for human
αl(I), α2, αl(II), and αl(III) chains.

It is apparent from these data that the α chains of the
interstitial collagens represent a set of closely related,
but clearly different, polypeptide chains. Although information on the primary structure of the α2, αl(II), and
αl(III) chains is at this time fragmentary, the available
comparative sequence data conclusively support the concepts
that these four polypeptide chains are homologous and that
synthesis of each of the chains is controlled by a distinct
genetic locus (3-15). The chemical differences between the
interstitial collagen α chains as well as the unique primary
structure of each chain are undoubtedly responsible for
certain distinctive characteristics of the Type I, Type I-
trimer, Type II, and Type III collagen molecules. These
include unique solubility properties and the occurrence of
apparently specific antigenic determinants. As noted above,
exploitation of their different solubility properties in
aqueous solvents has proven extremely useful in achieving
the separation and isolation of these collagens in native
form. Moreover, the availability of specific antibodies to
the different collagens is of immense value in studies
designed to precisely localize the respective antigens in
tissue sections by employing the conventional techniques of
indirect immunofluorescence and light microscopy.

The Basement Membrane Collagens.

Although fibrous cross-striated elements are not readily
identifiable in basement membranes, it is clear that such
structures contain a relatively high proportion of molecules
with collagenous sequences. Studies on the molecular organization of basement membranes, however, have led to considerable disagreement concerning the nature and number of their
collagenous constituents as well as the relationship of these
components to other macromolecular constituents of the membranes. On the one hand, collagen recovered following limited

TABLE II

*Compositional Features of the
Human Interstitial Collagen α Chains*

	Residues/1000 Residues			
	α1(I) [1]	α2 [1]	α1(II) [2]	α1(III) [3]
3-Hydroxyproline	1	1	2	0
4-Hydroxyproline	96	83	99	125
Glutamic acid	70	66	90	71
Proline	132	117	121	107
Glycine	334	336	333	350
Alanine	115	108	100	96
Half-cystine	0	0	0	0
Leucine	20	32	26	22
Phenylalanine	12	12	13	8
Hydroxylysine	5	9	15	5
Lysine	30	22	22	30
Arginine	49	51	51	46

[1] *Data from reference 11.*
[2] *Data from reference 12.*
[3] *Data from reference 7.*

proteolysis with pepsin from several basement membranes including glomerular basement membrane was reported to yield
on denaturation only a single type of chain (16). The chain
derived from basement membrane collagen appeared to have a
molecular weight somewhat in excess of that for the constituent α chains of the interstitial collagens and differed markedly from the latter chains with respect to overall compositional features. Based on these results, the collagen in
basement membrane structures has often been regarded as a
single molecular species, Type IV collagen, comprised of
three apparently identical α1(IV) chains. Additional studies
on acid soluble collagen from anterior lens capsule basement
membrane suggested that basement membrane collagen molecules

were associated in the tissue with rather extensive non-
collagenous sequences thus forming a subunit which resembled,
in many respects, a procollagen molecule (17). Subsequent
experiments on collagen biosynthesis with organ cultures of
parietal yolk sac endoderm essentially confirmed this view
with respect to the basement membrane collagen molecule by
indicating that basement membrane procollagen chains are
not converted to smaller molecular weight species at the
time collagen is deposited in the tissue (18). Since
cysteinyl residues are apparently present in basement membrane
collagenous sequences, and since lysyl-derived cross-linking
compounds can be identified in hydrolysates of basement membranes, it has been further proposed that the collagenous subunits are stabilized in the tissue by intermolecular disulfide
bonds and lysyl-derived cross-links as well as by disulfide
linkages with noncollagenous glycoproteins (19).

On the other hand, an entirely different view for the
molecular organization of basement membranes has been generated in studies demonstrating a high degree of molecular
heterogeneity for components solubilized in detergent-contain-
ing solvents from reduced and alkylated glomerular basement
membrane (20-22). Although it is likely that at least some
of the observed heterogeneity results from physiological
degradation of the basement membrane macromolecules (22), the
latter model visualizes basement membranes as comprised of a
meshwork of disulfide-linked chains which contain alternating
collagenous and noncollagenous sequences with the former
sequences being, for the most part, substantially shorter
than the sequence represented by an interstitial collagen
α chain (23). With respect to this model, it is noteworthy
that relatively short collagenous sequences are known to
exist in certain biological macromolecules such as Clq, a subcomponent of the first component of complement (24), and in
some forms of the enzyme acetylcholinesterase (25).

General Hypothesis for Basement Membrane Collagens.

The views discussed above for the molecular organization
of basement membranes and the nature of the collagenous components therein are quite divergent. It is possible, however,
that the approaches used by the different investigators (19,
22) for solubilizing and characterizing the basement membrane
constituents have served to emphasize different aspects of
the basement membrane molecular structure. Thus, it may be
argued that both views are essentially correct and that basement membranes contain at least two additional classes of

collagen molecules: one in which the polypeptide chains
comprising the molecules consist of collagenous sequences
which are somewhat longer than the α chains of the interstitial collagens; and another class of molecules containing
relatively short collagenous sequences.

Some evidence for the occurrence of such molecules in
the basement membranes of several tissues has recently been
obtained (26). The latter studies were initiated with the
assumption that triple-helical collagenous sequences occurring
in basement membranes would remain intact, but be rendered
soluble, under the condition of limited proteolysis commonly
employed to solubilize interstitial collagen molecules. It
was further assumed that collagens of basement membrane
origin could be separated from interstitial collagen molecules by differential salt precipitation under appropriate
conditions. This general approach was considered highly
advantageous as it does not involve prior isolation of an
anatomically defined basement membrane. At the same time,
it has the disadvantage of requiring further studies to
unequivocally establish that the collagens presumed to be
derived from basement membranes are actually derived from
such structures. Thus, the evidence for the origin of these
molecules in basement membranes remains largely indirect and
consists of data on amino acid composition and carbohydrate
content which indicate a close relationship to components
previously derived directly from isolated basement membranes.

In any event, experiments on pepsin-solublized collagen
from whole infant skin revealed that significant quantities
of collagen remain in solution under conditions leading to
the precipitation of Type I, II, and III collagens from the
original extract. Characterization of this more soluble
collagen showed that it was comprised of two chromatographi-
cally distinct chains (designated A chain and B chain), each
of which exhibits an apparent molecular weight of about
110,000 daltons. At this time, the molecular organization of
these chains has not been definitively evaluated and it is
possible that they occur either in a single molecular species
or that they are derived from separate molecules.

Far different results were obtained in studies leading
to the characterization of the more soluble collagen fraction
derived from specimens of vessel wall intima. In this case,
the collagen was recovered on denaturation as extremely high
molecular weight components which on reduction with sulfhydryl
reagents were quantitatively converted to subunits with an
apparent molecular weight of 55,000 daltons.

Table III lists a partial amino acid composition for A
chain, B chain, and the 55,000-molecular weight component

(55K) from human tissues and compares these data with those obtained in previous studies on a collagen chain isolated from human glomerular basement membrane (GBM).

TABLE III

Compositional Features of a Collagenous Component from Human Glomerular Basement Membrane and Related Components from Human Skin and Aortic Intima

| | Residues/1000 Residues | | | |
	GBM[1]	A Chain[2]	B Chain[2]	55K[2]
3-Hydroxyproline	11	7	10	0
4-Hydroxyproline	130	113	105	65
Glutamic acid	84	86	95	104
Proline	61	98	120	92
Glycine	310	346	334	318
Alanine	33	54	45	41
Half-cystine	8	0	0	20
Leucine	54	33	38	24
Phenylalanine	27	10	11	15
Hydroxylysine	45	22	39	48
Lysine	10	12	13	18
Arginine	33	48	40	64
Glc-Gal-Hydroxylysine	(34)	(5)	(29)	(41)
Gal-Hydroxylysine	(2)	(3)	(5)	(4)
Glucosamine	0	0	0	(22)

[1]*Data from reference 16.*
[2]*Data from reference 26.*

From these data, it is reasonably clear that the components listed above exhibit several compositional features which distinguish them, as a group, from the interstitial collagen α chains for which similar data is presented in Table II.

Although these distinguishing features are not apparent in
all components to the same extent, they may be summarized as
follows: a relatively high content of 3-hydroxyproline; a
moderately elevated content of glutamic acid; a striking
decrease in alanine content; and a relatively large amount
of lysine plus hydroxylysine with a high proportion of the
latter amino acid present as the disaccharide derivative.
In addition, the 55,000 molecular weight component contains
appreciable quantities of glucosamine suggesting the presence
of more complex heteropolysaccharide units.

The current data are, then, apparently compatible with
the notion that basement membrane structures collectively
contain several kinds of collagen molecules, the constituent
chains of which are comprised of collagenous sequences
approximately 15-20% longer than the collagenous sequences
in the α chains derived from interstitial collagens. Indeed,
two chains similar to A chain and B chain (in terms of size
and compositional features) have recently been isolated from
the more soluble collagen fraction of human amniotic and
chorionic membranes, and it was suggested that they could
originate in the collagen molecules of the morphologically
prominent basement membranes of these tissues (27). In
addition, the prevalence of disulfide-bonded high molecular
weight aggregates of the 55,000 molecular weight component
in extracts of certain tissues strongly suggests that
collagen-like molecules containing the latter sequences in
triple-helical form comprise at least a portion of the protein
in some basement membrane structures. It is possible,
although not as yet established, that these relatively short
collagenous sequences occur as portions of much larger chains
in which the collagenous and noncollagenous sequences alternate. At the moment, there are virtually no data concerning
the actual size of such chains or their genetic diversity.
However, by analogy with what appears to be the case for other
classes of collagenous sequences, it is tempting to speculate
that at least the relatively short collagenous sequences
exhibit a certain degree of genetic diversity, and thus
collectively constitute an additional class of collagen
sequences adapted for a specific biological function. The
above considerations are summarized in Figure 1 which diagram-
matically depicts the molecular conformation of an interstitial collagen molecule in relation to several parameters
proposed for collagen molecules originating in basement membrane structures.

It is reasonable to presume that further work will eventually lead to a more complete biochemical characterization
of the collagenous components present in basement membranes
of several tissues and provide crucial information in several

NH$_2$-TERMINAL
REGION

A.

CENTRAL
REGION   COOH-TERMINAL
                REGION

B.

C.

Fig. 1.   A schematic representation of the mole-
cular configuration for the proposed classes of colla-
gen molecules: (A) an interstitial collagen molecule
in which the non-helical NH$_2$-terminal region, the
central helical region, and the non-helical COOH-
terminal region account for about 2%, 95%, and 3% of
the molecular length, respectively: (B) a basement
membrane collagen molecule in which the constituent
chains contain collagenous sequences about 15-20%
longer than the α chains of the interstitial collagens.
As noted in the text, such molecules may exist in
tissues in a form resembling procollagen molecules
with extensive non-helical sequences at each extremity
and these latter sequences are not depicted; and (C)
a basement membrane collagen or "collagen-like" mole-
cule in which relatively short triple-helical collagen-
ous sequences alternate with noncollagenous sequences.

critical areas in which definitive data are currently lacking.
This is particularly true with respect to macromolecular interactions on the part of molecules containing the larger
collagenous sequences as well as the origin and macromolecular
organization of molecules containing the relatively short
collagenous sequences.   Information of this nature can be expected to considerably enhance our understanding of the molecular organization and function of these most important
biological structures.

REFERENCES

1. Miller, E.J., *Mol. Cell. Biochem.*, in press.
2. Chandrakasan, G., Torchia, D.A. and Piez, K.A., *J. Biol. Chem.* 251:6062-6066.
3. Mayne, R., Vail, M.S. and Miller, E.J., *Develop. Biol.* 54:230-240, 1976.
4. Narayanan, A.S. and Page, R.C., *J. Biol. Chem.* 251:5464-5471, 1976.
5. Miller, E.J., *Biochemistry* 11:4903-4909, 1972.
6. Trelstad, R.L., Kang, A.H., Toole, B.P. and Gross, J., *J. Biol. Chem.* 247:6469-6473, 1972.
7. Chung, E. and Miller, E.J., *Science* 183:1200-1201, 1974.
8. Gay, S., Fietzek, P.P., Remberger, K., Eder, M. and Kühn, K., *Klin. Wschr.* 53:205-208.
9. Timpl, R., Glanville, R.W., Nowack, H., Wiedemann, H., Fietzek, P.P. and Kühn, K., *Hoppe-Seyler's Z. f. Physiol. Chem.* 356:1783-1792, 1975.
10. Nowack, H., Gay, S., Wick, G., Becker, U. and Timpl, R., *J. Immunol. Meth.* 12:117-124, 1976.
11. Epstein, E.H., Jr., Scott, R.D., Miller, E.J. and Piez, K.A., *J. Biol. Chem.* 246:1718-1724, 1971.
12. Miller, E.J. and Lunde, L.G., *Biochemistry* 12:3153-3159, 1973.
13. Butler, W.T., Miller, E.J. and Finch, J.E., Jr., *Biochemistry* 15:3000-3006, 1976.
14. Fietzek, P.P. and Rexrodt, F.W., *Eur. J. Biochem.* 59:113-118, 1975.
15. Fietzek, P.P. and Rauterberg, J., *FEBS Letters* 49:365-368, 1975.
16. Kefalides, N.A., *Biochem. Biophys. Res. Commun.* 45:226-234, 1971.
17. Olsen, B.R., Alper, R. and Kefalides, N.A., *Eur. J. Biochem.* 38:220-228, 1973.
18. Minor, R.R., Clark, C.C., Strause, E.L., Koszalka, T.R., Brent, R.L. and Kefalides, N.A., *J. Biol. Chem.* 251:1789-1794, 1976.
19. Kefalides, N.A., *J. Invest. Derm.* 65:85-92, 1975.
20. Hudson, B.G. and Spiro, R.G., *J. Biol. Chem.* 247:4229-4238, 1972.
21. Hudson, B.G. and Spiro, R.G., *J. Biol. Chem.* 247:4239-4247, 1972.
22. Sato, T. and Spiro, R.G., *J. Biol. Chem.* 251:4062-4070, 1976.

23. Spiro, R.G., *New Eng. J. Med.* 288:1337-1342, 1973.

24. Reid, K.B.M. and Porter, R.R., *Biochem. J.* 155:19-23, 1976.

25. Lwebuga-Mukasa, J.S., Lappi, S. and Taylor, P., *Biochemistry* 15:1425-1434, 1976.

26. Chung, E., Rhodes, R.K. and Miller, E.J., *Biochem. Biophys. Res. Commun.* 71:1167-1174, 1976.

27. Burgeson, R.S., El Adli, F.A., Kaitila, I.I. and Hollister, D.W., *Proc. Natl. Acad. Sci. USA,* 73:2579-2583, 1976.

PROBLEMS IN THE ULTRASTRUCTURAL ASSEMBLY

OF BASEMENT MEMBRANE COLLAGEN MICROFIBRILS

Arthur Veis and David Schwartz

*Northwestern University Medical School
Chicago, Illinois*

ABSTRACT: *Bovine lens capsule basement membrane collagen was isolated following pepsin digestion. Electron microscopic observations of this collagen in the
non-aggregated state as well as under conditions which
would have formed native 700Å-periodic fibers in Type I
interstitial collagen showed that only non-striated
fibers would form in the lens capsule basement membrane
collagen. Segment-long-spacing precipitates were
observed although these were less well organized than
typical interstitial collagen preparations. Since
non-collagenous proteins had been removed it appears
that the organization of basement membrane collagen
into non-periodic filaments in vivo is related to some
intrinsic property of the basement membrane collagen
molecules.*

INTRODUCTION

The typical 700Å cross-striated fibrils of the Type I,
II and III interstitial collagens have not been observed in
intact basement membrane, although the membranes appear to
have a filamentous fine structure (1). As a part of our
study of the mechanism of fibril formation in collagens, we
have begun to examine the interactions between basement membrane collagen molecules in order to determine which factors

rule out the ordered native collagen fibril packing arrangements which give rise to the 700Å (D-stagger) periodicity of
the interstitial collagens.

It is fairly well established that basement membranes
(BM) are constructed from a collagen component of unique
composition and sequence and one or more glycoprotein components covalently linked to the collagen (2). The collagen
portion of those basement membranes which have been characterized, such as the BM of the anterior lens capsule (3) and the
parietal yolk sac endoderm (4), have molecular sizes equivalent to the pro α- or α-chains of the interstitial collagens.
The compositions of the BM collagens are somewhat different
from those of the Type I, II and III collagens in that the
BM collagens all contain 3-hydroxyproline in increased
amount, 9 residues/1000 residues compared with 1 residue/-
1000 residues in Type I, and more hydroxylysine, 58 residues-
/1000 residues as compared with ∿5 residues/1000 residues in
Type I (5). However, the remainder of the structure maintains the glycine triplet repeat and is capable of forming
typical collagen triple helix (6). In view of the strong
homologies between chain sequences in the interstitial
collagens (7) it seems unlikely that, in spite of their
genetically distinct origins, the distribution of side chain
functional groups other than the excess hydroxylysine (which
may replace interstitial collagen lysine positions in large
measure) will differ markedly in the BM and interstitial
collagens.

The relationship between the non-collagen glycoprotein
components of BM and the collagen is not clearly established,
but they are attached by disulfide bridges (2). It also
appears that BM collagen may be incorporated directly into
the membrane structure without loss of the pro-peptide extensions (4). Thus, one might hypothesize that it is either
the pro-peptide extensions or the non-collagen protein components which prevent the assembly of striated native collagen
fibrils. To examine the validity of this hypothesis we have
attempted to isolate the collagenous portion of a typical BM
collagen and examine its fiber forming capacity free of non-
collagenous proteins and pro-peptide extensions in contrast
to the comparable behavior of Type I collagen.

EXPERIMENTAL PROCEDURES

Anterior lens capsules were dissected from 200 bovine
eyes and cleaned and washed by the method of Kefalides and
Denduchis (8). One portion was extracted with citrate buffer
by the procedure of Olsen et al. (9), the remainder was

digested directly with pepsin (10:1 BM:Enzyme) in 0.1M acetic acid. The suspension was centrifuged and the soluble super-natant fraction collected. Soluble interstitial collagen was obtained by extraction of rat skins with either 1.0M NaCl or 0.5M acetic acid. These collagens were purified by repeated salt precipitation (10).

The citrate soluble and pepsin solubilized BM collagens were precipitated from acetic acid solution by dialysis against 17% NaCl. The collagen precipitates were collected and dissolved in 0.1N HAc. Very little citrate soluble collagen was obtained from the bovine lens capsule.

The pepsin soluble collagen was examined in the electron microscope in several ways:

A. *Monomers and Aggregates under Non-Aggregating Conditions*. In a procedure described in detail elsewhere (11), the monomers and aggregates present in a solution may be visualized with minimal artifactual aggregation. In essence, droplets of an acetic acid solution are placed on a carbon coated formvar film and then washed immediately with collagen free buffer solution at the same concentration. After final washing, the grids are wicked, and the specimens negatively stained with fresh 2% phosphotungstic acid solutions, at pH 7.0

B. *Native Fibrils*. Native fibers were produced, in Type I collagens, by dialysis of acid solutions against distilled water or dilute NaCl solutions at neutral pH. The precipitates were collected and examined in negatively stained preparations. BM collagen solutions were treated in identical fashion.

C. *Segment-Long-Spacing (SLS) Precipitates*. Precipitates of Type I collagen were readily produced by addition of adenosine triphosphate (ATP) in acid form to acetic acid solutions or by dialysis of the acid collagen solutions against ATP solution. The SLS precipitates were collected and viewed in negatively stained preparations. Pepsin solubilized BM collagen was treated in identical fashion.

RESULTS AND DISCUSSION

The pepsin soluble BM collagen is seen as extended rod-like molecules approximately 300 nm in length when examined under minimal aggregating conditions, Figure 1a. The aggregates which do form show no apparent preferential intermolecular alignments. At this stage, however, the BM collagen rodlets are less distinct and more curved than comparable preparations of Type I collagen, Figure 1b. There is no evidence

 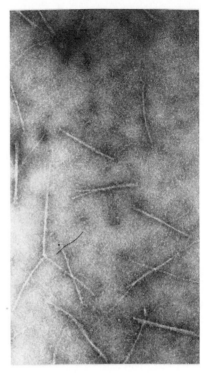

Fig. 1a. Pepsin-treated
BM collagen soluble in 0.1M
acetic acid examined under
non-aggregating conditions.
Negatively stained with 2%
PTA. Magnification X64,000.

Fig. 1b. Rat skin (inter-
stitial, Type I) collagen solu-
ble in 0.1M acetic acid and ex-
amined under conditions compa-
rable to 1a. Negatively stain-
ed with 2% PTA. X64,000.

in Figure 1a for the presence of residual globular procollagen extensions or globular non-collagenous proteins associated
with the BM collagen as seen in the micrographs of Olsen et al.
(9) for non-pepsin treated citrate extracted lens capsule BM
collagen.

When the BM collagen is brought to conditions under which
Type I collagen molecules form native 700 Å repeat or D-stagger, the BM collagen does not precipitate readily. At concentrations of collagen on the order of 5-10 mg/ml a filamentous
precipitate does form but it does not exhibit any axial
periodicity, Figure 2. The aggregates do not indicate any
preferential molecular stagger.

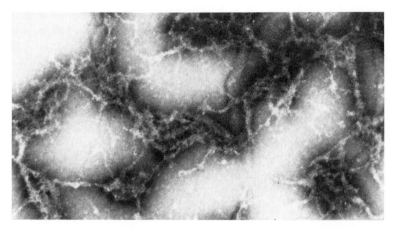

Fig. 2. Fibrous precipitate obtained upon dialysis of the pepsin treated BM collagen of Fig. 1a against H_2O. Under these conditions inter- stitial collagen would have formed D-periodic fibers. Negatively stained with 2% PTA. Magnification X20,000.

Although the attempts to form native D-stagger filaments were not successful segment-long-spacing precipitates, at zero-D were obtained. The SLS were poorly organized as com- pared to rat skin collagen, Figure 3, but did show typical collagen banding and asymmetry. It appears that the amino terminal ends can be well registered at zero-stagger but at the COOH-terminal region the spools are less well packed.

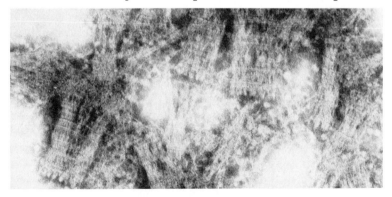

Fig. 3. Segment-long-spacing precipitate of the pepsin-treated BM collagen of Fig. 1a produced by dialysis against ATP. Negatively stained with 2% ATP. Magnification X100,000.

It has been assumed in earlier studies by Kefalides and co-workers (2,5,8) that pepsin digestion removes the major part of pro-peptide extensions and non-collagen glycoproteins from BM collagens. This appears to be the case in the present study since, as noted above, the amino terminal molecular end regions are well defined. That is, amino terminal extensions appear to be absent in micrographs such as Fig. 1a and Fig. 3. Hence, we can conclude that the interactions controlling aggregation in the pepsin-treated BM collagen system relate to the helical parts of the BM collagen molecules or a non-pepsin susceptible carboxyl terminal peptide extension.

Fiber formation begins with assembly of a limiting micro-fibril unit (12,13) of constant diameter and these micro-fibrils then associate to form fibrils of varying size. In the initial assembly, two molecules can interact in either of two ways, depicted schematically in Figure 4. The two mecha-nisms differ in an important aspect. In scheme A which

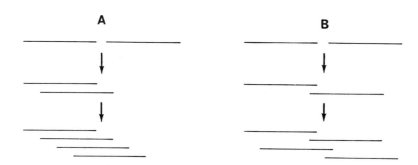

A   B

Models for assembly of collagen molecules into micro-fibrils.

Fig. 4a. Stepwise microfibril growth, molecules added with 1D-stagger. Microfibril growth could pro-ceed in either direction but by single molecule addi-tions.

Fig. 4b. Thin filament formation, followed by filament association. In this scheme molecules add at 4D based on specific end-region interactions. The filaments pack in D-stagger because of side-by-side interactions in helical regions.

emphasizes an initial D-stagger favored by charge pairing interactions (14) ionic helical region-helical region inter-actions provide the primary driving force for ordering. In scheme B end-region interactions of specific nature favor the initial association of 4D-stagger molecules, pairs which then form entropically driven microfibrils (15). Comper and Veis (16) have demonstrated the importance of molecular end-regions in assembly of microfibrils but also showed that partial end-region removal by pepsin did not prevent microfibril formation in Type I collagen *in vitro*. Even in process B, however, thin filaments ordering to the 1D-stagger must depend on compensating charge pairing interactions.

Although pepsin treated BM molecules are rod-like in character and similar in length to Type I collagens and will form filamentous structures or disorganized SLS, their failure to form D-periodic microfibrils and fibrils means that D-periodic intermolecular electrostatic charge pairing interactions cannot be established in the helical regions. In the absence of sequence data, we can speculate that either the increased numbers of hydroxylysines, to a total lysine plus hydroxylysine of about 70 residues/1000 (5), or the reduction of arginine content to 19 residues/1000 from the ∿50 residues/1000 value in interstitial collagens, drasti-cally inhibit interaction. A second possibility for inhibi-tion of charge pairing interactions is the high degree of hydroxylysine glycosylation. A final possibility is that some residual carboxyl-terminal extensions, hinted at in Figure 3, inhibits the interaction. It is probable, however, that we can exclude the other non-collagenous protein com-ponents of BM as the primary factor in the failure of forma-tion of D-periodic collagen fibrils in the lens capsule base-ment membranes. Thus, the organization of the basement membrane is related to some intrinsic properties of the BM collagen molecules.

REFERENCES

1. Fisher, R.F. and Wakely, J., *Proc. Roy. Soc. Lond.* B-193: 335-358, 1976.
2. Kefalides, N.A., *Int. Rev. Conn. Tiss. Res.* 6:63-104, 1973.
3. Grant, M.E., Schofield, D., Kefalides, N.A. and Prockop, D.J., *J. Biol. Chem.* 248:7432-7437, 1973.
4. Minor, R.R., Clark, C.C., Strause, E.L., Koszalka, T.R., Brent, R.L. and Kefalides, N.A., *J. Biol. Chem.* 251: 1789-1794, 1976.

5. Kefalides, N.A., *Biochem. Biophys. Res. Comm.* 47:1151-1158, 1972.
6. Gilman, R.A., Blackwell, J., Kefalides, N.A. and Tomichek, E., *Biochim. Biophys. Acta* 427:492-496, 1976.
7. Fietzek, P.P. and Kühn, K., *Int. Rev. Conn. Tiss. Res.* 7:1-60, 1976.
8. Kefalides, N.A. and Denduchis, B., *Biochemistry* 8:4613-4621, 1969.
9. Olsen, B.R., Alper, R. and Kefalides, N.A., *Eur. J. Biochem.* 38:220-228, 1973.
10. Clark, C.C. and Veis, A., *Biochemistry* 11:494-502, 1972.
11. Schwartz, D. and Veis, A. *(in preparation).*
12. Veis, A., Anesey, J. and Mussell, S., *Nature* 215:931-934, 1967.
13. Smith, J.W., *Nature* 219:157-158, 1968.
14. Doyle, B.B., Hukins, D.W.L., Hulmes, D.J.S., Miller, A., Rattew, C.J. and Woodhead-Galloway, J., *Biochem. Biophys. Res. Comm.* 60:855-864, 1974.
15. Veis, A. and Yuan, L., *Biopolymers* 14:895-900, 1975.
16. Comper, W. and Veis, A., *Biopolymers* (1977) *in press.*

B. Biosynthesis

COMPARISON OF

NEWLY SYNTHESIZED BASEMENT MEMBRANE PROCOLLAGEN

TO INTERSTITIAL TYPE I PROCOLLAGEN

Charles C. Clark and Nicholas A. Kefalides

University of Pennsylvania
Philadelphia, Pennsylvania

ABSTRACT: Basement membrane (BM) procollagen (PC) from rat embryo parietal yolk sac and interstitial (I) PC from chick embryo tendon cells were purified by DEAE cellulose chromatography. The PC's were each subjected to bacterial collagenase or chymotrypsin digestion, and the products examined by SDS gel filtration before and after disulfide (S-S) bond reduction.

Prior to enzymic digestion and reduction, all species eluted as S-S bonded aggregates; after reduction the MW for IPC pro-α chains was ∿125,000 and BMPC pro-α chains was ∿160,000. Chymotrypsin digestion without reduction converted IPC to α-chains with MW ∿95,000 while BMPC remained aggregated; subsequent reduction of BMPC gave chains with MW ∿135,000. Collagenase digestion of IPC without reduction yielded two fragments with MW 80-90,000 and 23,000, respectively. Upon reduction, the former now eluted with MW ∿35,000. A more complex pattern was obtained with BMPC, but most significant was a hydroxyproline-containing peptide (MW ∿70,000) which moved to a lower MW upon reduction.

The results suggest that BMPC contains S-S bonds in a protease-resistant hydroxyproline-containing portion of the molecule. A model of BMPC will be compared to that of IPC.

287

INTRODUCTION

Evidence in recent years has shown that interstitial Type I collagen is initially synthesized in a precursor form, procollagen, which has properties quite different from those of the collagen molecule (for reviews, see Schofield and Prockop, 1973; Bornstein, 1974; Martin et al., 1975). The molecular structure of Type I procollagen which emerges from these studies is that of a triple-helical collagen molecule with non-triple-helical precursor-specific appendages (propeptides) at both the amino- and carboxy-termini (Tanzer et al., 1974; Byers et al., 1975; Monson et al., 1975; Fessler et al., 1975). The carboxy-terminal propeptides contain tryptophan, are linked together by intermolecular disulfide bonds and appear to have a larger molecular weight than the amino-terminal propeptides which contain no tryptophan or intermolecular disulfide bonds (Tanzer et al., 1974; Murphy et al., 1975; Monson et al., 1975; Morris et al., 1975; Fessler et al., 1975; Olsen et al., 1976; Uitto et al., 1976). Although gel filtration studies in sodium dodecyl sulfate (SDS) yield molecular weights of $\sim$125,000 for pro-α(I) (Jimenez et al., 1971), more recent evidence suggests a molecular weight of 150,000 - consistent with a collagen portion of 1000 amino acid residues, an amino-terminal propeptide of 200 residues, and a carboxy-terminal propeptide of 340 residues (Monson et al., 1975).

Relative to interstitial procollagen, the structure of newly-synthesized basement membrane (BM) procollagen has not been studied extensively. Initial studies employing gel filtration in SDS suggested that the molecular weight of BM pro-α chain was 140,000 (Grant et al., 1972 and 1975), but more recent studies give a molecular weight of 150,000 to 160,000 (Kefalides et al., 1976; Minor et al., 1976). Although it was originally suggested that there was a time-dependent conversion of BM pro-α chains to BM α chains with a molecular weight of $\sim$120,000 analogous to that seen with interstitial procollagen (Grant et al., 1972 and 1975), more recent evidence suggests that there is no such conversion (Minor et al., 1976; Heathcote et al., this volume, 1976). Thus, intact basement membranes appear to contain a "procollagen-like" molecule (Minor et al., 1976).

The object of this study is to compare some properties of rat embryo parietal yolk sac BM procollagen to those of chick embryo tendon cell Type I interstitial procollagen. These properties include susceptibility to bacterial collagenase and to chymotrypsin, and analysis of the resultant fragments by SDS-gel filtration both before and after disulfide bond reduction.

EXPERIMENTAL PROCEDURE

The source of Type I interstitial procollagen was matrix-free, 17-day chick embryo tendon cells (Dehm and Prockop, 1971); the source of BM procollagen was whole, 14-day rat embryo parietal yolk sac (Clark et al., 1975). The cells or tissues were incubated with [^{14}C]proline for 4 hrs, centrifuged, and only the extracellular medium proteins were used. To prevent non-specific proteolysis, the medium was immediately made 10 mM in EDTA and N-ethylamaleimide, and 1 mM in benzamidine hydrochloride and phenylmethyl sulfonylfluoride (Monson and Bornstein, 1973). After dialysis against 0.4M NaCl-0.1M Tris·HCl (pH 7.4) containing the above inhibitors, the medium proteins were precipitated by the addition of solid ammonium sulfate to 30% of saturation and harvested by centrifugation. Our results show that this pellet contains >90% of the 4-hydroxy[^{14}C]proline originally present in the medium.

Both interstitial and BM procollagen samples were then treated before disulfide bond reduction with either bacterial collagenase (Dehm et al., 1974) or chymotrypsin (Schofield et al., 1974) with appropriate controls. The products of these digestions were monitored by SDS-gel filtration before and after disulfide bond reduction. Carrier Type I collagen was included in every run. Chromatographic fractions were analyzed for total radioactivity, and labeled (Juva and Prockop, 1966) and unlabeled (Switzer and Sammer, 1971) 4-hydroxyproline. Some samples were also monitored by SDS-polyacrylamide gel electrophoresis (Clark, 1976).

RESULTS

Initial analysis of the purified medium proteins by SDS-polyacrylamide gel electrophoresis is shown in Figure 1. In contrast to Type I interstitial procollagen which shows pro-α1 and pro-α2 in a ratio of 2:1 (Fig. 1A), BM procollagen shows only a single component (Fig. 1B). This is consistent with the original finding of only a single type of α chain in BM collagen (Kefalides, 1971). Based on the included collagen standards, pro-α1(I) has a molecular weight of $\sim$125,000 while pro-α(IV) has a molecular weight of $\sim$160,000.

The medium proteins were also chromatographed on SDS-agarose before and after disulfide bond reduction. As shown in Figure 2A, BM procollagen elutes in the void volume as one might expect for a disulfide-bonded aggregate. An identical pattern was obtained with Type I procollagen (not shown, see Schofield et al., 1974). On the other hand, after disulfide

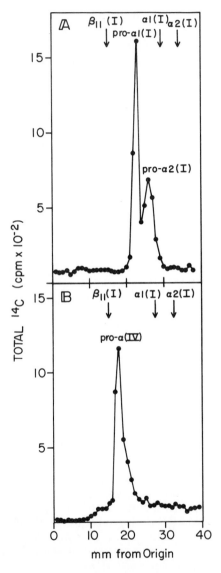

Fig. 1. SDS-polyacryla-mide gel electrophoresis of A) reduced Type I procollagen and B) reduced Type IV pro-collagen. The migration of added rat tail tendon collagen standards are shown. The tracking dye migrated to 78mm.

bond reduction, the BM sample elutes as a pro-α(IV) chain with a molecular weight of ∿160,000 (Fig. 2B). The elution position of pro-α(I) with a molecular weight of ∿125,000 is indicated.

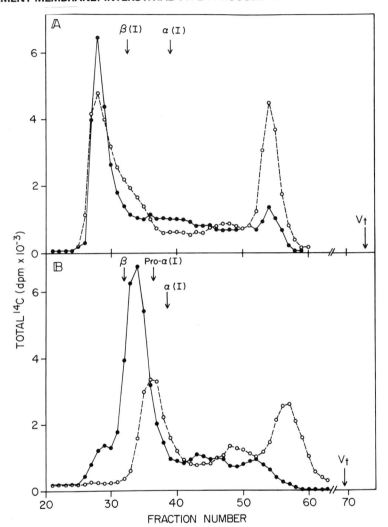

Fig. 2. SDS-gel filtration on 6% agarose (Bio
Gel A-5M, 200-400 mesh) of BM procollagen. A) Unreduced
and B) reduced. The same number of counts were initially
present in all samples. The elution position of Type I
collagen and procollagen standards are shown. All of the
4-hydroxy[^{14}C]proline (not shown) was associated with the
respective peaks in fractions 25-40. Chymotrypsin
control (●——●), chymotrypsin digest (o---o).
V_t indicates position of ^3H_2O.

The results after digestion with chymotrypsin are also shown in Fig. 2. Before disulfide bond reduction, BM procollagen still elutes as a disulfide-bonded aggregate (Fig. 2A). This finding has also been reported by Williams et al. (1976) for kidney glomerular BM procollagen. These results are also similar to those reported for Type III collagen (Chung and Miller, 1974; Epstein, 1974). After disulfide bond reduction, the void volume peak now elutes with an apparent molecular weight of 130,000-140,000 (Fig. 2B). In contrast to these observations with BM procollagen, Type I procollagen samples eluted with α(I) chains both before and after disulfide bond reduction as expected (not shown). These results suggest that similar to Type I, Type IV procollagen is partially susceptible to proteolytic digestion; however, in contrast to Type I, disulfide bonding in Type IV procollagen is present in a protease-resistant region of the molecule and may account for the larger molecular size of the resultant α chain (130,000 vs. 95,000).

Samples were next digested with bacterial collagenase. Greater than 90% of the [14C]proline and essentially all of the 4-hydroxy[14C]proline of Type I procollagen became dialyzable. When the collagenase resistant fraction of Type I procollagen was examined on SDS-agarose before reduction (Figure 3A), the major peak of radioactivity eluted with an apparent molecular weight of 80,000 to 90,000; upon reduction (Fig. 3B) this peak now eluted in a position corresponding to a molecular weight of 35,000 to 40,000. The smaller peak (fractions 65-73) was not sensitive to reduction (M <20,000). Based on the similarities of these results to those of Olsen et al. (1976), the major peak presumably represents the disulfide-bonded carboxy-terminal propeptide region, while the smaller peak presumably represents the non-disulfide-bonded amino-terminal propeptides. The origin of the peak eluting in fractions 65-73 is not known as yet.

A significantly different pattern was obtained when BM procollagen was treated with bacterial collagenase (Figure 4). In contrast to the results with Type I, only ∿50% of the 14C and ∿90% of the 4-hydroxy[14C]proline became dialyzable. The peak of major interest is that which elutes between fractions 34-45. This fraction contains a 4-hydroxy[14C]proline total 14C ratio of 0.30, and elutes with an apparent molecular weight of ∿70,000. After reduction, this peak now eluted as indicated by the dashed arrow (Fig. 4). These observations suggest that in contrast to Type I, in Type IV procollagen a disulfide bond is present in a hydroxyproline-containing collagenase-resistant peptide.

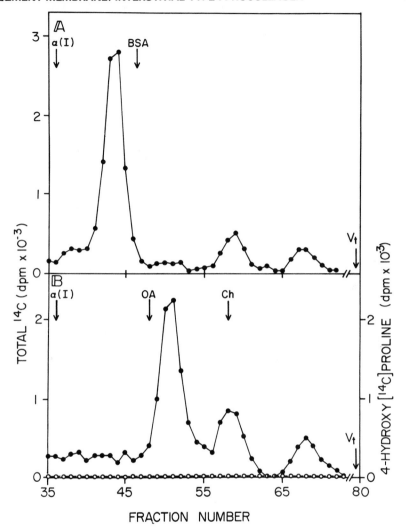

Fig. 3. SDS-gel filtration on 8% agarose (Bio Gel A-1.5M, 200-400 mesh) of bacterial collagenase digest of Type I procollagen. A) Unreduced and B) reduced. The elution positions of α(I), bovine serum albumin (BSA), ovalbumin (OA) and chymotrypsinogen (Ch) are shown. Total ^{14}C (●——●), 4-hydroxy-[^{14}C]proline (○---○).

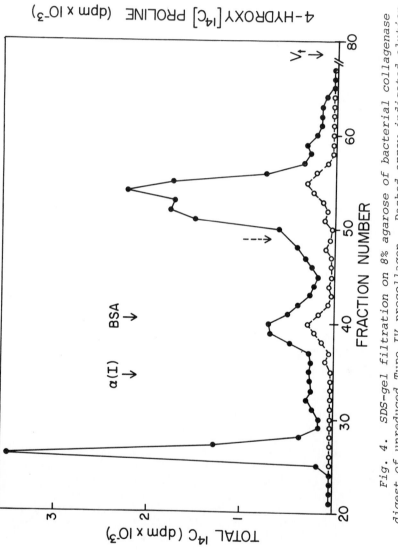

Fig. 4. SDS-gel filtration on 8% agarose of bacterial collagenase digest of unreduced Type IV procollagen. Dashed arrow indicated elution position of peak in fractions 34-45 after disulfide bond reduction. Total 14C (●———●), 4-hydroxy[14C]proline (o---o).

The small molecular weight peptides present after collagenase digestion (Fig. 4, fractions 50-57) are not sensitive
to disulfide bond reduction and presumably arise from the pro-
peptide regions of Type IV procollagen. On the other hand,
the major peak in the void volume (Fig. 4) is probably not
derived from procollagen and may represent non-collagen BM
glycoproteins.

DISCUSSION

From the results of these experiments, we can speculate
upon the structure of BM procollagen as compared to interstitial Type I procollagen. In Figure 5 we schematically
represent the structures of both Type I and Type IV procollagens, and the major products arising after either bacterial collagenase or chymotryptic digestion.

In the upper panel is shown the model for tendon cell
Type I procollagen which is consistent with that described
by others (Olsen et al., 1976; Uitto et al., 1976). The
amino- and carboxy-termini are indicated. The latter is
disulfide-bonded so the procollagen molecule will elute as an
aggregate before reduction. Chymotrypsin digestion removes
the propeptides (and disulfide-bonds) to give rise to $\alpha(I)$
chains upon denaturation both with and without reduction.
Bacterial collagenase digestion yields the disulfide-bonded
carboxy-terminal propeptide trimer and the non-disulfide-
bonded amino-terminal propeptides (Fig. 3A). Upon reduction,
the carboxy-terminal propeptides are generated (Fig. 3B).

In the lower panel is a model for Type IV procollagen
which is consistent with our data. Note that the amino-
and carboxy-termini are not indicated. It must be emphasized,
however, that this model is only one of many which could be
drawn and still be consistent with the data. In spite of
this shortcoming, we can use this model to point out the
differences we observed between Type I and Type IV procollagens:

1. Type IV procollagen behaves as a molecule larger
   than Type I procollagen both by SDS-polyacrylamide
   gel electrophoresis (Fig. 1) and SDS-gel filtration
   (Fig. 2B). Although this differential behavior in
   SDS may be artifactual, we have drawn Type IV
   bigger than Type I.

2. Type I has a chain composition of $(\text{pro-}\alpha 1)_2$
   $\text{pro-}\alpha 2$ while Type IV has a chain composition of
   $(\text{pro-}\alpha)_3$ (Fig. 1)

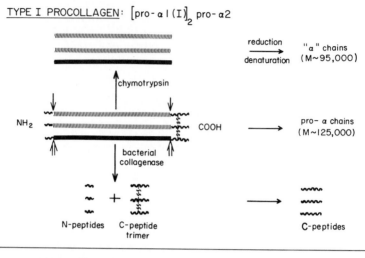

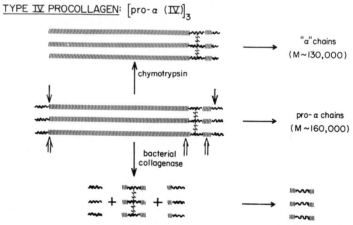

Fig. 5. Schematic representation of the structures of Type I procollagen (upper panel) and Type IV procollagen (lower panel), and the products arising after either chymotrypsin or bacterial collagenase digestion. The collagenous domain is indicated by the broad straight lines and the propeptide domains by the squiggle lines. The single arrows show the presumed cleavage sites of chymotrypsin and the double arrows show the presumed cleavage sites of bacterial collagenase.

3. Upon limited proteolysis (e.g., chymotrypsin digestion) prior to reduction, both molecules are susceptible. However, Type IV procollagen remains disulfide bonded whereas Type I does not (Fig. 2A). Upon reduction and comparison of the "α" chains generated, it appears as if each pro-α chain has lost 30,000 molecular weight (Type I: ∿125,000 to 95,000; Type IV: ∿160,000 to 130,000) (Fig. 2B).

4. After bacterial collagenase digestion, Type IV procollagen yields a hydroxyproline-containing, disulfide-bonded peptide (Fig. 4), whereas Type I procollagen yields a disulfide-bonded peptide which does not contain hydroxyproline (Fig. 3).

In order to account for the observed differences in susceptibility to enzymic digestion, we have proposed (Fig. 5) that the propeptide region on one end of the Type IV procollagen molecule may contain a collagen-like domain which protects a disulfide-containing, non-collagenous domain from being digested. This collagenous domain, however, is susceptible to bacterial collagenase and thus yields a disulfide-bonded, hydroxyproline containing peptide.

Future experiments will determine the viability of this model.

ACKNOWLEDGEMENTS

We are indebted to Marjorie Albrecht and Jacqueline Zavodnick for their expert technical help. This work was supported by NIH Grants AM-14526, AM-20553 and HL-15061. C.C.C. is a recipient of NIH Research Career Development Award AM-00063.

REFERENCES

Bornstein, P., Annu. Rev. Biochem. 43:567, 1974.
Byers, P.H., Click, E.M., Harper, E. and Bornstein, P.,
    Proc. Nat. Acad. Sci., U.S.A., 72:3009, 1975.
Chung, E. and Miller, E. J., Science 183:1200, 1974.
Clark, C.C., In: "Methodology of Connective Tissue Research",
    (D.A. Hael, ed.) Oxford, Joynson-Bruvvers Ltd., pp. 205-226,
    1976.
Clark, C.C., Tomichek, E.A., Koszalka, T.R., Minor, R.R. and
    Kefalides, N.A., J. Biol. Chem. 250:5259, 1975.

Dehm, P., Olsen, B.R. and Prockop, D.J., *Eur. J. Biochem.*
  46:107, 1974.
Dehm, P. and Prockop, D.J., *Biochim. Biophys. Acta* 240:358,
  1971.
Epstein, E.H., Jr., *J. Biol. Chem.* 249:3225, 1974.
Fessler, L.I., Morris, N.P. and Fessler, J.H., *Proc. Nat.
  Acad. Sci., U.S.A.*, 72:4905, 1975.
Grant, M.E., Harwood, R. and Williams, I.F., *Eur. J. Biochem.*
  54:531, 1975.
Grant, M.E., Kefalides, N.A. and Prockop, D.J., *J. Biol.
  Chem.* 247:3545, 1972.
Heathcote, J.G., Sear, C.H.J. and Grant, M.E., (*this volume*).
Jimenez, S.A., Dehm, P. and Prockop, D.J., *FEBS Letters*
  17:245, 1971.
Juva, K. and Prockop, D.J., *Anal. Biochem.* 15:77, 1966.
Kefalides, N.A., *Biochem. Biophys. Res. Comm.* 45:226, 1971.
Kefalides, N.A., Cameron, J.D., Tomichek, E.A. and Yanoff, M.,
  J. Biol. Chem. 251:730, 1976.
Martin, G.R., Byers, P.H. and Piez, K.A., *Adv. Enzymol.* 42:
  167, 1975.
Minor, R.R., Clark, C.C., Strause, E.L., Koszalka, T.R.,
  Brent, R.L. and Kefalides, N.A., *J. Biol. Chem.* 251:1789,
  1976.
Monson, J.M. and Bornstein, P., *Proc. Nat. Acad. Sci., U.S.A.*,
  70:3521, 1973.
Monson, J.M., Click, E.M. and Bornstein, P., *Biochemistry*
  14:4088, 1975.
Morris, N.P., Fessler, L.I., Weinstock, A. and Fessler, J.H.,
  J. Biol. Chem. 250:5719, 1975.
Murphy, W.H., von der Mark, K., McEneany, L.S.G. and Bornstein,
  P., *Biochemistry* 14:3243, 1975.
Olsen, B.R., Hoffmann, H.P. and Prockop, D.J., *Arch. Biochem.
  Biophys.* 175:341, 1976.
Schofield, J.D. and Prockop, D.J., *Clin. Orthop.* 97:175, 1973.
Schofield, J.D., Uitto, J. and Prockop, D.J., *Biochemistry*
  13:1801, 1974.
Switzer, B.R. and Summer, G.K., *Anal. Biochem.* 39:487, 1971.
Tanzer, M.L., Church, R.L., Yaeger, J.A., Wampler, D.E. and
  Park, E.D., *Proc. Nat. Acad. Sci., U.S.A.*, 71:3009, 1974.
Uitto, J., Lichtenstein, J.R. and Bauer, E.A., *Biochemistry*
  15:4935, 1976.
Williams, I.F., Harwood, R. and Grant, M.E., *Biochem. Bio-
  phys. Res. Comm.* 70:200, 1976.

SYNTHESIS OF AN EPITHELIAL BASEMENT MEMBRANE

GLYCOPROTEIN BY CULTURAL HUMAN AND MURINE

CELLS

Lewis D. Johnson,

Janice Warfel and Judith M. Megaw

Emory University
Atlanta, Georgia

ABSTRACT: A line of cells, termed parietal yolk sac carcinoma cells, isolated from murine testicular teratomas, and a cell line, termed AF cells, isolated from human amniotic fluids, both secrete an epithelial basement membrane glycoprotein when grown in vitro. The murine cells were grown in Dulbecco and Vogt's medium supplemented with 10% fetal bovine serum, and the human cells were grown in the same medium supplemented with 25% serum. The glycoproteins were isolated by precipitation with $(NH_4)_2SO_4$, dissolution in 0.05M phosphate buffer, pH 7.2 and chromatography on controlled pore glass (CPG-10-350). Elution from CPG-10-350 resulted in two major peaks, a smaller one at the void volume and a larger peak which was retarded. The void volume peak was extensively dialyzed against distilled water and lyophilized. Antisera produced against the protein in this peak reacted specifically with renal tubular and glomerular basement membranes of sections of frozen kidney assayed by indirect immuno-fluorescent microscopy. Analyses of amino acids and sugars indicated an acidic glycoprotein containing 13%

carbohydrate. Component sugars were mannose, galac-
tose, glucose, fucose, hexosamines and sialic acid.
SDS-agarose chromatography and SDS-acrylamide gel
electrophoresis of reduced, denatured glycoprotein
indicated a molecular weight of 32,000. Modifica-
tion of the growth medium resulted in marked differ-
ences in the carbohydrate composition of the murine
basement membrane glycoprotein.

INTRODUCTION

In addition to unique types of collagen, basement membranes also contain glycoproteins. These glycoproteins have
received far less attention than collagen, perhaps because
they contain no unusual amino acids such as hydroxyproline
and hydroxylysine, the presence of which provide ready markers
for identifying collagen during isolation procedures. The
observation several years ago (1) that a murine neoplasm
synthesized abundant amounts of basement membrane material
provided the basis for the studies of murine basement membrane glycoprotein which will be described. The finding that
epithelial basement membrane glycoproteins of human and murine
origin are antigenically quite similar provided the means for
identifying the glycoprotein associated with cultured human
cells (2).

ISOLATION AND CHARACTERIZATION OF MURINE
    BASEMENT MEMBRANE GLYCOPROTEIN

The experimental procedures described herein have been
published in further detail elsewhere (3,4).
The murine neoplastic epithelial cells were grown in
Dulbecco and Vogt's medium supplemented with 10% fetal bovine
serum. The medium was harvested from the cultures and
centrifuged to remove cellular material. Solid $(NH_4)_2SO_4$ was
added to the supernatant to 50% saturation, and the protein
was allowed to precipitate overnight at 4°. The precipitate
was collected by centrifugation, redissolved in 0.05 M phosphate buffer, pH 7.2, and dialyzed extensively against the
same buffer. Denatured or undissolved protein was removed by
centrifugation, and aliquots of the supernatant were chromatographed on a 90 x 2.5 cm column of CPG-10-350 which is a
controlled pore glass molecular sieve with an exclusion limit
of 4×10^6 daltons.
The column was eluted with 0.05 M phosphate buffer, pH
7.2, the eluate was monitored at 280 nm and collected in

5 ml fractions. The elution pattern is shown in Figure 1 and
indicates a peak in the void volume, designated Peak I, with
the bulk of the protein being retarded and eluting as two
poorly-defined peaks of lower molecular weight.   These latter
peaks are predominantly serum proteins, none of which chroma-
tographically elute in the void volume.
    The void volume protein was concentrated and rechromato-
graphed on phosphocellulose. As shown in Figure 2, the
protein chromatographed as a single peak when the column was
eluted with a linear gradient of phosphate buffer, pH 7.2.
The slope of the gradient is indicated by the broken line.

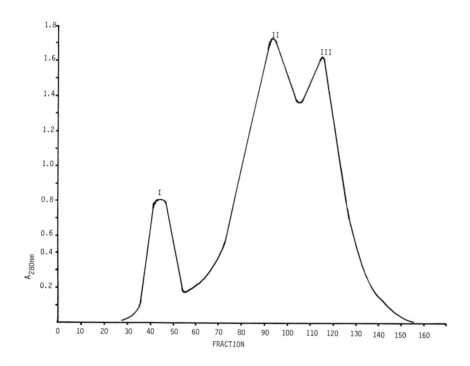

*Fig. 1. Elution of final supernatant from
CPG-10-350.   From Johnson and Warfel, 1976 (4).*

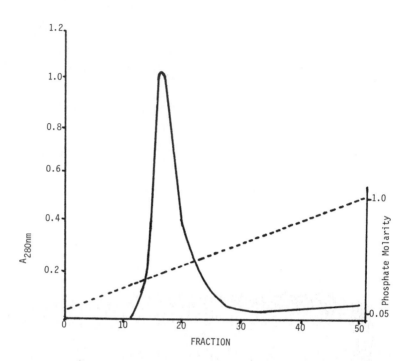

Fig. 2. *Elution pattern of void volume protein*
from P-cellulose. From Johnson and Warfel, 1976 (4).

This protein was identified as a component of epithelial
basement membrane by indirect immunofluorescence (5). Antiserum was produced in rabbits by intramuscular injection of
the glycoprotein emulsified in complete Freund's adjuvant (4).
Frozen sections of mouse kidney were incubated initially with
the antiserum and subsequently with commercial goat anti-
rabbit IgG. Control sections were incubated with normal
rabbit serum. As seen in Figure 3, linear fluorescence is
present in the area of the tubular basement membrane and was
also present, but less intense, in the glomeruli.
     The elution position of the glycoprotein when chromato-
graphed on molecular sieve columns, suggested either a very
large molecule or highly aggregated small molecules. To
resolve this point, the void volume protein was rechromato-
graphed following denaturation. The protein was extensively
dialyzed against 0.1 M Tris buffer, pH 8.2, containing 1%
sodium dodecyl sulfate (SDS). 2-Mercaptoethanol was added to
a concentration of 5%, and the mixture was heated at 100° for
1 minute followed by incubation at 37° for two hours. The
reduced, denatured protein was applied to a 90 x 2.5 cm

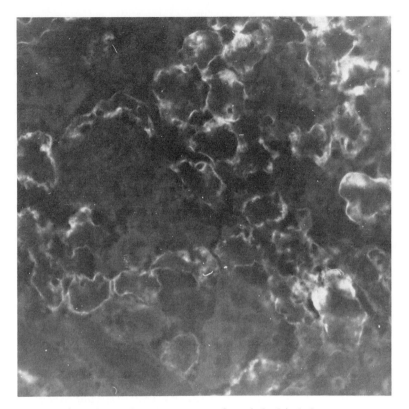

Fig. 3. Fluorescence of epithelial basement membranes after reaction of murine kidney with antiserum to neoplastic basement membrane glycoprotein.

column of Bio-gel A5 M equilibrated with Tris buffer containing 1% SDS. When the column was eluted with the same buffer, the pattern shown in Figure 4 resulted. A small amount of protein eluted with the void volume, designated Peak A, and with an apparent molecular weight of 60,000, designated Peak B. However, the bulk of the protein eluted with an apparent molecular weight of 30,000. The elution positions of marker proteins of known molecular weight are indicated for reference. The broken curve is the elution pattern when the initial heating time at 100° was increased to 5 minutes. These results indicate the presence of only one molecule and were confirmed when the reduced, denatured glycoprotein was electrophoresed on SDS acrylamide gels (4).

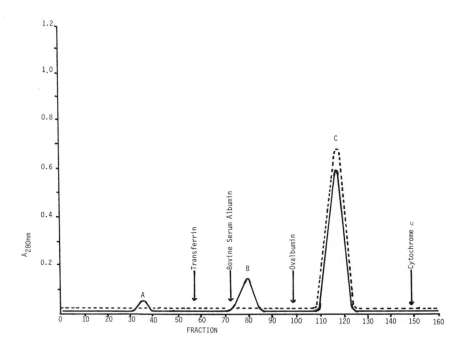

Fig. 4. *Elution of reduced, denatured basement membrane glycoprotein from SDS-agarose gel. The marker proteins used were transferrin (MW 90,000), bovine serum albumin (60,000), ovalbumin (45,000) and cytochrome c (12,400). The elution pattern indicated by the solid curve resulted when the glycoprotein was initially heated at 100° for 1 minute, and that indicated by the broken curve resulted when initial heating at 100°C was increased to 5 minutes (4).*

The amino acid composition of the glycoprotein was determined by analysis on a Jeolco amino acid analyzer after hydrolysis of the glycoprotein in 6N HCl for 24 hours at 100°. Analyses were performed according to the method of Miller and Piez (6) in order to rule out the presence of collagenous protein. As shown in Table I, subsequent analyses, performed by the method of Moore and Stein (7) indicated that the glycoprotein was acidic with glutamic and aspartic acids constituting approximately 20% of the amino acid residues.

The sugar composition of the glycoprotein was determined by gas chromatographic analysis of trimethylsilyl derivatives of the methyl glycosides. By this approach, sialic acids and glucose were detected and their presence was confirmed respectively by the thiobarbituric acid assay (8) and digestion with glucose oxidase. The total sugar accounts for approximately 13% of the weight of the glycoprotein with the composition shown in the first column of Table II.

TABLE I

Amino Acid Composition of Epithelial Basement Membrane Glycoprotein
The content of each amino acid is reported as residues per 1000 residues of total amino acid.

Amino acid	Residues
Lysine	70.5
Histidine	21.3
Arginine	44.0
Aspartic Acid	102.5
Threonine	62.5
Serine	73.3
Glutamic Acid	89.8
Glycine	81.5
Alanine	77.6
Valine	57.5
Methionine	8.8
Isoleucine	47.5
Leucine	99.6
Tyrosine	36.6
Phenylalanine	51.1
Proline	56.7
Half-cystine	20.0

TABLE II

Sugar Composition of Murine Epithelial Basement Membrane Glycoprotein (Column 1) and the Effects of Sequential Reduction of Fetal Bovine Serum From the Tissue Culture Medium

Sugar (µg/mg glycoprotein)	Days in Culture				
	3	6	15	21	30
Fucose	17.40	16.50	0	0	0
Mannose	18.00	17.50	12.00	5.65	0
Galactose	20.60	20.50	18.50	16.21	0
Glucose	10.00	9.00	4.00	2.25	0
N-acetylgalactosamine	20.90	16.50	8.50	6.38	6.00
N-acetylglucosamine	40.30	37.00	20.00	15.34	13.50
Sialic acids	1.84	1.25	0	0	0
TOTAL	129.04	122.25	63.00	45.63	19.50

The sugar composition of the glycoprotein can be markedly altered by the amount of supplemental fetal bovine serum
in the growth medium. The data shown in the subsequent
columns resulted when the amount of serum was altered in the
following manner. Cultures were established with 10%
supplemental serum. On day 3, the medium was decanted, the
glycoprotein extracted and identified as basement membrane
glycoprotein by Ouchterlony double diffusion. Sugar analysis
was performed. Fresh growth medium containing 5% serum was
added to the cultures. On day 6, the serum was decreased
to 2%, again decreased to 1% on day 12, and maintained at
that level. Sugar analyses of isolated glycoprotein indicated
initial loss of fucose and sialic acids, presumably terminal
sugars, with subsequent loss of all sugars to the point that
only reduced amounts of hexosamines remained. We have not
progressed far enough in these experiments to venture an
explanation for the observed changes.

HUMAN BASEMENT MEMBRANE GLYCOPROTEIN

Finally, I would like to describe preliminary results
of studies of basement membrane glycoprotein synthesis by
human cells grown *in vitro*. When cells obtained for diagnos-
tic amniocentesis are cultured, three morphologically distinct
types of cells can be identified (9). The predominant type,
designated AF-type by these investigators, have been cultured
as colonies or clones by Dr. Jean Priest at Emory. These
cells were grown in slide chambers in Dulbecco and Vogt's
medium supplemented with 15% fetal bovine serum. Once sufficient growth occurred, the medium was decanted and the cell
layer was washed with balanced salt solution. The slides were
then incubated with antiserum against murine basement membrane
glycoprotein followed by fluorescein labelled goat antirabbit
IgG. As shown in Figure 5a, strands of positively stained
extracellular material are scattered throughout the cell
layer. By contrast, incubation with normal rabbit serum resulted in no staining of extracellular material as demonstrated in Figure 5b.
AF-cells were also grown in large containers for the
isolation of glycoprotein from the growth medium. The isolation procedure was identical to that described for the
isolation of murine glycoprotein. The void volume peak from
molecular sieve chromatography was allowed to react with
anti-murine glycoprotein in an Ouchterlony double diffusion
system. As shown in Figure 6, the center well contained antiserum against murine basement membrane glycoprotein. Well A

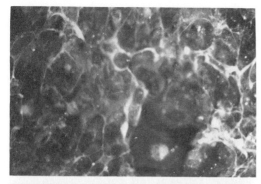

Fig. 5a. Reaction of antiserum directed against epithelial basement membrane glycoprotein with Type AF amniotic fluid cells.
From Megaw et al. 1977 (10).

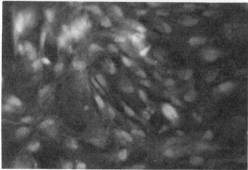

Fig. 5b. Control showing lack of reaction when incubation was carried out using normal rabbit serum. From Megaw et al. 1977 (10).

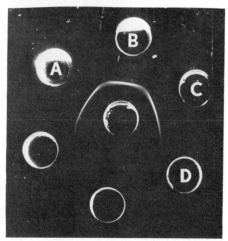

Fig. 6. Immunodiffusion of anti-EBM glycoprotein (center well) against glycoprotein isolated from the medium in which human AF-type amniotic fluid cells were grown (Well A), murine neoplastic EBM glycoprotein (Well B), murine kidney EBM glycoprotein (Well C) and protein isolated from the culture medium in which human fibroblasts were grown. Lines of identity are present between the glycoproteins of human and murine origin (10).

contained the glycoprotein of murine neoplastic origin, Well B, the human glycoprotein, and Well C, a soluble glycoprotein extracted from murine kidney. Well D contained culture medium from human fibroblast cultures. As this Figure indicates, lines of identity are present between human and murine epithelial basement membrane glycoprotein.

In conclusion, these studies indicate the presence of soluble basement membrane glycoprotein in the culture medium of both murine and human epithelial cell cultures. The glycoproteins are markedly similar antigenically. In addition, the sugar composition of the murine glycoprotein appears to depend on some factor present in fetal bovine serum and can be markedly decreased by decreasing the amount of supplemental serum. This phenomenon is quite intriguing in terms of understanding the controls of protein glycosylation and the role of sugars in the transport of glycoproteins out of the cell.

ACKNOWLEDGEMENTS

This work was supported by Grant No. CA 14797 from the National Cancer Institute.

REFERENCES

1. Pierce, G.B., Jr., Beals, T.F., Sri Ram, J. and Midgeley, A.R., Jr., *Amer. J. Pathol.* 45:929-961, 1964.
2. Johnson, L.D., Smith, J.J. and Kennedy, L.J., *Clin. Immunol. and Immunopath.* 2:178-184,1974.
3. Johnson, L.D. and Starcher, B.C., *Biochim. et Biophys. Acta* 290:158-167, 1972.
4. Johnson, L.D. and Warfel, J., *Biochim. et Biophys. Acta* 455:538-549, 1976.
5. Coons, A.H., *IN:* "General Cytochemical Methods", (J.F. Danielli, editor), Academic Press, New York, pp. 399-435, 1958.
6. Miller, E.J. and Piez, K.A., *Anal. Biochem.* 16:320-326, 1966.
7. Moore, S. and Stein, W.H., *J. Biol. Chem.* 211:907-913, 1954.
8. Warren, L., *J. Biol. Chem.* 234:1971-1975, 1959.
9. Hoehn, H., Bryant, E.M., Fantel, A.G. and Martin, G.M., *Humangenetik* 29:285-290, 1975.
10. Megaw, J.M., Priest, J.H., Priest, R.E. and Johnson, L.D., *J. Med. Genet. (1977) in press.*

SECRETION WITHOUT DEPOSITION OF

NON–HYDROXYLATED BASEMENT MEMBRANE COLLAGEN

M. E. Maragoudakis, H. J. Kalinsky

and J. Wasvary

CIBA-GEIGY Corporation
Ardsley, New York

SUMMARY: The rate of secretion of basement membrane
(BM) collagen synthesized by the rat parietal yolk
sac system (PEMT)* was essentially the same in the
presence or absence of either dipyridyl or GPA 1734
(8,9-dihydroxy-7-methylbenzo[b]quinolizinium bromide),
with no evidence of intracellular accumulation of
unhydroxylated BM collagen. This was shown by measuring collagenase-digestible radioactivity and [^{14}C]-
hydroxyproline in tissues and medium after 3 and 6
hour of incubation of the PEMT with [^{14}C]proline or
[^{14}C]proline plus [^{3}H]glycine. The identity of the
non-dialyzable radioactivity found in the medium in
the presence of dipyridyl or GPA 1734 with non-hydroxylated BM collagen was established by incubation with
purified prolyl hydroxylase from rat skin; this converted the material to hydroxyproline-containing protein
with a ratio of total [^{14}C]proline to [^{14}C]hydroxy-
proline close to that of control BM collagen.
Non-hydroxylated BM collagen was not deposited on

*Abbreviations used: PEMT, rat parietal yolk sac;
SDS, sodium dodecyl sulfate; TPCK, L-1-tosylamide-2-phenyl-
ethylchloromethylketone; PMSF, phenylmethylsulfonylfluoride.

Reichert's membrane, as evidenced by measurements of
total [^{14}*C]proline and [*^{14}*C]hydroxyproline in collagen-*
ase digests of membranes isolated following incubation
and freed of trophoblast and epithelial cells.

INTRODUCTION

Our interest in basement membrane (BM) collagen is in
relation to diabetic microangiopathy and our objective is the
development of inhibitors of BM collagen synthesis. Such
inhibitors could be useful experimental tools and potential
therapeutic agents in diabetic capillary disorders and other
collagen diseases.

We used the rat parietal yolk sac (PEMT) (1) as a BM
collagen synthesizing system. In our search for inhibitors
of BM collagen biosynthesis we found that 8,9-dihydroxy-7-
methylbenzo[b]quinolizinium bromide (GPA 1734) was a potent
inhibitor of both hydroxyproline and hydroxylysine formation
(2). In the present report, the rate of secretion of unhydroxylated BM collagen into the medium and its deposition on
Reichert's membrane was assessed in comparison to fully
hydroxylated BM collagen. The identity of the material
synthesized in the presence of GPA 1734 or dipyridyl with
unhydroxylated BM collagen was also established.

MATERIALS AND METHODS

Parietal yolk sacs were obtained from rats at the 14th
day of gestation as described previously (1). The entire
structure (PEMT) consisting of the parietal endoderm (PE),
Reichert's membrane (M) and trophoblast (T) was used. At
least three flasks were set up for each parameter, using 5-10
PEMT per flask in 5 ml Krebs-glucose buffer, pH 7.4, containing 10% dialyzed fetal calf serum. GPA 1734 (50 µM) or
dipyridyl (0.3 mM) were added prior to a 1-hour preincubation
at 37° to deplete endogenous substrates, after which 2.5 µCi
of [^{14}C]proline or [^{14}C]lysine were added. The flasks were
incubated for 3 hr and further synthesis stopped by the addition of 1 mg of cycloheximide and 10 µmoles of dipyridyl.
Tissues and medium were generally processed together; in the
experiments indicated, tissue and medium were separated after
incubation and the tissue dissected further into PE, M and T.
The trophoblast layer was discarded and the epithelial cells
removed by repeated shaking in buffer, leaving a clean
Reichert's membrane as checked by inspection (50X). After
exhaustive dialysis and acid hydrolysis, [^{14}C]hydroxyproline
and [^{14}C]hydroxylysine were determined (3,4). Lyophilized
[^{14}C] proteins from tissue or medium were separated by SDS-
Agarose chromatography as described previously (5).

RESULTS AND DISCUSSION

Table I shows the inhibitory effect on hydroxyproline formation of GPA 1734, with hydroxyl groups in the 8,9-positions, as compared with a closely related analog GPA 1967, with the hydroxyl groups in the 9,10-positions. At 30 µM, GPA 1734 inhibited hydroxyproline formation by more than 90% without appreciable effect on total protein synthesis. At comparable concentrations GPA 1967 was not inhibitory, and at higher concentrations had a non-specific inhibitory effect on both total protein synthesis and hydroxyproline formation. The formation of hydroxylysine was also inhibited by GPA 1734 (97% at 40 µM), and the effect could not be reversed by ascorbate. The relatively small inhibitory effect of GPA 1734 on total incorporation, at the concentrations used, suggested that it acts specifically on the hydroxylation of proline and lysine in the procollagen peptide. The inhibitory effect of GPA 1734 could be completely abolished by adding equimolar concentrations of Fe^{++}, suggesting that GPA 1734 may act as a chelator of Fe^{++}, which is required for the activity of purified prolyl and lysyl hydroxylase. Pulse-chase experiments also suggested that the mechanisms of action of GPA 1734 and dipyridyl are very similar. It should be mentioned, however, that other known chelators such as EDTA, desferrioxamine, 2,3-dihydroxynapthalene and others were without effect at comparable concentrations. Therefore, we must assume either that these agents cannot permeate the cell membrane or that GPA 1734 and dipyridyl bind to a specific site on prolyl and lysyl hydroxylase, which facilitates chelation of Fe^{++} from these enzymes.

TABLE I

Comparison of Inhibition by GPA 1734 and GPA 1967 in the PEMT System

Additions	Total Incorporation		Hydroxyproline formed	
	DPM x 10^{-3}	% of Control	DPM x 10^{-3}	% of Control
None	585.8	(100)	88.0	(100)
GPA 1734, 0.02 mM	551.6	94.2	67.6	77.0
GPA 1734, 0.03 mM	500.8	85.4	10.4	11.8
GPA 1734, 0.04 mM	496.4	84.7	3.1	3.5
GPA 1967, 0.05 mM	435.2	74.3	57.4	65.0
GPA 1967, 0.1 mM	367.7	62.8	46.2	53.0
GPA 1967, 0.5 mM	240.6	41.1	26.7	30.0

In the PEMT system, the BM collagen synthesized by the
epithelial cells only (1) can be secreted into the incubation
medium, be deposited on the preformed BM collagen of Reichert's membrane, or remain within the cell. This system is
suitable for studying these fractions separately. The finding that inhibition of hydroxylation with GPA 1734 or dipyridyl had no appreciable effect on total protein synthesis
in this system suggested that the unhydroxylated material
was accumulating intracellularly or was being secreted into
the medium at essentially the same rate as control BM collagen was extruded. Table II shows that the collagenase-
digests of the proteins synthesized from [U-^{14}C]proline and
recovered in tissue fractions contained about the same radioactivity whether or not GPA 1734 or dipyridyl were present.
There is no suggestion of intracellular accumulation of
unhydroxylated BM collagen. Similarly, collagenase-digestible [^{14}C] proteins secreted into the medium were about the
same.

TABLE II

*Effect of GPA 1734 and Dipyridyl on the
Extrusion of BM Collagen*

Fraction		Control	GPA 1734	Dipyridyl
		DPM x 10^{-3}*		
Tissue	3 hr	15.1	12.5	9.4
	6 hr	22.3	17.1	14.1
Medium	3 hr	2.3	2.4	2.4
	6 hr	5.5	5.3	3.8

*Values are total [^{14}C] collagenase-digestible proteins.

SDS-agarose chromatographs of reduced and denatured [^{14}C]
proteins from cells, medium and Reichert's membrane are shown
in Figure 1. Several [^{14}C]labelled protein peaks are present
in the intracellular material. There is only one hydroxyproline-containing protein peak, substantially larger than
lathyritic rat skin collagen (LC). Both the medium and the
Reichert's membrane demonstrated a single hydroxyproline

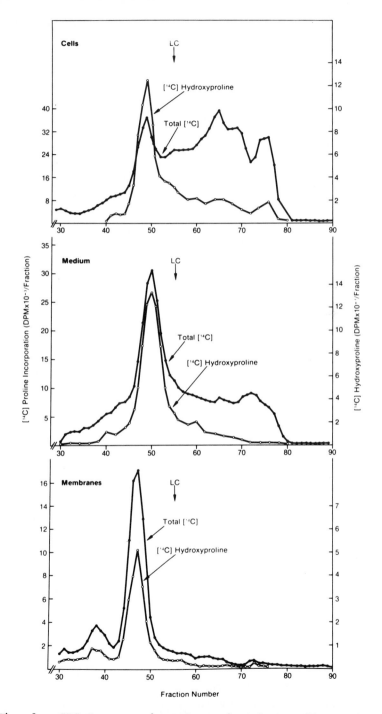

Fig. 1. SDS-Agarose chromatography of BM collagen from cells, medium and Reichert's membrane. A Bio-Gel-5m (200 – 400 mesh) column (3.8 x 58 cm) was equilibrated with phosphate buffer (0.1 M, pH 7.4 containing 0.1% SDS and 0.01% NaN_3). Fractions of 3.4 ml were collected. V_o = 99 ml and V_t = 320 ml.

containing protein peak of about the same size as that found
in the cells. This does not agree with the results obtained
previously with chick lens cells (5). There is no evidence
in the PEMT system for an intracellular precursor form of
BM collagen. It was also obvious that the major protein
secreted into the medium was BM collagen. Consequently, the
material secreted into the medium in the presence of GPA 1734
or dipyridyl was most probably the unhydroxylated form of the
same collagen.

Direct evidence for the identity of this material with
unhydroxylated BM collagen was obtained from the following
experiment: Collagenase-digests of medium proteins were
subjected to ion-exchange chromatography with authentic gly-
pro-hypro and gly-pro-pro tripeptides (6). With normally
hydroxylated proteins, a radioactive peak comprising about
12% of the total radioactivity was eluted with the gly-pro-
hypro peak. In the presence of GPA 1734 or dipyridyl, a
$[^{14}C]$-gly-pro-hypro peak was absent, and a peak not seen in
the control was eluted close to authentic gly-pro-pro. These
results suggest that the material secreted into the medium
when dipyridyl or GPA 1734 are present is unhydroxylated BM
collagen.

The medium proteins synthesized in the presence of hydroxylation inhibitors appear to be extremely susceptible to
extracellular protease, as shown by the multiplicity of peaks
on SDS-chromatography. However, when special precautions
were taken to inactivate proteases, the unhydroxylated
material could be obtained as a single peak on SDS-agarose,
with an apparent molecular size similar to that of hydroxylated BM collagen (Figure 2). An aliquot of each fraction
was subsequently dialyzed and precipitated with acetone to
remove SDS. Each sample was then incubated with prolyl
hydroxylase prepared from the skin of newborn rats (7), and
$[^{14}C]$hydroxyproline was determined after hydrolysis. This
protein was capable of serving as a substrate for prolyl
hydroxylase, thus identifying it as unhydroxylated BM collagen.

This is substantiated in Table III, which shows analyses
of total $[^{14}C]$ proteins and $[^{14}C]$hydroxyproline from lyophilized medium proteins synthesized with or without the hydroxylation inhibitors, both before and after incubation with
excess prolyl hydroxylase. No additional hydroxylation could
be obtained in the BM collagen synthesized under control
conditions. The material synthesized in the presence of GPA
1734, which contained less than 3% $[^{14}C]$hydroxyproline before
treatment with polyhydroxylase, was converted to material
with more than 30% of the total radioactivity as $[^{14}C]$hydroxy-
proline. Similarly, dipyridyl material was converted to a

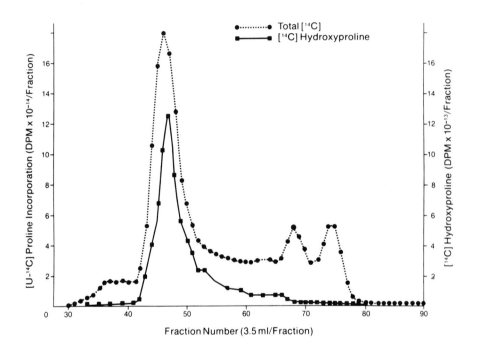

Fig. 2. SDS-Agarose chromatography of unhydroxylated BM collagen from medium of PEMT prepared under conditions inhibiting extracellular proteases. After 135 minutes of incubation the medium was centrifuged and heated at 100° for 5 minutes. Acetic acid, dipyridyl, N-ethylmaleimide, PMSF, TPCK were added to final concentrations of 0.5M, 0.3 mM, 10 mM, 0.15 mg/ml and 0.04 mg/ml, respectively, before dialyzing overnight against the same mixture. Following exhaustive dialysis against 0.5M acetic acid, the acetic acid-soluble material was lyophilized and chromatographed as described in Figure 1.

protein containing about 34% of the total radioactivity as hydroxyproline. This is comparable to the values obtained with the control, in which about 40% of the total activity was present as hydroxyproline.

TABLE III

Hydroxylation by Prolyl Hydroxylase (PH) of BM Collagen
Synthesized by the PEMT System and Secreted into the Medium
in the Presence or Absence of GPA 1734 and Dipyridyl

Starting Material	Incubated Without PH*			Incubated With PH	
	Total $[^{14}C]$	$[^{14}C]$ Hypro		Total $[^{14}C]$	$[^{14}C]$ Hypro
DPM X 10^{-3}					
Control	46.9	19.8		50.1	22.1
GPA 1734	65.3	1.7		71.9	22.6
Dipyridyl	51.4	1.7		50.3	17.5

Prolyl hydroxylase (PH) purified through the DEAE-
Sephadex stage (7) corresponding to 0.5 mg of protein
was used in 10 ml mixture.

Since the unhydroxylated BM collagen was secreted from the cells, it was of interest to learn whether this material was also deposited on Reichert's membrane. Experiments in which the cleaned membranes were obtained after incubation with $[^{14}C]$ proline showed only negligible activity on membranes from dipyridyl or GPA 1734 samples as compared with the controls, despite abundant activity in the cells and medium. It was, therefore, concluded that the unhydroxylated material cannot be deposited to any significant extent.

The secretion of unhydroxylated BM collagen, the fact that it cannot be deposited on pre-existing BM and its susceptibility to extracellular proteases are suggestive of potential therapeutic value in the use of compounds which inhibit hydroxylation, to prevent the pathological deposition of BM collagen in diabetic microangiopathy.

REFERENCES

1. Clark, C.C., Tomichek, E.A., Koszalka, T.R., Minor, R.R. and Kefalides, N.A., *J. Biol. Chem.* 250:5259-5267, 1975.
2. Maragoudakis, M.E., Kalinsky, H.J., Wasvary, J. and Hankin, H., *Fed. Proc.* 35:679, 1976.
3. Blumenkranz, N. and Prockop, D.J., *Anal. Biochem.* 15:77-83, 1966.
4. Juval, K. and Prockop, D.J., *Anal. Biochem.* 15:77-38, 1966.

5. Grant, M.E., Kefalides, N.A. and Prockop, D.J., *J. Biol. Chem.* 247:3545-3551, 1972.
6. Paz, M.A. and Gallop, M., *In Vitro* 11:302-312, 1975.
7. Rhoads, R.E. and Udenfriend, S., *Arch. Biochem. Biophys.* 139:329-339, 1970.

IN VITRO BASAL LAMINA SYNTHESIS BY HUMAN

GLOMERULAR EPITHELIAL AND MESANGIAL CELLS,

EVIDENCE FOR POST-TRANSLATIONAL HETEROGENEITY

G. E. Striker, P. D. Killen,

L. C. Y. Agodoa, V. Savin

and L. D. Quadracci

University of Washington
Seattle, Washington

SUMMARY: The morphology of outgrowing cells from ex-
planted glomeruli changes from a monolayer of epithelial
cells immediately post explantation to a multilayer of
mesangial cells late in culture. Correlated with this
change was a decrease in the ratio of 3-hydroxyproline
to total hydroxyproline (0.132 to 0.06) of the basal
lamina collagen secreted by these cells. Epithelial
and mesangial cells were isolated with collagenase.
Both cells were found to synthesize a basal lamina col-
lagen. The basal lamina collagen synthesized by epi-
thelial cells had a 3-hydroxyproline to total hydroxy-
proline ratio of 0.177 while that of mesangial cells
was 0.085. By SDS-PAGE both proteins migrated as high
molecular aggregates which on reduction co-migrated

Supported in part by NIH grants AM-15867, HL-03174 and
AM-05221.

Paul D. Killen is a recipient of a predoctoral fellow-
ship funded by NIH grant GM-02103.

319

with beta chains of collagen. These data indicate
that both epithelial and mesangial cells synthesize
a basal lamina collagen and that these cell products
differ with respect to post-translational processing.

INTRODUCTION

There are few sites in the body from which basal lamina can be isolated without significant contamination by other structural proteins. The glomerulus has been thought to be one such source and there are several reports on its chemical composition (1).

The glomerular basal lamina is a homogeneous, electron dense structure without demonstrable fibrils. It is bounded on each side by different cell types both of which are known to synthesize a basal lamina in other locations. Hence the contribution, if any, that each of these cells make to the total basal lamina is unknown. Previous studies of the composition of the basal lamina of intact glomeruli have also included the mesangial matrix, the material surrounding the third glomerular cell type. Mesangial cells resemble smooth muscle cells, which are surrounded by a basal lamina and have been shown to synthesize types I and III collagen (2,3,4). Thus, the composition and cellular sites of synthesis of basal lamina proteins, even when restricted to collagenous proteins may be quite complex. Therefore, we have attempted to isolate the individual glomerular cells and study the structural proteins they synthesize.

METHODS

Isolation of Glomerular Cells

Kidneys were obtained under aseptic conditions from adults at nephrectomy or autopsy. Glomeruli were isolated by mincing the cortex and sieving the tissue through increasingly fine stainless steel meshes (5).

Isolated glomeruli were either explanted immediately or incubated with 1.5 mg/ml collagenase (Worthington CLSPA 11) in complete Waymouth's medium to remove the epithelial cells. Incubation times from 10-40 minutes at 37°C were found to be effective and not to adversely affect cell viability. The glomerular suspension was allowed to settle to remove whole glomeruli. The supernate containing partially digested glomerular segments and cells was then centrifuged at low speed

to sediment the segments while keeping the cells in suspension.
The supernate was examined microscopically to ensure complete
sedimentation of glomerular fragments.

The pelleted glomeruli and the suspended cells were
separately cultivated in complete Waymouth's medium containing 20% or 30% fetal calf serum. Control cultures consisted
of explanted undigested glomeruli.

At various times after initiation of the cultures, 1 µCi
of ^{3}H-thymidine was added to the cultures and the incubation
was continued for 18 hours. The glomeruli and cells were
lightly trypsinized, collected on glass-fiber filters and the
thymidine incorporation into DNA assessed by liquid scintillation techniques.

Labeling of Cell Proteins

Glomerular cell cultures were incubated for two hours in
serum and proline free Waymouth's medium supplemented with
β-amino proprionitrile and sodium ascorbate (50 µg/ml). Fresh
medium containing 20 µCi/ml 2,3^{3}H L-proline, or 7 µCi/ml of
^{14}C L-proline (New England Nuclear Corp.) was added. After
24 hours the medium was decanted and protease inhibitors were
added to achieve final concentrations of 25 mM EDTA, 10 mM N-
ethyl maleimide, and 1 mM phenylmethylsulfonyl fluoride. The
inhibited medium was concentrated 5 fold and aliquots for
^{3}H amino acid analysis and SDS gel electrophoresis were
dialyzed exhaustively against 1 mM ammonium bicarbonate and
lyophilized. The remainder was immediately used for salt
precipitation. The cell layer was extracted with 0.1 M Tris,
0.5 M NaCl, pH 7.4 1% β-mercaptoethanol and processed for
^{3}H amino acid analyses.

^{3}H Amino Acid Analysis

Acid hydrolysates of culture media were chromatographed
on a 58 x 0.9 cm column containing sulfonated polystyrene
resin (Beckman UR-30) and were eluted with a 0.2 N sodium
citrate buffer pH 3.00 at ambient temperatures. 3-Hydroxy-
proline and 4-hydroxyproline were resolved from proline by
78 and 60 ml, respectively. Elution volumes were confirmed
on a stream split column using acid hydrolysates of human
glomerular basement membrane isolated as previously described
(6). Fractions were collected and ^{3}H or ^{14}C was assessed
using standard liquid scintillation counting techniques.
Corresponding medium from cells labelled with ^{14}C and ^{3}H were
pooled, hydrolyzed, chromatographed, and counted, using
standard double label counting techniques. This latter procedure was performed to assess the loss of ^{3}H from the

3 position of proline during biological oxidation. Unless
otherwise noted, all values reported have been corrected for
this loss.

Neutral Salt Fractionation of Medium

    Protein was precipitated from concentrated medium by the
addition of 20% NaCl at neutral pH. After stirring at 4°C
overnight, the medium was centrifuged at 105,000 X g for 1.5
hours resulting in a clear supernate. The NaCl concentration
in the supernate was increased in 5% increments following each
centrifugation until a 35% NaCl concentration was achieved.
All fractions were dialyzed and analyzed for hydroxyproline
isomers as described above. In other experiments, media from
multiple cultures were precipitated with 25% NaCl at neutral
pH and analyzed for hydroxyproline.

Sodium Dodecyl Sulfate (SDS) Polyacrylamide Gel
    Electrophoresis

    Polyacrylamide gel electrophoresis was performed on
labeled medium from glomerular cells, before and after short
collagenase digestion (7). Samples were incubated at 67°C
for 10 minutes in 0.10 M phosphate buffer pH 7.0 containing
1% SDS, 0.5 M urea with or without 1% β-mercaptoethanol and
were applied to 5% polyacrylamide gels (8).
    Gels were frozen, sliced in 0.9 sections, dissolved in
30% H_2O_2 at 70° and ^{3}H counted. Monkey skin collagen and
labeled fibroblast medium were used for collagen standards.

RESULTS AND DISCUSSION

Characterization of the Cells in Culture

Mixed glomerular cell culture

    As previously reported (9), many glomeruli attach to the
culture surface within 7-10 days. The cellular outgrowth has
two distinct phases. The first consists of an outgrowth of
cells which form a monolayer of flat, angular cells resembling epithelial cells (Figure 1). They have a low nuclear-
cytoplasmic ratio and appear to maintain a relatively uniform
size and shape. Incorporation of ^{3}H-thymidine peaks prior to
the time the epithelial cells reach confluence and rapidly
decreases thereafter until the second phase begins (Figure 2).
The onset of the second phase is recognized by the rapid rise

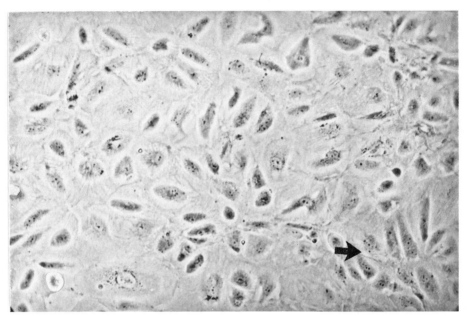

Fig. 1. *Phase photomicrograph of a confluent
cell layer of mixed glomerular cells demonstrating
epithelial cells and a cluster of mesangial cells
(arrow). Original magnification, X 100.*

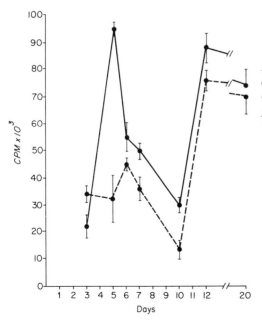

Fig. 2. *^{3}H Thymidine
incorporation following
explantation of intact
glomeruli. Replicate sam-
ples were analyzed at the
times indicated.*

in ^{3}H-thymidine incorporation into the cell layer and the appearance of a second cell type in the cultures. This cell population is characterized by the elongated shape of both nucleus and cytoplasm and their tendency to multilayer in an irregular fashion (Figure 3). This gives the cultures an appearance of ridges and depressions.

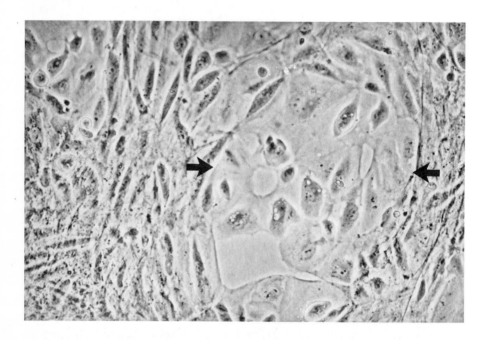

Fig. 3. Phase photomicrograph of the cellular outgrowth of intact glomeruli several weeks after visual confluence. Multilayered mesangial cells surround an area of confluent epithelial cells (arrows). Original magnification, X 100.

Epithelial Cell Cultures from Collagenase
  Digests of Glomeruli

The supernatant cells were found to be quite sensitive to vigorous pipetting or centrifuging. They attach to the culture surface within twelve hours and assume a shape identical to the early outgrowth from undigested glomeruli. The cells rapidly reach a stable confluent monolayer. Multilayering was not seen even at three weeks after confluence was reached. Fluorescence microscopy demonstrated that these cells did not have factor VIII antigen on their surface. Electron microscopic examination of the confluent monolayer revealed that their cytoplasm contained many free ribosomes, a small number of elongated mitochondria, sparsely distributed profiles of rough and smooth endoplasmic reticulum, a small number of vesicles and very few microfilbrils. No peripheral dense bodies were seen. There were well defined junctions between cells which resemble desmosomes (Figure 4). In addition, cytoplasmic fibrils were noted to traverse the cytoplasm to connect the junctions with adjacent cells, but were sparse elsewhere, mimicking the *in vivo* morphology of epithelial cells. Taken together, these features suggest that the cells are epithelial in origin, and likely derived from the visceral glomerular epithelium.

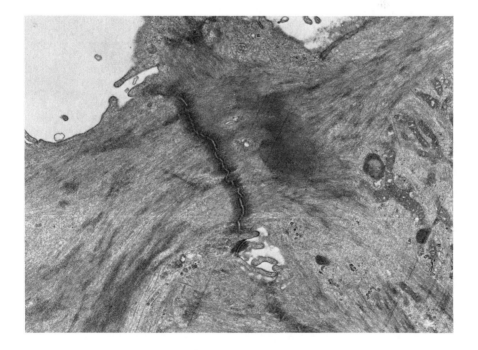

Fig. 4. En face electron micrograph of cell junctions between epithelial cells. These junctions are characterized by cytoplasmic filaments connecting adjacent junctions, increased electron density of the cytoplasm and cell membrane in these zones, and their limited number in any one cell. Original magnification, X 50,400.

Mesangial cell cultures from collagenase
  digests of glomeruli

The pellets contained both large segments of relatively
intact glomeruli and smaller fragments representing a few
capillary loops. By electron microscopy the epithelial cells
were not present, although the epithelial surface could often
be identified by the presence of adherent, irregular fragments
of pedicels. The glomerular segments adhered to one another
and were difficult to suspend. They also adhered more
rapidly to the culture surface than control cultures. Stellate cells appeared shortly after initiation of the cultures.
In contrast to the epithelial cells, the stellate cells grew
as cords of cells which multilayered early and quite extensively and closely resembled the cells from the mixed glomerular cell culture which grew out late. Before the cell layer
became confluent, there were areas of the culture where densely packed multilayers had retracted from the culture surface.
Electron microscopic examination of these cells demonstrated multiple peripheral dense bodies and microfilaments
and a characteristic, tangled array of extracellular fibrillar material (Figure 5). In addition, an interrupted basal
lamina surrounded many of these cells when they were multilayered. There cells, therefore, share many features with
smooth muscle cells *in vivo* and *in vitro* and also, therefore,
with mesangial cells. Since these cells were derived from
both segments of glomeruli and intact glomeruli, it was considered likely that they were mesangial, rather than vascular
pole cells.

Basal Lamina Collagen Characterization

3-Hydroxyproline Assay

Basal lamina collagen uniquely contains a high proportion
of its total hydroxyproline (T-Hyp) as the 3 isomer. The
ratio of 3-hydroxyproline (3-Hyp) to T-Hyp has been reported
for isolated human glomerular basement membrane and its pepsin solubilized species. The reported values vary but are
significantly different from interstitial collagens (1).
Under the labeling conditions employed, ^{3}H-proline incorporation into nondialyzable protein remained linear for
the entire 24-hour labeling period. Pulse chase experiments
revealed that 70% of the protein bound hydroxyproline was in
the medium and 30% was associated with the cell layer. ^{3}H-
amino acid analysis of cell layer extracts and medium from
post confluent mixed glomerular cells are compared in Table I.

Fig. 5. En face electron micrograph of mesangial cell culture. Extracellular material consists of a tangled array of fibrils without identifiable substructure. Original magnification X 8,000.

TABLE I

^{3}H Amino Acid Analyses of Post Confluent Mixed Glomerular Cell Cultures

Sample Glomerular Cells	^{3}H Amino Acid Analysis	
	3-Hyp/T-Hyp*	T-Hyp/Pro
Medium	.073	.087
Cell Layer	.071	.018
Medium	.071	.101
Cell Layer	.063	.029
Medium	.070	.133
Cell Layer	.065	.033

* *These values are not corrected for ^{3}H loss in 3-hydroxy-proline.*

These data show no significant difference between the 3-Hyp to T-Hyp ratio of the cell layer and the medium. However, the collagenous protein content was always higher in the medium as assessed by T-Hyp content.

The ability of mixed glomerular cells, isolated glomerular epithelial cells and mesangial cells to synthesize collagenous proteins containing 3-Hyp was compared with human
fascial fibroblasts. The results of multiple amino acid
analyses are shown in Table II.

TABLE II

*Ratio of ^3H Labeled 3-Hydroxyproline/Total Hydroxyproline
and Total Hydroxyproline/Proline of
Human Glomerular Cells and Fibroblasts in vitro*

Cell Type	3-Hyp/T-Hyp*	T-Hyp/Pro*
Confluent Mixed Glomerular Cells	.129 ± .013,3	.104 ± .026,3
Mixed Glomerular Cells	.101 ± .048,4**	.170 ± .027,4
Weeks Post Confluence		
1	.132**	.204
2	.081**	.178
3	.062**	.144
4	.060**	.152
Epithelial Cells	.177 ± .012,4	.086 ± .025,4
Mesangial Cells	.085 ± .008,3	.104 ± .006,3
Fibroblasts	.012 ± .004,4**	.286 ± .058,4

* Ratio of 3-Hydroxyproline/Total Hydroxyproline and
Total Hydroxyproline/Proline. Mean ± Standard Deviation.

** *These values are not corrected for ^3H loss in
3-hydroxyproline.*

Mixed glomerular cells labeled shortly after confluence
synthesize a soluble collagenous protein as assessed by T-Hyp
synthesis. The mean ratio of 3-Hyp to T-Hyp was approximately
13%, similar to values previously reported for basal lamina
collagen. This value was significantly different from the
ratio observed for human fascial fibroblast ($p < .001$). Thus,
there is among this mixed population of cells a cell type
capable of synthesizing a basal lamina collagen.

Neutral Salt Precipitation

Serial precipitation of concentrated medium with a neutral salt (20,25,30,35% w/v) at 4° demonstrated marked heterogeneity of the collagenous proteins with respect to 3-Hyp content (Table III).

TABLE III

*Serial Neutral Salt Precipitation of
Mixed Glomerular Cell Medium*

Sample	% T-Hyp* Recovered	3 Hyp/T-Hyp**	T-Hyp/Pro
Whole Medium	100%	.073	.087
20% NaCl Sediment	53%	.053	.190
25% NaCl Sediment	23%	.077	.098
30% NaCl Sediment	7%	.080	.047
35% NaCl Sediment	2%	.088	.039
35% NaCl Supernate	12%	.132	.027

*T-Hyp = Total Hydroxyproline.

**These values are not corrected for ^3H loss in 3-hydroxyproline.

The recovery of hydroxyproline was 97% and no differences were observed in the analysis of the starting material and the mathematical summation of all fractions. If the species of collagen were homogenous, then the ratio of 3-Hyp to T-Hyp should remain constant and equal the ratio observed for the whole medium. However, this ratio ranges from .053 to .132. Thus, the medium would appear to reflect a weighted average of several collagenous proteins. Precipitation of medium proteins with 25% NaCl yielded two fractions with mean ratios of .100 and .190 (n=19). The 25% salt supernate consistently had the highest 3-Hyp to T-Hyp ratio.

Control cultures were serially labeled over a 4-week
time period during which time the predominate cell morphology
shifted from epithelial to mesangial. The observed ratio of
3-Hyp to T-Hyp declined from .132 to .060 (see Table II).
These data suggested that the two cell types might synthesize
basal lamina collagens with marked differences in 3-Hyp content. Alternatively the older cultures could be contaminated
with fibroblasts or other cell types capable of synthesizing
interstitial collagens with a low 3-Hyp content.

SDS Polyacrylamide Gel Electrophoresis

Human glomerular basal lamina collagen and other types
of basal lamina collagen synthesized *in vitro* exist natively
as a high molecular weight disulfide bonded aggregate which
on reduction yields a subunit with a molecular weight significantly greater than that observed for types I, II and III
collagen (10,11). Thus, contamination of late culture
medium by non-basal lamina collagens which have a low 3-Hyp
content should be detectable by polyacrylamide gel electrophoresis.

SDS gel electrophoretic profiles of whole medium from
confluent epithelial cultures demonstrated high molecular
weight aggregates which on reduction co-migrated with β-chains
of skin collagen standards (Figure 6A). Significantly, little
radioactivity was seen in the region of pro-α chains obtained
from media of fibroblast cultures. Electrophoretic profiles
of medium from multilayered cultures are shown in Figure 6B.
Again, the predominate specie is a β migrating peak. Little
radioactivity is observed in the region of pro-α or α-chains.
After short collagenase digestion, 70% of the β migrating
protein was digested, establishing its collagenous nature. A
much smaller fraction of the pro-α migrating material was
digested, suggesting that only a small fraction was collagenous. These data suggest that the observed differences in the
3-Hyp ratios of early and late cultures was not due to dilution by contaminating interstitial collagens.

Basal Lamina Collagen Synthesis by Isolated
    Glomerular Cells

Cultures of isolated glomerular cells are capable of
synthesizing collagenous proteins as assessed by hydroxyproline isomer synthesis (Table II). Moreover, the isolated cell
populations remained homogeneous over the time periods shown
for control cells. Similarly the ratio of 3-Hyp to T-Hyp
showed little variation over this time period. The mean ratio
observed for the epithelial cells was 0.177. This value is in

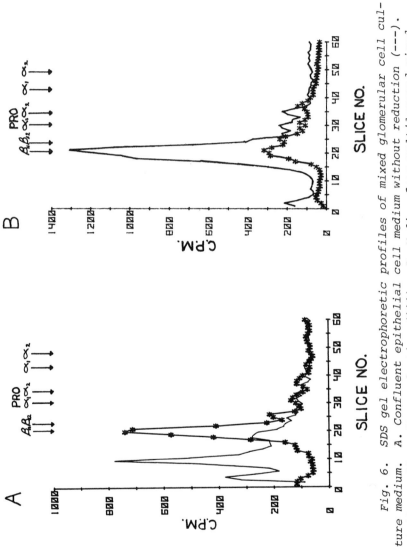

Fig. 6. SDS gel electrophoretic profiles of mixed glomerular cell culture medium. A. Confluent epithelial cell medium without reduction (---). Similar medium after reduction (***). B. Medium from multilayered mixed glomerular cell cultures after reduction (---). Similar medium following 2 hour collagenase digestion (***).

the same range as observed for the epithelial cell outgrowth
of mixed glomerular cell cultures. It was also significantly
different from both fibroblast and mesangial cell cultures
(p < .001).

The mean 3-Hyp to T-Hyp ratio from isolated mesangial
cell cultures was 0.085. This value approximated those
observed for multilayered mixed glomerular cell cultures and
was significantly different from epithelial cell and fibroblast medium (p < .001).

CONCLUSION

These data confirm previous speculations that glomerular
epithelial cells synthesize basal lamina proteins. The fact
that mesangial cells synthesize a collagen is not unexpected
but it is interesting to note that while these cells morphologically resemble vascular smooth muscle cells they do not
appear to synthesize collagen types I or III. Further, the
collagen they do synthesize resembles basal lamina collagen
with respect to 3-Hyp content but differs significantly from
that synthesized by glomerular epithelial cells. While these
differences are the result of post translational modification,
preliminary data from cyanogen bromide digests suggest there
may be differences in the primary structure as well. The
ultimate documentation of these differences awaits purification and analysis of the collagen from each of the glomerular
cells.

Previous determinations of the chemical composition of
glomerular basal lamina have included both the mesangial and
peripheral basal lamina. Since we have demonstrated that the
cells in these regions synthesize different basal lamina
collagens, analyses of whole glomeruli may represent a
weighted average of these collagen types. This fact becomes
more important in the studies of sclerosing renal diseases
where accumulation of matrix could be related to individual
or collective abnormalities in epithelial, endothelial or
mesangial cells. In these instances, the peripheral basal
lamina and mesangial regions should be isolated and analyzed
individually.

REFERENCES

1. Kefalides, N.A., *J. Invest. Derm.* 65:85, 1975.
2. Layman, D., et al., *Proc. Nat. Acad. Sci.*, 74:671, 1977.
3. Burke, J.M., et al., *Biochemistry (in press)*, 1977.
4. Macarak, E. and Kefalides, N.A., *This volume.*
5. Quadracci, L.J. and Striker, G.E., *Proc. Soc. Exp. Biol. and Med.* 135:147, 1970.
6. Striker, G.E., et al., *In:* "Proceedings of the International Symposium on the Morphology, Natural History and Treatment of Glomerulonephritis, (E.L. Becker, ed.), John Wiley & Sons, N.Y., 1:3, 1973.
7. Peterkovsky, B. and Diegelmann, R., *Biochemistry* 10: 988, 1971.
8. Goldberg, B., et al., *Proc. Nat. Acad. Sci.*, *USA,* 69:3655, 1972.
9. Striker, G.E., et al'., *Contributions to Nephrology* 2:25, 1976.
10. Minor, R.R., et al., *J. Biol. Chem.* 251:1789, 1976.
11. Killen, P.D., et al., *Fed. Am. Soc. Exp. Path.* 33:617, 1974.

BIOSYNTHESIS OF RAT LENS CAPSULE COLLAGEN

J. Godfrey Heathcote, Christopher H. J. Sear

and Michael E. Grant

University of Manchester
Manchester, England

SUMMARY: Isolated rat lens capsules were found to synthesize hydroxy[^{3}H]proline-containing polypeptides when incubated in the presence of [^{3}H]proline. The collagenous polypeptides produced during a 2 hr incubation were denatured, reduced, alkylated and found to have an apparent molecular weight of approximately 180,000 by sodium dodecyl sulfate-agarose gel filtration chromatography. No evidence was obtained for conversion of these polypeptides to a smaller molecular weight species in experiments where capsules were labelled for 2 hrs and chased with 'cold' proline for a further 4 hrs. Polyacrylamide gel electrophoresis showed that the collagen synthesized in vitro by rat lens epithelial cells consists of a single type of polypeptide chain with a lower electrophoretic mobility than either embryonic chick tendon or cartilage pro-α chains.

INTRODUCTION

Recent investigations suggest that basement membranes contain a collagenous component (designated type IV collagen) which is distinct from the three types of fibrillar collagen found in vertebrate tissues (1). The initial biosynthetic

335

precursor of basement membrane collagen is less clearly
defined than the precursors of collagen types I to III and
there is also considerable controversy regarding the nature
and inter-relationship of the subunits of the functional
basement membrane (1,2). However, the synthesis of type IV
collagen has been demonstrated *in vitro* in several systems
including chick lens (3,4), rat kidney glomerulus (5), rat
parietal yolk sac (6) and rabbit corneal endothelium (7).
Incubation of these tissues with [^{14}C]proline resulted in
the synthesis of large hydroxy[^{14}C]proline-containing polypeptides having many of the characteristics of basement membrane collagens. On the basis of gel filtration studies
using SDS-agarose columns the molecular weights of these
newly-synthesized collagen polypeptides have been reported to
fall in the range 140,000 - 160,000 (3-7). A time-dependent
conversion to a species of slightly lower molecular weight
has been observed with chick lenses (4) and rat glomeruli (5)
but in these experiments no steps were taken to avoid the
possibility of non-specific proteolysis; and it is noteworthy that in organ cultures of rat parietal yolk sac
endoderm no conversion to a smaller molecular weight species
was observed (6).

The present study was designed to investigate the
characteristics of the collagen synthesized *in vitro* by rat
lens epithelium. Evidence is presented suggesting that the
collagenous component of rat lens capsule is composed of a
single type of polypeptide chain having a molecular weight
of approximately 180,000. Subsequent pulse-chase studies
provided no evidence for a conversion to collagenous molecules of a smaller size when stringent precautions were
taken to avoid degradation of the sample during the analytical procedures.

MATERIALS AND METHODS

Male Sprague Dawley rats (130-160 g) were killed by a
blow on the head and the lenses were immediately isolated
and dissected free of adhering vitreous body. The lens capsules were carefully removed under a dissecting microscope
and washed in modified Krebs medium (8) at 4°C before being
pre-incubated for 15 min at 37°C in filter-sterilized Krebs
medium containing 10% (v/v) dialyzed fetal calf serum plus
penicillin G (100 IU/ml) and streptomycin (100 μg/ml).
Incubations with [5-^{3}H]proline (29 Ci/mmol) were continued
for up to 6 hrs at 37°C and terminated by the addition of
α,α'-bipyridyl (1 μM), cycloheximide (100 μg/ml), phenyl-
methylsulphonyl fluoride (1.7 mM), N-ethylmaleimide (10 mM)

and EDTA (25 mM) to the final concentrations indicated.   In
pulse-chase experiments the lens capsules were incubated for
2 hrs with [^{3}H]proline followed by a further 4 hrs in the
presence of 100 μg/ml of 'cold' proline.

The molecular sizes of the collagenous [^{3}H]polypeptides
synthesized by the isolated lens capsules were investigated
by gel filtration chromatography. To the incubation mixture
containing the proteinase inhibitors was added 50,000 cpm of
[^{14}C]proline-labelled chick tendon procollagen which served
as an internal marker in the column chromatographic analyses.
The samples were treated with SDS and urea, reduced with
dithiothreitol, alkylated with iodoacetamide and dialyzed
exhaustively against 0.02M Tris-HCl buffer, pH 7.4, containing 0.1% SDS, 1 mM EDTA and 0.02% NaN$_3$. These procedures
solubilized over 95% of the peptide-bound radioactivity. Gel
filtration chromatography on SDS-agarose was carried out as
described previously (9). Fractions (1.8 ml) were collected
and an aliquot taken for determination of ^{3}H and ^{14}C using
a triton-toluene based scintillation fluor. 4-Hydroxy[^{3}H]-
proline in the fractions was assayed by the method of Juva
and Prockop (10).

The polypeptide composition of the newly synthesized
collagenous molecules was investigated by SDS-polyacrylamide
gel electrophoresis (11) using a separating gel 10 cm long
and containing 5% (w/v) acrylamide. Samples were prepared as
for gel filtration analysis with additional dialysis against
electrophoresis sample buffer overnight in the presence of
proteinase inhibitors. The gels were stained with Coomassie
Brilliant Blue R and the distribution of ^{3}H-macromolecules
was determined by slicing and measurement of radioactivity.
Regions corresponding to the peaks of radioactivity were
removed from duplicate gels, hydrolyzed (12) and assayed for
4-hydroxy[^{3}H]proline (10). Embryonic chick tendon and cartilage procollagens labelled with [^{14}C]proline were prepared by
the methods of Harwood et al. (13) and electrophoresed con-
currently.

RESULTS AND DISCUSSION

The collagenous nature of the lens capsule is well
documented (1) and preliminary studies in which isolated rat
lenses were incubated for 2 hrs with [^{3}H]proline indicated
that intact lenses were capable of the synthesis of hydroxy-
[^{3}H]proline-containing peptides. The ratio of 4-hydroxy[^{3}H]-
proline to total ^{3}H present in non-diffusible peptides varied
from 1.5 to 3.5% and was, therefore, similar to that obtained

in experiments with embryonic chick lens cells (3,4). This basement membrane is synthesized by the lens epithelium (14) and by careful dissection techniques it was found possible to peel off the capsule from the lens to yield a membranous structure supporting a single layer of epithelial cells (Fig. 1). When this tissue was incubated with [³H]proline 3 to 6% of the total radioactivity incorporated was present as 4-hydroxy[³H]proline and protein synthesis was found to continue in an approximately linear manner for up to 6 hrs

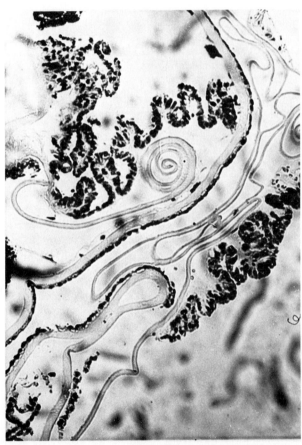

Fig. 1. Photomicrograph of isolated rat lens capsule stained with haematoxylin and eosin. Over much of its area the anterior lens capsule shows an attached single layer of epithelial cells whereas the thinner posterior capsule is devoid of adherent cells.

The molecular size of the collagenous peptides synthesized by isolated lens capsules was determined by gel filtration chromatography under denaturing conditions. Since
previous studies have indicated that procollagen is markedly
susceptible to proteolysis (for review see ref. 15) particular attention was given to the inclusion of proteinase
inhibitors during the procedures subsequent to incubation.
Lens capsules were labelled for 2 hrs with [^{3}H]proline and
the ^{3}H-proteins, after denaturation, reduction and alkylation,
were co-chromatographed on an SDS-agarose column with [^{14}C]-
proline-labelled procollagen secreted by embryonic chick
tendon cells (13). The column had previously been calibrated
with standard proteins and it was found that most of the
newly synthesized polypeptides were retarded by the agarose
matrix and exhibited apparent molecular weights in the range
12,000 to 100,000 (Fig. 2). However, a significant peak of
radioactivity, accounting for 18% of the total cpm recovered
from the column, eluted two fractions ahead of the internal
marker of type I pro-α chains [molecular weight of approx.
150,000 (13)]. This component had an apparent molecular
weight of 180,000 and contained all the 4-hydroxy[^{3}H]proline
synthesized by the lens capsules during the incubation period.
Analysis of a hydrolysate of the peak fraction by ion-exchange
chromatography indicated that 50% of the radioactivity in the
peak was present as hydroxy[^{3}H]proline and, of this, 5% was
present as the 3-isomer. These values confirm that the polypeptides are collagenous in nature and the presence of a
significant amount of 3-hydroxyproline is characteristic of
basement membrane collagen preparations (1).
     In subsequent experiments capsules were incubated for
2 hrs with [^{3}H]proline, followed by a 4 hr chase period in
the presence of unlabelled proline. The samples were treated
as described above for gel filtration chromatography and an
elution profile was obtained almost identical to that shown
in Fig. 2. No evidence was obtained for the presence of
hydroxy[^{3}H]proline in a species of molecular weight lower
than 180,000 suggesting that cleavage of the basement membrane
collagen polypeptides did not occur under the experimental
conditions employed. Nevertheless, longer term tissue culture
experiments with isolated lens epithelium will be necessary
before the possibility of cleavage can be completely excluded.
     A preliminary characterization of the collagen polypep-
tides synthesized was undertaken using SDS-polyacrylamide gel
electrophoresis. The major peak of radioactivity electrophoresed at or near the dye front but a second prominent
symmetrical peak representing high molecular weight material
and migrating slightly ahead of collagen β-chains, was also

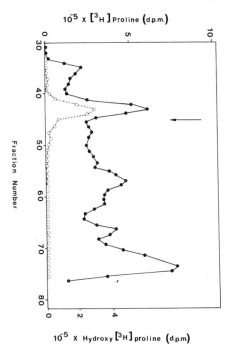

10^{-5} x [³H] Proline (d.p.m.)

Fraction Number

10^{-5} X Hydroxy [³H] proline (d.p.m.)

Fig. 2. Gel filtration on SDS-agarose (Bio Gel A-5m) of ³H-proteins synthesized by isolated rat lens capsules during a 2 hr incubation period. The sample was denatured, reduced, alkylated and chromatographed as described in the text. The arrow indicates the elution position of the internal standard, [¹⁴C]proline-labelled chick tendon pro-α chains. [³H]proline (o——o); 4-hydroxy[³H]-proline (o - - - o).

observed (Fig. 3). The fast-moving peak was shown not to contain hydroxy[³H]proline after acid hydrolysis of this segment of the gels and was considered to represent newly synthesized crystallins (14). Similar experiments established the presence of hydroxy[³H]proline in the high molecular weight species demonstrating that these polypeptides were collagenous. The detection of a single, symmetrical peak of radio-labelled collagenous molecules in these experiments suggests that the collagen synthesized *in vitro* by rat lens epithelium is composed of a single polypeptide species, presumably pro-α1 (IV) chains. These observations are consistent with previous studies indicating that the collagen extracted from bovine anterior lens capsule contains three identical poly-peptide chains (16).

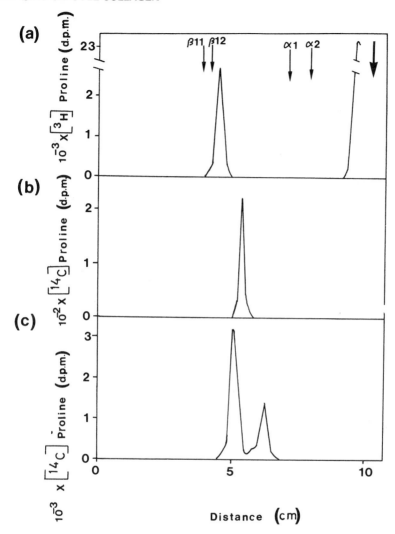

Fig. 3. Electrophoretic analysis of the ^3H-proteins synthesized by isolated rat lens capsules. Samples were treated as described in the text and electrophoresed for 10 hrs at 1.5 mA per gel. The migration of (a) the newly synthesized lens proteins was compared with that of [^{14}C]proline-labelled standards of (b) chick cartilage procollagen and (c) chick tendon procollagen. The mobilities of rat tail tendon collagen components are indicated and the heavy arrow marks the position of the dye front.

ACKNOWLEDGEMENTS

    J. Godfrey Heathcote and Christopher H. J. Sear are
Research Fellows financed by the National Kidney Research
Fund and Medical Research Council, respectively.

REFERENCES

1. Kefalides, N.A., *Int. Rev. Connect. Tissue Res.* 6:63-
    104, 1973.
2. Spiro, R.G., *Diabetologia* 12:1-14, 1976.
3. Grant, M.E., Kefalides, N.A. and Prockop, D.J., *J. Biol.
    Chem.* 247:3539-3544, 1972.
4. Grant, M.E., Kefalides, N.A. and Prockop, D.J., *J. Biol.
    Chem.* 247:3545-3551, 1972.
5. Grant, M.E., Harwood, R. and Williams, I.F., *Eur. J.
    Biochem.* 54:531-540, 1975.
6. Minor, R.R., Clark, C.C., Strause, E.L., Koszalka, T.R.,
    Brent, R.L. and Kefalides, N.A., *J. Biol. Chem.* 254:
    1789-1794, 1976.
7. Kefalides, N.A., Cameron, J.D., Tomichek, E.A. and
    Yanoff, M., *J. Biol. Chem.* 254:730-733, 1976.
8. Dehm, P. and Prockop, D.J., *Biochim. Biophys. Acta* 240:
    358-369, 1971.
9. Harwood, R., Bhalla, A.K., Grant, M.E. and Jackson, D.S.,
    Biochem. J. 152:291-302, 1975.
10. Juva, K. and Prockop, D.J., *Anal. Biochem.* 15:77-83,
    1966.
11. Laemmli, U.K., *Nature* 227:680-685, 1970.
12. Houston, L.L., *Anal. Biochem.* 44:81-88, 1971.
13. Harwood, R., Merry, A.H., Woolley, D.E., Grant, M.E. and
    Jackson, D.S.,.*Biochem. J. (in press).*
14. Clayton, R.M.,·*In:* "Current Topics in Developmental
    Biology", (Moscona, A.A. and Monroy, A., Eds.), Academic
    Press, New York, Vol. 5, pp. 115-180, 1970.
15. Grant, M.E. and Jackson, D.S., *Essays in Biochem.* 12:77-
    113, 1976.
16. Kefalides, N.A., *Biochem. Biophys. Res. Commun.* 45:226-
    234, 1976.

BIOSYNTHESIS OF PROCOLLAGEN BY ENDOTHELIAL

AND SMOOTH MUSCLE CELLS *IN VITRO*

Edward J. Macarak and Nicholas A. Kefalides

University of Pennsylvania
Philadelphia, Pennsylvania

ABSTRACT: The synthesis of collagen in cultures of
calf endothelial cells and in cultures of smooth
muscle cells has been studied. Calf endothelial
cultures have been characterized by the presence of
factor VIII antigen and Weible-Palade bodies.
Weible-Palade bodies are endothelial-specific organelles not found in other components of the vascular
wall such as smooth muscle or adventitial connective
tissue. Factor VIII antigen was localized in endothelial cells using immunofluorescence microscopy.
When cultured bovine endothelial cells are incubated
with bovine factor VIII antibody, a strong fluorescence
is observed. If cultures of endothelial cells and
bovine smooth muscle cells are labelled with $[^{14}C]$-
proline, both types of cells synthesize hydroxy$[^{14}C]$-
proline. Gel filtration chromatography (Agarose A5M)
in the presence of sodium dodecyl sulfate indicate
major differences in the collagen and noncollagen
proteins synthesized by these cells. Endothelial cells
synthesize a component which, after reduction, contains
approximately 50% of the total hydroxy$[^{14}C]$proline and
elutes in a peak close to the exclusion volume. This
exclusion volume peak contains significant quantities
of the isomer 3-hydroxyproline (3-hydroxy$[^{14}C]$proline =
14%). The chromatogram has two other components with
apparent molecular weights of 250,000 and 140,000.

*Only the 140,000 component contains hydroxy[^{14}C]-
proline. Gel filtration chromatography of [^{14}C]-
labelled material from smooth muscle cultures treated
similarly to that produced by endothelial cells shows
only one peak eluting with an apparent molecular weight
of 130,000. The percent hydroxylation of the peak
fractions was determined to be 44. The quantity of
3-hydroxy[^{14}C]proline present in the peak fractions
was determined to be 2.4%. These data indicate that
endothelial cells and smooth muscle cells are most
likely producing different kinds of collagen.*

Endothelial cells perform a variety of important
physiological functions such as the selective passage of
materials from the blood to the interstitium (1), the
maintenance of non-thrombogenic surface (2), the metabolism
of various vasoactive compounds such as serotonin (3), the
control of granulocyte diapedesis (4), and the synthesis of
specialized proteins such as the basement membrane (5).

The importance of vascular endothelial cells in the
pathogenesis of thrombosis and atherosclerosis has been
recognized for some time, but experimental studies in the
past have been limited by the relative inaccessibility of
this tissue. Because of this difficulty, few meaningful
biochemical studies have been carried out on endothelial
cells. Recently, methodological advances in tissue culture
have enabled several groups (6,7,8,9) to isolate endothelial
cells from a variety of sources and to grow them *in vitro*.
Tissue culture affords one the opportunity to isolate tissues
away from physiologic controls and to study, in detail,
various aspects of cell behavior, such as collagen biosynthesis.

One important aspect of endothelial cell metabolism is
its ability to synthesize and secrete the collagen and non-
collagen proteins which together constitute the basement
membrane (10). It has been shown that endothelial cells in
culture synthesize a collagen whose physical and chemical
properties most closely resemble those of basement membrane
or Type IV collagen (5). Other components of the vascular
wall such as smooth muscle cells are known to synthesize
collagen Types I and III (11). For this reason, we have
carried out studies to compare the physical and chemical
properties of the collagens and procollagens synthesized by
endothelial and smooth muscle cells *in vitro* with the purpose
of demonstrating the unique properties of the endothelial
cell procollagen.

METHODS

Cell Culture

Endothelial cultures were initiated from cells isolated from calf aortas according to the methods of Macarak et al. (9). Cells were grown in Medium 199 as modified by Lewis et al. (12), supplemented with 20% fetal bovine serum (FBS, Rehatuin, Armour Chemical Company) and antibiotics (Amphotericin B, 2.5 µg/ml, and Gentamicin, 50 µg/ml). Smooth muscle cells were grown from segments (1 mm^2) of calf aorta in Dulbecco-Vogt modified Eagle's medium according to the method of Ross (13). For subculture, smooth muscle cells were treated briefly with 0.1% trypsin and 0.05% ethylenediaminetetraacetic acid (EDTA) in calcium- and magnesium-free saline. Released cells were split in a 1 to 4 ratio. All cultures were grown in a humidified atmosphere of 5% CO_2 in air.

Immunofluorescence

For indirect immunofluorescence, calf endothelial cells were grown on round cover slips (15 mm in diameter) and subsequently treated with bovine Factor VIII antibody kindly supplied by Dr. Earl W. Davie, University of Washington, Seattle, Washington. Staining and microscopy procedures were carried out as previously described (9).

Collagen Biosynthesis

For the study of collagen biosynthesis, both endothelial and smooth muscle cells were incubated in Eagle's minimal essential medium (MEM) supplemented with ascorbic acid (50 µg/ml), β-aminopropionitrile (50 µg/ml) and [U-^{14}C]-proline (2 µCi/ml; 200 mCi/mMole). After a 24 to 48 hr labeling period, the culture medium was removed from the cells and the cell and medium fractions treated separately. The medium is centrifuged at 400 x g to remove any cells and subsequently chilled to 4°C. After addition of rat tail tendon collagen to serve as a "carrier", the collagen and other [^{14}C]proteins in the medium were precipitated using a 40% of saturation amount of ammonium sulfate, in a phosphate buffer (0.01M) containing several protease inhibitors, EDTA (4mM), N-ethylmalemeimide (10mM) and phenylmethylsulfonylfluoride (0.9 mM) (14). The precipitate was centrifuged into a pellet which, after extensive dialysis against 0.01 M phosphate with the above inhibitors, serves as a source of [^{14}C]labeled proteins for analysis.

Characterization of Synthesized Collagen

The material precipitated by ammonium sulfate was
prepared for chromatography by reduction and denaturization
in 5% 2-mercaptoethanol, 1% sodium dodecyl sulfate in 0.1%
sodium phosphate buffer, pH 7.4 for 2 hrs at 43°C. Gel
filtration was carried out on a calibrated column of 6%
Agarose (Bio-Gel A-5m) (5).

The content of 3- and 4-hydroxy[^{14}C]proline, was
determined in selected chromatographic fractions. For
determination of 4-hydroxy[^{14}C]proline, the method of
Juva and Prockop (15) was used. The quantity of the 3-isomer
of hydroxyproline was determined as described by Grant et al.
(16).

For pepsin digestion, the ammonium sulfate precipitate
was dialyzed against 0.05M acetic acid and treated with
pepsin (100 µg/ml) for 6 hrs at ·15°C, as described by
Howard et al. (5).

RESULTS

Since the methods used to isolate endothelial cells can
also lead to the removal of other cellular components of the
vascular wall, we sought to overcome the problem of cell
identification by using both morphological and immunological
criteria to identify endothelial cells. At confluency, one
can easily distinguish between endothelial and smooth muscle
cells; however, one can easily be deceived if cell cultures
are not pure but rather a mixture of different kinds of cells.
We have used two criteria to identify endothelial cells:
1) the presence of Weible-Palade bodies and 2) the presence
of Factor VIII antigens. Weible-Palade bodies are only
found in endothelial cells (14) and although they are
present in calf endothelium, their frequency appears to be
much less than in human endothelium. However, a small petri
dish may contain several million cells, and it is difficult
to use the presence of these organelles to monitor culture
purity. For this reason, we have sought additional criteria
to assure ourselves that the cultures do, in fact, contain
endothelial cells. Factor VIII antigen, which is produced
by endothelial cells and not by other components of the
vascular wall has been localized in endothelial cell cultures by immunofluorescence microscopy (9). When Factor
VIII antibody (kindly supplied by Dr. Earl W. Davie) conjugated to fluorescein-isothiocyanate was incubated with
cultured calf endothelial cells and viewed by immunofluorescence microscopy, a strong positive fluorescence was

observed (9). Control cultures using pre-immunization serum
rather than antisera did not stain. The presence of Factor
VIII antigen in virtually all cultured cells indicates that
our cultures contain almost exclusively endothelial cells.
     It has been shown (5) that endothelial cells synthesize
collagen. It was of interest, however, to compare the material synthesized by endothelial cells with that produced by
smooth muscle cells. For these experiments, both calf endothelial and smooth muscle cells were labeled with [^{14}C]proline.
Figure 1 shows a chromatogram of endothelial [^{14}C]proteins.
Endothelial [^{14}C]proteins exhibit a peak at or near the
exclusion volume, a large peak eluting with an apparent
molecular weight of 250,000 and a minor peak with an apparent
molecular weight of 150,000. When these peaks were analyzed
for the presence of hydroxy[^{14}C]proline, 55% of the total
hydroxy[^{14}C]proline was found in the void volume fractions
and 45% in the peak with an apparent molecular weight of
150,000. Little or no hydroxy[^{14}C]proline was found in the
250,000 molecular weight peak. Further analysis, showed that
the exclusion volume peak contained 13% 3-hydroxy[^{14}C]proline.
The 150,000 molecular weight peak contained approximately 7%
of the 3-isomer of hydroxyproline. When [^{14}C]labeled
proteins synthesized by smooth muscle cell in culture were
treated and analyzed similarly, only one peak with an apparent
molecular weight of 135,000 was found (Figure 2). When this
peak was analyzed for 3-hydroxy[^{14}C]proline, it was found to
contain approximately 2.4% of this isomer. For pepsin digestion, ammonium sulfate precipitates from endothelial and
smooth muscle cells were treated as described in the Methods.
After limited pepsin digestion [^{14}C]labeled proteins were
reduced and analyzed by SDS-agarose gel filtration.
Figure 3 shows the chromatogram of pepsin digested endothelial
cell [^{14}C]labeled proteins. The pepsin resistant material
elutes with an apparent molecular weight of 115,000 and has a
total hydroxyproline:total hydroxyproline + proline ratio of
0.59. When smooth muscle cell [^{14}C]labeled proteins were
analyzed after limited pepsin digestion, the chromatogram
shown in Figure 4 was obtained. The pepsin resistant hydroxy-
[^{14}C]proline peak elutes with an apparent molecular weight of
95,000 daltons.

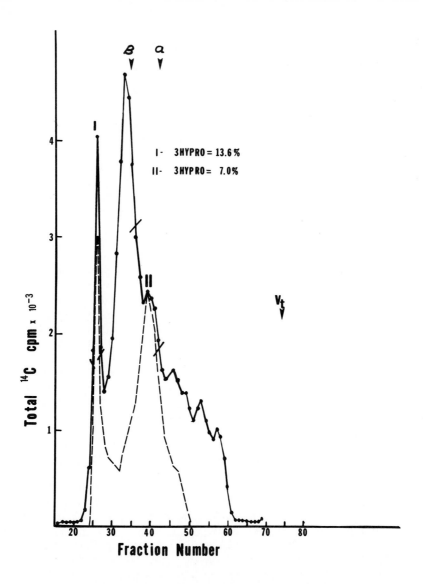

Fig. 1. Gel filtration chromatography on SDS-
agarose A-5m of [^{14}C]protein from endothelial medium.
Material was precipitated with ammonium sulfate and
reduced in 1% SDS-5% mercaptoethanol. Solid line indicates elution of total [^{14}C]protein. Dashed line
indicates elution of 4-hydroxy[^{14}C]proline. α and β
indicate elution of peptides from calfskin collagen.
Slash marks on peaks I and II indicate samples analyzed for 3-hydroxy[^{14}C]proline as described in text.

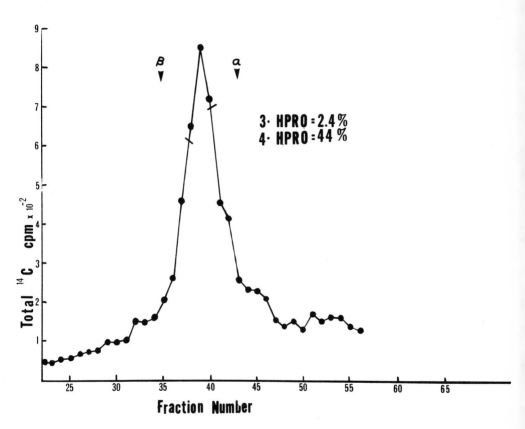

Fig. 2. Gel filtration chromatography on
SDS-agarose A-5m of [^{14}C]protein from smooth muscle
medium. Material was precipitated with ammonium
sulfate and reduced in 1% SDS-5% mercaptoethanol.
α and β indicate elution of peptides from calfskin
collagen.

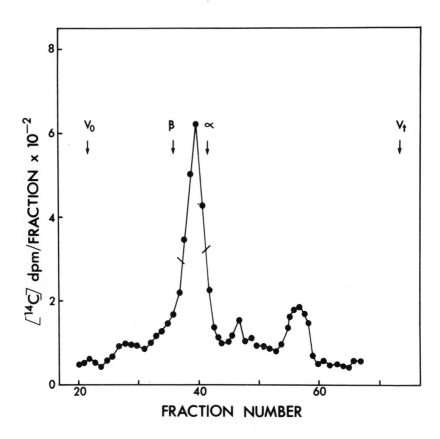

Fig. 3. Gel filtration chromatography on SDS-agarose A-5m of $[^{14}C]$protein from endothelial medium. Material was precipitated with ammonium sulfate and digested with pepsin. Pepsin-resistant $[^{14}C]$protein was reduced in 1% SDS-5% mercaptoethanol as described in the text. α and β indicate elution of peptides from calfskin collagen.

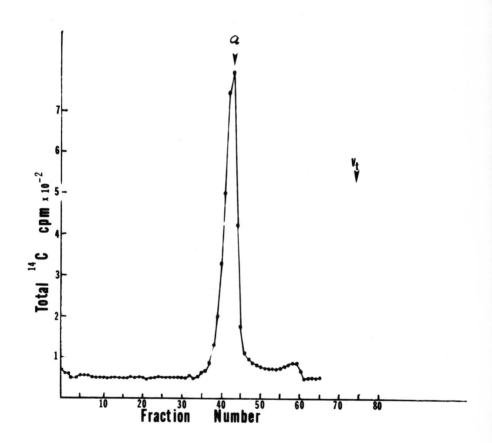

Fig. 4. Gel filtration chromatography on SDS-
agarose A-5m of [^{14}C]protein from smooth muscle
medium. Material was precipitated with ammonium
sulfate and digested with pepsin. Pepsin-resistant
[^{14}C]protein was reduced in 1% SDS-5% mercaptoethanol
as described in the text. α and β indicate elution
of peptides from calfskin collagen.

DISCUSSION

We have compared the collagens and procollagens synthesized by endothelial and smooth muscle cells in culture. A
major problem in the use of cultured cells is the absolute
identification of cells in culture. This is an especially
difficult problem for those who isolate endothelial cells
from blood vessels because these cells are anatomically
juxtaposed to other vascular components, such as smooth
muscle. To ensure that endothelial cultures do, in fact,
contain endothelial cells, we have used the presence of
Factor VIII antigen, which is associated only with endothelial
cells. Using immunofluorescence microscopy, we were able to
examine large numbers of cells and determine that our culture
methodology isolates endothelial cells with minimal contamination by smooth muscle.

Studies of collagen synthesis by cultured endothelial
and smooth muscle cells show that endothelial cells:
1) secrete a procollagen whose behavior upon gel filtration
is quite different from that of smooth muscle cells, 2)
secrete a procollagen which, after limited pepsin digestion,
has a higher molecular weight than smooth muscle collagen(s)
and 3) secrete a procollagen whose chemical properties (total
hydroxylation and the amount of the 3-isomer of hydroxyproline) are greater than that of smooth muscle. These results as well as those previously reported (5,16) indicate
that smooth muscle cells and endothelial cells are producing
different kinds of collagen. Further, these data lend further
support to the hypothesis that endothelial cells in culture
synthesize and secrete a basement membrane collagen.

ACKNOWLEDGEMENTS

This work was supported by NIH Grants AM-14526 and
AM-20553.

REFERENCES

1. Scow, R.O., Blanchette-Mackie, E. Joan and Smith, L.D.,
    Clinical Res. 39:149, 1976.
2. Johnson, S.A., *In:* "The Circulating Platelet" (S. A.
    Johnson, ed.) Academic Press, New York, (Abs.) 1971.
3. Small, R., Macarak, E.J. and Fisher, A., *J. Cell
    Physiol*. 90:225, 1977.

4. MacGregor, R.R., Macarak, E.J. and Kefalides, N.A.,
    Comparative adherence of granulocytes to endothelial
    monolayers and nylon fiber. Submitted for publication
    to *J. Clin. Inves.*, 1977.
5. Howard, B.V., Macarak, E.J., Gunson, D. and Kefalides,
    N.A., *Proc. Natl. Acad. Sci., U.S.A.*, 73:2361, 1976.
6. Jaffe, E.A., Nachman, R.L., Becker, C.G. and Minick,
    C.R., *J. Clin. Invest.* 52:2745, 1973.
7. Gimbrone, M.A., Jr., Cotran, R.S. and Folkman, J.,
    J. Cell Biol. 60:673, 1974.
8. Slater, D.N. and Stone, J.M., *Atherosclerosis* 21:259,
    1975.
9. Macarak, E.J., Howard, B.V. and Kefalides, N.A., *Lab.
    Invest.* 36:62, 1977.
10. Kefalides, N.A., *Int. Rev. Conn. Tiss. Res.* 6:63, 1973.
11. Gay, S., Balleisen, L., Remberger, K., Fietzek, P.P.,
    Adelmann, B.C. and Kuhn, K., *Klin. Wochenschr.* 53:899,
    1975.
12. Lewis, L.J., Hoak, J.C., Maca, R.D. and Fry, G.L.,
    Science 181:453, 1973.
13. Ross, R., *J. Cell Biol.* 50:172, 1971.
14. Fessler, L.I., Morris, N.P. and Fessler, J.H., *Proc.
    Nat. Acad. Sci., U.S.A.*, 72:4905, 1975.
15. Juva, K. and Prockop, D.J., *Anal. Biochem.* 15:77, 1966.
16. Grant, M.E., Kefalides, N.A. and Prockop, D.J., *J. Biol.
    Chem.* 247:3539, 1972.
17. Macarak, E.J., Howard, B.V. and Kefalides, N.A., *Ann.
    N.Y. Acad. Sci.* 275:104, 1976.

SYNTHESIS OF BASEMENT MEMBRANE COLLAGEN BY

CULTURED HUMAN ENDOTHELIAL CELLS

Eric A. Jaffe, C. Richard Minick,

Burt Adelman, Carl G. Becker

and Ralph Nachman

Cornell University Medical College
New York, New York

SUMMARY: Studies were performed to determine if
cultured human endothelial cells synthesized basement
membrane collagen. In culture, endothelial cells were
attached to grossly visible membranous structures
which on light microscopy were composed of ribbons of
dense, amorphous material. On transmission electron
microscopy, these membranous structures consisted of
amorphous basement membrane, and material morphologically similar to microfibrils and elastic fibers.
By immunofluorescence microscopy, these membranous
structures stained brightly with antisera to human
glomerular basement membrane. Cultured endothelial
cells incorporated ^3H-proline into protein; 18% of
the incorporated ^3H-proline was solubilized by purified collagenase. When endothelial cells were cultured
with ^{14}C-proline, 7.1% of the incorporated counts were
present as ^{14}C-hydroxyproline. Cultured endothelial
cells were labeled with ^3H-glycine and ^3H-proline and
digested with pepsin. The resulting fractions on
analysis by SDS-polyacrylamide gel electrophoresis
contained two radioactive protein peaks of Mol. wt.
94,200 and 120,500. Both these peaks disappeared

*after digestion with purified collagenase. The peak
of Mol. wt. 120,500 corresponds to that of α1(IV)
collagen; the peak of Mol. wt. 94,200 may represent
α1(III) collagen.*

   *Thus, cultured human endothelial cells synthesize
material which is morphologically and immunologically
like amorphous basement membrane and biochemically
like basement membrane collagen. Cultured endothelial
cells probably also synthesize material which is
morphologically similar to microfibrils and elastic
fibers.*

INTRODUCTION

   Endothelial cells *in vivo* rest on the subendothelium,
a complex array of acellular, connective tissue structures.
Ultrastructural studies of the subendothelium *in vivo* have
shown that it is composed of at least five different components; amorphous basement membrane, microfibrils, elastic
fibers, fibrillar collagen, and mucopolysaccharides (1).
The subendothelium is thought to have several important
physiologic functions. It provides most of the rigidity of
capillary walls (2), serves as an anchor for the overlying
endothelial cell layer, and may act as a filter (3). The
subendothelium is also involved in pathologic processes.
Loss of endothelial cells and exposure of the subendothelium
to blood is an important step in the development of thrombosis
since the subendothelium can initiate platelet adhesion and
aggregation (1). Alterations in the structure and function
of the subendothelium may contribute significantly to the
pathogenesis of atherosclerosis (1) and the vascular disease
of diabetes mellitus. For example, one of the earliest
changes seen in diabetes mellitus is thickening of capillary
basement membranes (4).

   We show here that cultured human endothelial cells
synthesize material which is morphologically and immunologically like amorphous basement membrane and biochemically like
basement membrane collagen. Cultured endothelial cells
probably also synthesize material which is morphologically
similar to microfibrils and elastic fibers.

RESULTS

Cell Culture of Human Endothelial Cells

Human endothelial cells were derived from umbilical
cord veins and cultured using methods and material previously
described (5,6). Endothelial cells were cultured in plastic
35 x 10 mm plastic Petri dishes or plastic T-25 or T-75
flasks. The culture medium consisted of Medium 199 containing either (a) 20% heat-inactivated fetal calf serum or
horse serum; or, (b) 20% heat-inactivated horse serum plus
10% heat-inactivated human serum. The culture media also
contained penicillin (100 U/ml), streptomycin (100 μg/ml),
and L-glutamine (2mM). Ascorbic acid (50 μg/ml) was added
daily to the culture media because of rapid turnover of the
vitamin in tissue culture (7).

Immunofluorescence Studies

When sections of umbilical cord were treated with goat
antiserum to human glomerular basement membrane and then with
fluorescein-conjugated rabbit antiserum to goat IgG, fine,
often coalescent immunofluorescent lines were observed in the
subendothelium. Basement membrane of smooth muscle cells was
also immunofluorescent (Fig. 1). In sections of kidney (not
shown) the same antisera stained glomerular basement membrane,
smooth muscle basement membrane, all endothelial cell basement membrane, and renal tubular basement membrane. The antiserum did not stain collagen in the adventitia of blood
vessels or in the renal medulla (6).
When endothelial cells which had been cultured for 1 week
were treated with the same antiserum, intracellular and extracellular staining was observed (Fig. 2). The intracellular
staining was granular and probably represents basement membrane components in vesicles awaiting extracellular release.
The extracellular staining appeared as a delicate meshwork;
in some areas the extracellular matrix had condensed and
appeared as brightly staining ribbons. Immunofluorescent
staining of endothelial and smooth muscle basement membranes
in sections of umbilical vein was completely inhibited by
previous absorption with human glomerular basement membrane
as was immunofluorescent staining of the cytoplasm and extracellular matrix of cultured endothelial cells. Absorption of
the antisera with human dermal collagen reduced the staining
only slightly. Control studies using normal rabbit serum
was negative (6).

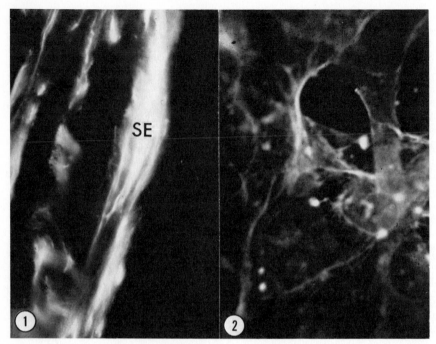

Fig. 1. Section of human
umbilical vein treated with
goat antiserum to glomerular
basement membrane followed by
fluoresceinated rabbit antibody to goat IgG. Delicate,
often coalescent immunofluorescent lines are seen in the subendothelium (SE). The basement
membranes of smooth muscle
cells are also immunofluorescent. Lumen is to the right.
X 500.

Fig. 2. Cultured endo-
thelial cells treated with
the same reagents as in Fig.
1. A meshwork of fine and
coarse immunofluorescent extracellular fibrils is seen
as well as less intense immunofluorescent staining of
the cytoplasm. X 500.
(From Jaffe et al. (6) by
permission of the publishers
of the J. Exp. Med.)

Morphological Studies of Subendothelium

Ultrastructural studies of the subendothelium *in vivo* have shown that it is composed of five different components. These are amorphous basement membrane, microfibrils, elastic fibers, fibrillar collagen, and mucopolysaccharides. These components are all present in the subendothelium of umbilical veins.

When cultured endothelial cells were detached from the surface of a Petri dish with 0.5 mM EDTA, the cells rounded up and floated in the buffer attached to a membranous

structure visible to the naked eye. The cells and attached
membranous structures were pelleted by centrifugation and
prepared for electron microscopy. Thick sections (0.5 μm) of
the embedded cells were stained with toluidine blue and examined by light microscopy. Endothelial cells were embedded
in ribbons of dense, amorphous material (Fig. 3, inset).

   The bands of dense, amorphous material seen in Fig. 3
(inset), when examined by electron microscopy, were composed
largely of granular amorphous basement membrane. Microfilamentous structures varying from 40 to 120 A in size were
embedded in the amorphous basement membrane and were present
both as single fibrils and as meshworks (Figs. 3 and 4).

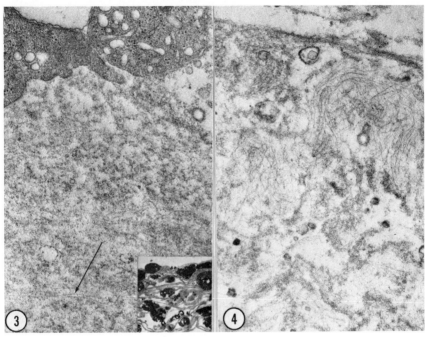

Fig. 3. Electron photo-
micrograph of pellet of cul-
tured endothelial cells. In
toluidine blue-stained thick
sections (inset), endothelial
cells are surrounded by bands
of dense, amorphous staining
material. By electron micros-
copy, endothelial cells adhere
to bands of dense granular
material containing micro
filamentous structures (arrow).
X 33,000

Fig. 4. Electron photo-
micrograph at higher magnifi-
cation of extracellular matrix
similar to that in Fig. 3.
Granular material resembling
amorphous basement membrane
surrounds tangles of micro-
filamentous structures re-
sembling microfibrils.
X 42,000

Some of the larger fibrils (100 A) had a "hollow core" appearance on cross section and resembled elastic microfibrils.
Amorphous elastin and structures resembling elastic fibers
were also seen and were composed of a central area of
relatively electron-lucent amorphous elastin bordered by
microfibrils (Fig. 5).

When semipurified extracellular matrix synthesized by
cultured endothelial cells was treated with purified collagenase, the amorphous basement membrane component was digested,
leaving exposed an extensive meshwork of microfibrils (Fig. 6).

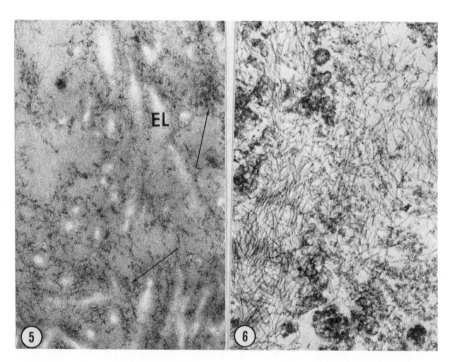

*Fig. 5. Electron photo-
micrograph of extracellular
matrix of cultured endothelial
cells revealing electron-
lucent islands of amorphous
elastin (EL) bordered by
microfibrils cut in cross
section (arrows).
X 78,000*

*Fig. 6. Semipurified
extracellular matrix prepared
from cultured endothelial
cells and treated with purified collagenase for 20 hrs.
resulting in digestion of
amorphous basement membrane
and revealing tangles of approximately 100 A micro-
fibrils. X 50,000
From Jaffe et al. (6) by permission of publishers of J.
Exp. Med.*

When pellets of cells and their surrounding extracellular matrix were treated with trypsin, the amorphous elastin component was left intact but the microfibrils were digested. When similar pellets were treated with elastase, the elastic fibers were no longer identified (6).

To determine the orientation of the extracellular matrix, cultured endothelial cells were fixed and embedded for electron microscopy *in situ* in Petri dishes. The block with its embedded cells was then rotated 90° and sectioned perpendicular to the surface of the Petri dish. Extracellular matrix was invariably present between the cells and the bottom of the dish or between overlapping cells. The extracellular matrix was never seen on the upper surface of the cells (6).

Biochemical Studies

In preliminary studies, cultured endothelial cells were shown to incorporate ^{3}H-proline into TCA-precipitable cellular and insoluble extracellular protein as a linear function of incubation time and dose of ^{3}H-proline (6). When ^{3}H-proline labeled endothelial cells and their attached extracellular matrix were digested with purified collagenase, 18% of the ^{3}H-proline incorporated into TCA-precipitable material was digested and solubilized. When ^{3}H-proline labeled endothelial cell postculture medium was digested with purified collagenase, 2.2% of the ^{3}H-proline incorporated into TCA-precipitable material was digested and solubilized (6).

To demonstrate that endothelial cells can form hydroxyproline presumably by synthesizing collagen and/or elastin, the following experiment was performed. Cultured endothelial cells were incubated with ^{14}C-proline for 24 hrs, washed extensively, and the cells and extracellular matrix hydrolyzed in 6 N HCl. ^{14}C-Proline and ^{14}C-hydroxyproline were separated with an amino acid analyzer. When these fractions were analyzed for radioactivity, 7.1% of the ^{14}C-proline incorporated into proteins was present as ^{14}C-hydroxyproline (6).

To characterize the synthesized collagen, cultured endothelial cells were incubated with radioactive amino acids, digested with pepsin, and analyzed by SDS-polyacrylamide gel electrophoresis. The digested cell pellet contained two major peaks in the collagen region (Fig. 7). These peaks had a molecular weight of 120,000 ± 3,300 and 94,200 ± 1,500 (average of six experiments ± SEM, each performed on a different endothelial cell culture). When the pepsin-treated cell pellet was digested with purified collagenase, both peaks disappeared. In some experiments, small amounts of

collagenase-digestible polypeptides of approximate molecular
weight 159,000 and 180,000 were seen (6).

When labeled postculture medium was digested with pepsin
and analyzed by SDS-polyacrylamide gel electrophoresis, peaks
of molecular weight 183,000, 152,000, and 115,000 were
observed (Fig. 8). All three peaks, but not the rest of the
radiolabeled protein, were digested by purified collagenase
(6).

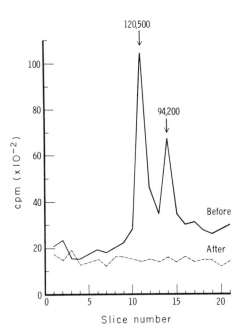
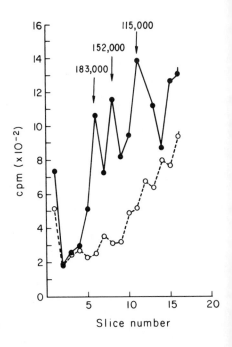

Fig. 7. SDS-polyacryla-
mide gel electrophoresis of
pepsin-digested cell pellet
before and after digestion
with purified collagenase. Two
collagenase-digestible
polypeptides were detected.
The upper portion of the
sliced gel is shown.

Fig. 8. SDS-polyacryla-
mide gel electrophoresis of
pepsin-digested postculture
medium. Three collagenase-
digestible polypeptides were
detected. Before collagenase
digestion (●), after collagenase digestion (o). The upper
portion of the sliced gel is
shown. (From Jaffe et al. (6)
by permission of publishers of
J. Exp. Med.).

DISCUSSION

The studies reported here demonstrate that cultured human endothelial cells synthesize and release collagen which is incorporated into the extracellular matrix. *In vivo*, the extracellular matrix is a component of the subendothelium. Cultured endothelial cells synthesized two collagen polypeptides, one with a molecular weight of 120,500 and the other with a molecular weight of 94,100. Since these molecular weights were obtained after pepsin digestion of the proteins, they probably represent two different types of collagen chains rather than one type of chain and one of its higher molecular weight precursor (8). Four different types of collagen are known; types I, II, III, and IV (8,9). Using antisera specific for types I, II and III collagen, Gay et al. showed that the subendothelium in human aorta contained type III collagen but not types I or II (10). Type IV collagen has been identified in whole human aorta (11) as well as in basement membrane (9). It is thus probable that subendothelium *in vivo* contains types III and IV collagen but lacks types I and II collagen. Type IV collagen, as found in glomerular basement membrane and chick anterior lens capsule, is composed of $\alpha 1(IV)$ chains which have a molecular weight of 120,000 (12,13). Type III collagen is composed of polypeptide chains with a molecular weight of 95,000 (14). It is likely, therefore, that the larger collagen chain (molecular weight 120,500) synthesized by cultured endothelial cells is $\alpha 1(IV)$ collagen. The smaller collagen chain (molecular weight 94,200) may be $\alpha 1(III)$ collagen but this is not clear.

The pepsin digested, radiolabeled postculture medium contained three collagenase-digestible polypeptides. The polypeptide of molecular weight 183,000 may represent either a dimer, B_{11}, of the smaller molecular weight (94,200) collagen molecule or a large precursor of the $\alpha 1(IV)$ collagen chains. These dimers have been described in purified preparations of type III collagen (10,15,16). The polypeptide of molecular weight 152,000 may represent a smaller precursor of the $\alpha 1(IV)$ chain, or a large precursor of the smaller (molecular weight 94,100) collagen chain (17). This molecular weight collagen polypeptide was infrequently detected in collagen preparations obtained from cultured cells and extracellular matrix presumably due to its susceptibility to pepsin digestion with resultant conversion to a smaller molecular weight collagen chain. The polypeptide of molecular weight 115,000 probably represents the $\alpha 1(IV)$ collagen chains (molecular weight 120,500) seen in the particulate fractions.

It has previously been shown for the other systems
(Descemet's membrane, glomerular basement membrane) that
amorphous basement membrane is a complex made up of collagen
and one or more glycoproteins (9,18). Cultured endothelial
cells and extracellular matrix stained weakly with an antisera to human dermal collagen and this staining was completely abolished by absorption of the antisera with human
dermal collagen. Thus, endothelial cells and extracellular
matrix contained proteins immunologically related to collagen.
Endothelial cells were stained brightly with antiserum to
glomerular basement membrane. Absorption of the anti-
glomerular basement membrane sera with dermal collagen reduced the staining only slightly, whereas absorption with
glomerular basement membrane completely inhibited the
staining. These immunofluorescent observations can be explained by the presence of additional antigens associated with
basement membrane collagens and/or by differences between
dermal and basement membrane collagens. By analogy with
glomerular basement membrane, these other components could be
high and low molecular weight glycoproteins (9,18).

Extracellular matrix synthesized by cultured endothelial
cells also contained two other morphologically recognizable
components, microfilamentous structures varying in diameter
from 40 to 120 A and amorphous elastin. Microfilamentous
structures measuring approximately 100 A in diameter had a
tubular profile and a "hollow core" in cross section. These
fibrils resemble those described by Palade and Bruns in the
outer portion of capillary basement membranes (19) and those
described as elastic microfibrils (20). Similar microfibrils
often bordered areas of amorphous elastin thus forming structures like elastic fibers. Like their *in vivo* counterparts,
the amorphous elastin was digestible by elastase but not by
trypsin or purified collagenase whereas the microfibrils were
digested by trypsin but not by purified collagenase. Despite
the absence of direct biochemical evidence for synthesis, the
cell culture technique used in these studies suggests that
these proteins were probably synthesized by the cultured
endothelial cell. Endothelial cells were initially isolated
and later passed using a crude preparation of collagenase
which contains, in addition, a variety of proteases other
than collagenase. The enzyme preparation is capable of
digesting the endothelial cell extracellular matrix, leaving
behind bare cells which are then passed. An initial isolation
and two passages by this technique plus a 10- to 20-fold
increase in cell number and area covered by the cells make it
unlikely that much of the extracellular matrix was carried
along with the cells from the umbilical vein wall.

Type IV collagen has been found in other tissues in structures that morphologically resemble amorphous basement membrane. It is thus probable that the type IV collagen synthesized by endothelial cells is present in the amorphous basement membrane component of the subendothelium. The fact that the amorphous basement is digestible by purified collagenase strongly suggests that at least one of its components is a collagen and it may be that both the collagens are associated with this particular structure. Fibrillar collagen was not seen in the subendothelium synthesized by cultured endothelial cells. This may be due either to the fact that endothelial cells do not synthesize fibrillar collagen *in vivo* or to lack of the proper conditions in culture. The last component of the subendothelium, mucopolysaccharides, is not identifiable by electron microscopy using standard techniques and thus was not observed in our studies. It is reasonable to expect that cultured human endothelial cells synthesize mucopolysaccharides and incorporate them into the extracellular matrix since it has been shown by Buonassisi that cultured rabbit aortic endothelial cells synthesize cellular mucopolysaccharides (21,22).

REFERENCES

1. Stemerman, M.B., *In:* "Progress in Hemostasis and Thrombosis", (Spaet, T.H., Ed.), Grune and Stratton, Inc., New York, 2:1, 1975.
2. Murphy, E.M. and Johnson, P.C., *Microvasc. Res.* 9:242, 1975.
3. Caulfield, J.P. and Farquhar, M.G., *J. Cell Biol.* 63:883, 1974.
4. Siperstein, M.D., Unger, R.H. and Madison, L.L., *J. Clin. Invest.* 47:1973, 1968.
5. Jaffe, E.A., Nachman, R.L., Becker, C.G. and Minick, C.R., *J. Clin. Invest.* 52:2745, 1973.
6. Jaffe, E.A., Minick, C.R., Adelman, B., Becker, C.G. and Nachman, R.L., *J. Exp. Med.* 144:209, 1976.
7. Peterkofsky, B., *Arch. Biochem. Biophys.* 152:318, 1972.
8. Martin, G.R., Byers, P.H. and Piez, K.A., *Adv. Enzymol. Relat. Areas Mol. Biol.* 42:167, 1975.
9. Kefalides, N.A., *Int. Rev. Connect. Tissue Res.* 6:63, 1973.
10. Gay, S., Balleisen, L., Remberger, K., Fietzek, P.P., Adelmann, B.C. and Kühn, K., *Klin. Wochenschr.* 53:899, 1975.
11. Trelstad, R.L., *Biochem. Biophys. Res. Commun.* 57:717, 1974.

12. Grant, M.E., Schofield, J.D., Kefalides, N.A. and Prockop, D.J., *J. Biol. Chem.* 248:7432, 1973.
13. Grant, M.E., Harwood, R. and Williams, I.F., *Eur. J. Biochem.* 54:531, 1975.
14. Chung, E. and Miller, E.J., *Science (Wash., D.C.)* 183:1200, 1974.
15. Piez, K.A., Eigner, E.A. and Lewis, M.S., *Biochemistry* 2:58, 1963.
16. Byers, P.H., McKenney, K.H., Lichtenstein, J.R. and Martin, G.R., *Biochemistry* 13:5243, 1974.
17. Minor, R.R., Clark, C.C., Koszalka, T.R., Brent, R.L. and Kefalides, N.A., *J. Cell Biol.* 67:287a, 1975.
18. Kefalides, N.A., *J. Invest. Dermatol.* 65:85, 1975.
19. Palade, G.E. and Bruns, R.R., *In:* "Small Blood Vessel Involvement in Diabetes Mellitus", (Siperstein, M.D., Colwell, A.R. and Meyer, K., editors), American Institute of Biological Sciences, Washington, D.C., 39, 1964.
20. Ross, R. and Bornstein, P., *J. Cell Biol.* 40:366, 1969.
21. Buonassisi, V., *Exp. Cell Res.* 76:363, 1973.
22. Buonassisi, V. and Root, M., *Biochim. Biophys. Acta* 385:1, 1975.

TUNICAMYCIN EFFECTS ON PROCOLLAGEN

BIOSYNTHESIS, GLYCOSYLATION AND SECRETION

Marvin L. Tanzer, Fred Rowland,

Louann Murray and Jerry Kaplan

University of Connecticut Health Center
Farmington, Connecticut

SUMMARY: Chick embryo cells, grown in third passage
culture, and originally obtained from fourteen day-old
calvaria, were briefly exposed to the antibiotic, tunicamycin. These cells, when compared to control cultures,
showed a severe, progressive inhibition of the incorporation of glucosamine and mannose into total cellular
macromolecules. Inhibition of the incorporation of
glycine, leucine and proline was also progressive but
not as marked as for the carbohydrates. Cellular
secretion of all macromolecules was severely impaired
but no selective effects were noted. The small amount
of procollagen which did appear in the medium of tunicamycin treated cells was identical to control procollagen, in terms of subunit sizes and their degree of
glycosylation. The intracellular content of collagenous
proteins was unchanged, comparing treated and control
cells. It was noted that the migration of intracellular
procollagen polypeptides during SDS-polyacrylamide
electrophoresis was much slower than the extracellular
polypeptides, although both contained glucosamine. In
vitro studies, using subcellular fractions which contain
procollagen, showed that tunicamycin selectively inhibited glucosamine, but not mannose, incorporation into macromolecules. The composite results indicate that tunicamycin effectively inhibits protein synthesis, protein glycosylation and protein secretion in these avian cells.

367

INTRODUCTION

The biosynthetic precursor of collagen, a large procollagen protein, has several unique features. It contains
both NH$_2$- and COOH-terminal procollagen extensions which
differ from the collagenous region in amino acid composition
(1-3). Although the exact mass of these extensions is not
known, current estimates suggest that 120-200 amino acids are
at the NH$_2$-terminal end while 250-300 amino acids are at the
COOH-terminal end (4-6). The central collagenous region has
been sequenced, for one genetic type of collagen polypeptide,
and is approximately 1000 amino acids. The procollagen
appendages at both ends are similar to each other in that they
resemble most globular proteins in amino acid composition and
in their susceptibility to ordinary proteases (6). They
differ, however, in that only the COOH-terminal extensions are
bridged by disulfide bonds while the NH$_2$-terminal extensions
have intrachain disulfide loops (2,7); also tryptophan is
unique to the COOH-terminal extensions (1,8).

The central collagenous region, in addition to containing
a unique amino acid sequence and conformation, has monosaccharide and disaccharide units linked to hydroxylysine. The
chemical structure and biosynthesis of these units has been
well delineated (9) and it has recently been shown that
nascent procollagen chains are the substrate for such glycosylation (10). The initial step is transfer of galactose to
the hydroxyl group of hydroxylysine, followed by transfer of
glucose. The glucosyl transferase has recently been purified
and partially characterized (11).

In contrast to glycosylation of the collagenous region,
we know relatively little about glycosylation of the procollagen regions. Very recent studies have shown that there
is incorporation of mannose and glucosamine into secreted
procollagen polypeptides as well as the presence of galactosamine in them (12, 13). The data also show that these sugars
are limited to the procollagen regions of the polypeptides,
presumably attached as oligosaccharide chains.

The antibiotic, tunicamycin, interferes with glycosylation of glycoproteins in some eucaryotic organisms (14, 15).
One site of inhibition by tunicamycin *in vitro* is in the
synthesis of N-acetylglucosaminylpyrophosphoryldolichol, an
early intermediate in the glycosylation pathway of some glycoproteins (16). Tunicamycin will reduce the secretion of
glycoprotein enzymes into the medium, when yeast spheroplasts
are exposed to the antibiotic, but intracellular protein
synthesis is not impaired (14). Thus, some selectivity of
tunicamycin action occurs in yeast cells, with a primary
effect as an inhibitor of glycoprotein synthesis, perhaps by
interfering with a specific transfer reaction. We have

studied the effects of tunicamycin on procollagen producing cells, to ascertain if any selective inhibitory effects can be detected in procollagen biosynthesis and secretion.

MATERIALS AND METHODS

Avian cells were obtained from the calvaria of 14 day-old chick embryos by some modifications of the method of Peterkofsky (17). These cells were subcultured and studied when they just became confluent in the third passage. Incorporation of radioactive substances into total cellular macromolecules was measured as previously described (18). The total macromolecules in the culture medium were obtained following dialysis; procollagen was partially purified by established methods (19). Collagenase digestion was done by modifications of the Peterkofsky and Diegelmann method (20). SDS-polyacrylamide gel electrophoresis was done as previously reported (19). Tunicamycin was a generous gift of Professor G. Tomura, University of Tokyo.

RESULTS AND DISCUSSION

Incorporation of the radioactive precursors, glucosamine and mannose, into control cell macromolecules progressively increased during a two hour incubation period. In contrast, cells which had been previously exposed to tunicamycin for 30 minutes, followed by several rapid changes of fresh media, showed a markedly reduced level of incorporation during a subsequent two hour incubation period. At the end of two hours these cells incorporated only 20% of the radioactivity found in control cells. Similar results were seen when either radioactive glycine, proline or leucine incorporation was studied, although the degree of inhibition by tunicamycin was not as pronounced as in the case of the monosaccharide precursors.

The accumulation of radioactive macromolecules in the culture medium was also measured during the two hour incubation period; this accumulation progressively increased in the case of the control cells. In contrast, the accumulation of extracellular macromolecules in the culture media of treated cells was markedly impaired and reached a constant value by two hours. There was a 5-6 fold difference at two hours, comparing control and treated cells, using either amino acid precursors or monosaccharide precursors.

The proportion of radioactivity in collagenase digestible proteins, which had been labeled with radioactive proline, was examined. Initially, 44% of the proline labelled proteins in the cell were digestible by collagenase. By two hours this had progressively diminished to 26% of the proline-labelled

proteins. No significant differences were found, comparing
control and tunicamycin treated cells. In the case of the
culture media from these cells, approximately 8-10% of the
proline labelled proteins initially secreted were digestible
by collagenase and this increased to 50-60% for the two hour
samples. The tunicamycin treated cells showed similar results
with the exception that 20-30% of the initially secreted
proline labelled proteins were digested by collagenase.

SDS-polyacrylamide electrophoresis of the partially purified procollagens obtained from the media of control and
treated cells showed similar patterns. Three major polypeptide components were detected, and each component contained
radioactive proline and glucosamine. The ratio of radioactivity in each component was similar and was unchanged by
prior treatment with tunicamycin. No significant differences
were found in the relative abundance of each polypeptide,
comparing control and treated cells. Prior incubation with
bacterial collagenase completely eliminated the polypeptides
from their characteristic locations in the gels. Comparison
of the gel profiles of the secreted procollagen polypeptides
with the intracellular procollagen polypeptides showed that
the latter migrated considerably slower than the secreted
forms. Both intracellular and extracellular procollagens
contained glucosamine and proline; the reason for the difference in their migration behavior is not known.

A subcellular fraction of these cells, which contains
almost all of the procollagen polypeptides (21, 22), was
isolated and incubated with the "active" nucleotide sugar
precursors of mannose and glucosamine. Rapid incorporation
into macromolecules was obtained, reaching a plateau within
15 minutes. Addition of tunicamycin at the start of the
reactions completely suppressed glucosamine incorporation
but did not affect mannose incorporation. These results are
similar to those noted when a beef liver microsome fraction
was studied (16).

These composite results indicate that tunicamycin is a
powerful inhibitor of the incorporation of precursor amino
acids and carbohydrates into the cellular macromolecules of
avian cells in culture. The mechanism(s) of the inhibition
cannot be ascertained from these studies but may be related
to the very strong binding of tunicamycin to plasma membranes
(23). Such binding could interfere with transport processes
into the cell as well as with membrane associated reactions
such as protein synthesis and protein glycosylation. The
secretion of macromolecules from the treated cells is also
impaired, but seems to be a generalized effect because the
relative abundance of procollagen in the secreted material
was unaltered, compared to controls. Moreover, collagenous
proteins did not accumulate intracellularly following

tunicamycin treatment, indicating that the balance between biosynthesis of these proteins and their secretion was unaltered. Thus tunicamycin had a general inhibitory effect on fundamental cell processes such as macromolecular synthesis and secretion. In contrast, more selectivity was observed using *in vitro* preparations, in which tunicamycin had a preferential effect upon glucosamine incorporation. Although the reason for this discrepancy is not apparent, tunicamycin may prove to be more useful in studying procollagen glycosylation *in vitro* than in whole cell systems.

ACKNOWLEDGEMENTS: This investigation was supported by USPHS (AM-17220 and AM-12683), the NSF (GB-31077), the American Heart Association (75-773) and the University of Connecticut Research Foundation.

REFERENCES

1. Tanzer, M.L., Church, R.L., Yaeger, J.A., Wampler, D.E. and Park, E., *Proc. Nat. Acad. Sci., USA,* 71:3009-3013, 1974.
2. Fessler, L.I., Morris, N.P. and Fessler, J.H., *Proc. Nat. Acad. Sci., USA,* 72:4905-4909, 1975.
3. Byers, P.H., Click, E.M., Harper, E. and Bornstein, P., *Proc. Nat. Acad. Sci., USA,* 72:3009-3013, 1975.
4. Lenaers, A., Ansay, M., Nusgens, B.V. and Lapiere, C.M., *Eur. J. Biochem.,* 23:533-543, 1971.
5. Monson, J.M., Click, E.M. and Bornstein, P., *Biochemistry* 14:4088-4092, 1975.
6. Becker, U., Timpl, R., Hjelle, O. and Prockop, D.J., *Biochemistry* 15:2853-2862, 1976.
7. Fietzek, P.P. and Kühn, K., *Intl. Rev. Conn. Tiss. Res.* 7:1-60, 1976.
8. Uitto, J., Lichtenstein, J.R. and Bauer, E.A., *Biochemistry* 15:4935-4942, 1976.
9. Spiro, R.G., *In* "Glycoproteins" (A. Gottschalk, Ed.) Part B, pp. 964-999. Elsevier Publishing Co., N.Y., 1972.
10. Brownell, A.G. and Veis, A., *Biochem. Biophys. Res. Commun.* 63:371-377, 1975.
11. Myllylä, R., Risteli, L. and Kivirikko, K.I., *Europ. J. Biochem.* 61:59-67, 1975.
12. Clark, C.C. and Kefalides, N.A., *Proc. Nat. Acad. Sci., USA,* 73:34-38, 1976.
13. Oohira, A., Kusakabe, A. and Suzuki, S., *Biochem. Biophys. Res. Commun.* 67:1086-1092, 1975.
14. Kuo, S.C. and Lampen, J.O., *Biochem. Biophys. Res. Commun.* 58:287-295, 1974.

15. Frisch, A., Levkowitz, H. and Loyter, A., *Biochem. Biophys. Res. Commun.* 72:138-145, 1976.

16. Tkacz, J.S. and Lampen, J.O., *Biochem. Biophys. Res. Commun.* 65:248-257, 1975.

17. Peterkofsky, B., *Arch. Biochem. Biophys.* 152:318-328, 1972.

18. Kaplan, J. and Moskowitz, M., *Biochem. Biophys. Acta.* 389:290-305, 1975.

19. Church, R.L., Yaeger, J.A. and Tanzer, M.L., *J. Mol. Biol.* 86:785-800, 1974.

20. Peterkofsky, B. and Diegelmann, R., *Biochemistry* 10:988-994, 1971.

21. Diegelmann, R.F., Bernstein, L. and Peterkofsky, B., *J. Biol. Chem.* 248:6514-6521, 1973.

22. Park, E., Tanzer, M.L. and Church, R.L., *Biochem. Biophys. Res. Commun.* 63:1-10, 1975.

23. Kuo, S.C. and Lampen, J.O., *Arch. Biochem. Biophys.* 172:574-581, 1976.

PROCESSING OF PROCOLLAGEN

John H. Fessler, Deborah G. Greenberg

and Liselotte I. Fessler

University of California
Los Angeles, California

ABSTRACT: *The sequential modifications of procollagen pose the following problem: is the reason for one modification preceeding another that they occur in successive physical locations of a pathway, or do a faster and a slower enzyme action proceed simultaneously and independently of cellular location? Pulse-chase experiments in short term chick organ cultures show that in type I and II procollagens the amino extension peptides are mostly removed before the carboxyl extensions, and autoradiography indicates that secretion from cells occurs during the same time interval as this first cleavage. Extraction of cultured tendon cells has so far yielded only amino cutting enzyme activity and not carboxyl cutting enzyme, while both activities are present in culture media. This suggests that (1) separate enzymes exist for the two cleavages and (2) the enzyme which acts first is physically located up-stream in the procollagen processing pathway. Tendon organ cultures rapidly cut off amino extensions, and the model of simultaneous fast and slow cleavages cannot be excluded. Although amino extension cleavage may accompany secretion from cells it is not a prerequisite for secretion in cell culture. Limited association of procollagen molecules can occur without cleavage.*

*Variations in extent and in relative speed of cutting
could influence tissue specific fiber morphology.*

   *Biosynthesis of type III chick embryo blood vessel
procollagen shows that inter-chain disulfide bonds
are formed within three parts of each molecule:
amino extensions, carboxyl portion of final collagen
chain and carboxyl extensions. Initial processing
retains these bridge regions and full processing to
$(\alpha III)_3$ is much slower than for types I and II, if
it occurs at all.*

   Two of the outstanding characteristics of basement membrane are its shape and the maintenance of this shape, i.e.,
the amazing structural stability documented in this symposium
by Drs. Meezan, Brendel and Carlson (1). A path of assembly
must have led to such an end result, but one knows very little
about this in basement membrane. We are studying the
assembly path of some other collagenous structures and report
here about the types of problems which we have encountered,
and some of our results on types I, II and III collagen.
   By pulse-chase studies with radioactive label we proved
(12,13) that the procollagen type I synthesized by chick
embryo skull bones in short term organ culture is converted
via at least one physiologically intermediate form, which we
called altered procollagen, to the final collagen. Subsequently, we showed (4) that this was due to the sequential
removal of first the amino propeptides, giving altered procollagen, and the cleavage of the carboxyl propeptides.
   Goldberg and associates (5) had observed various cleaved
procollagens in fibroblast cultures. A potentially significant difference between their experimental arrangement and
in vivo conditions was that while the culture medium surrounding a monolayer of cells represents a relatively huge extracellular space, in the tissues the fibroblasts and the fibers
are in close confinement. Cells in culture could have
released into the medium both specific procollagen cutting
enzyme(s) and non-specific proteases and therefore, it was
not known which proteolytic cleavages were physiological.
But in our short-term organ cultures we did not disturb the
spatial relationship of cells and fibers. Also we observed
a temporal sequence of changes and as materials were isolated
in the same way at all time points, all were affected equally
by any artifactual proteolysis during isolation. We have
observed the same steps of sequential cleavage in the

physiological processing of type I procollagen in rat calvarial bones and in embryonic chick tendon (6) and of type II
procollagen in embryonic chick sternal cartilage (6). This
suggests that the removal of first the amino and then the
carboxyl propeptide may be related to the transport and/or
fiber assembly mechanisms of these collagens, and poses the
following problem: is the reason for one modification preceeding the other that they occur in successive physical
locations of a pathway, or do a faster and a slower enzyme
action proceed simultaneously and independently of cellular
location?

       An approximate correlation of the biochemical time
studies with autoradiographic time studies of calvarial
bones labeled with [^{3}H]proline (3) indicates that secretion
of labeled material from cells occurs during the same time
interval as the first cleavage, the removal of the amino-
propeptides. This leads to the prediction that cells should
contain active aminopropeptidase activity and that the extracellular cell culture medium should contain carboxyl propeptidase activity, although both enzymatic activities
presumably are due to proteins synthesized by the cells, and
also eventually both enzymes might leach out into the
culture medium.

       To test this, tendon fibroblast cultures were derived
from embryonic chick tendon. Secondary passage cells and
media conditioned by them were assayed for propeptidase
activities in the following way (see Figure 1). Radioactive
proline or leucine labeled procollagen and altered procollagen
were isolated from chick calvaria by DEAE-cellulose chromatography (4). Each was used separately as substrate to test
for propeptidase activities in extracts of the washed cells
and in aliquots of the medium conditioned by cells. The
digestion products were analyzed, after reduction, by
electrophoresis on sodium dodecyl sulfate polyacrylamide slab
gels followed by quantitative fluorography (4,7). Figure 1
shows that while the cell extract actively cut off the amino-
propeptides from procollagen, thereby converting it to altered
procollagen, it only cut off few carboxyl propeptides either
from procollagen or from separately supplied altered procollagen (the amount may vary slightly). In contrast, the
conditioned medium contained both cutting activities and
there was somewhat faster cleavage of carboxyl propeptides
than of amino propeptides. The final cleavage products were
identified as α1 and α2 chains and the corresponding amino
and carboxyl propeptides.

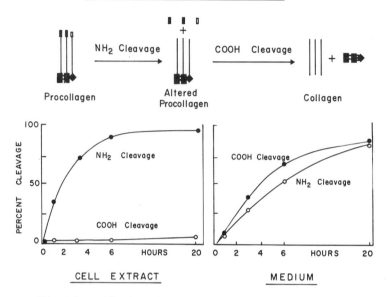

Fig. 1. *Cleavage of procollagen and altered pro-collagen by extracts of tendon cells and tendon cell culture medium. The rate of cleavage of propeptides from procollagen and altered procollagen is shown.*

This finding supports the concept that removal of the amino propeptides is associated with cells and that carboxyl propeptide cleavage is an extracellular event. It also strongly suggests that there are at least two enzymatic activities rather than a single procollagen peptidase (8). Note, however, that we have not eliminated the possibility that both enzymes attach to substrate at the same time but, independently, act at different rates, so that s substantial portion of the amino propeptides could be released first due to trivial empirical conditions. Also, freshly isolated cells and cells in culture can release procollagen. Therefore, while cleavage of the amino propeptides probably normally occurs in association with cells it appears not to be a prerequisite for release of procollagen from cells into culture medium. Removal of the propeptides is also <u>not</u> an essential prerequisite for association of procollagen molecules, according to our findings both by electron microscopy and by cross-linking of procollagen (9), though it is necessary for the formation of wide, native type I fibers. At present we do not know at which stages of the coming together of collagenous molecules the propeptides are excised. In basement membrane

they are presumably not removed (4,10,12) and their presence
will have its own morphogenic effects, at least in preventing
the formation of wider fibrils.

We have investigated the processing of type III procollagen in embryonic chick blood vessels (6) and this may
present a situation in between that of types I and II
collagen on the one hand and basement membrane on the other.
We have found sedimentation analysis in denaturant, 6 M urea
sucrose gradients rather useful for isolating disulfide-
linked procollagen species (3,4,13,14). With this method
Dr. Burgeson observed that some of the procollagens made by
chick embryo body wall fibroblast cultures contained a
disulfide-linked component (15) that sedimented faster than
the type I procollagen, which has interstrand S-S bridges
only between the carboxylpropeptides (4). Dr. Burgeson also
showed that it was very likely that this material was made
up of type III chains.[1]

To examine the material further, to find the reason for
its faster sedimentation in 6 M urea and to observe its
physiological processing, we made short term *in vitro* labeling studies of a likely source of type III procollagen: the
principal blood vessels of embryonic chicks. Figure 2 shows
a typical sedimentation diagram of an extract of blood
vessels labeled with [^{3}H]proline for 40 minutes, and
Figure 3 shows electrophoretic analyses, after reduction, of
material recovered from the fastest sedimenting radioactive
peak A, and the next fastest peak B. Disulfide-linked type I
procollagen makes up much of peak B, as judged by comparison
with known chick bone procollagen I and shown in Figure 3
by proα1 and proα2 chains which co-migrate with markers in
the same sodium dodecyl sulfate acrylamide slab gel. In
contrast, peak A material, which sediments 1.2 times as
fast as peak B, contained only one proα chain. Peak A
material could readily be renatured and after pepsin digestion
gave an electrophoretic pattern of cyanogen bromide cleavage
peptides which was not inconsistent with it being made up
of type III collagen chains. By a scheme of analyses outlined
in Figure 4 and given elsewhere in detail[2] we proved that
the peak A material has interstrand S-S bridges in three
regions of this procollagen III molecule: (1) like type I
procollagen there are disulfide-linked carboxyl propeptides;
(2) the carboxyl terminal portion of the pepsin-resistant
collagen stem is disulfide-linked (consistent with known
properties of type III collagen); and (3) the amino propeptides

[1]*Personal communication and in preparation for
publication*

[2]*In preparation for publication.*

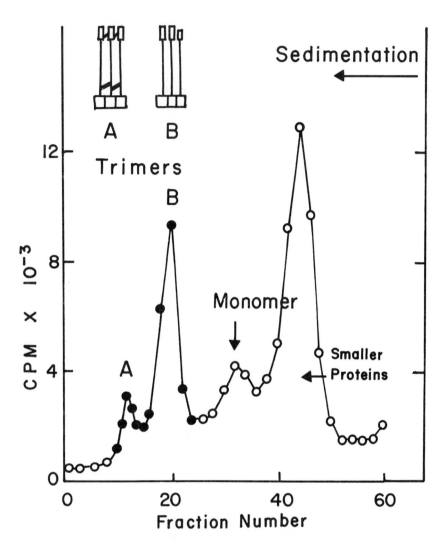

Fig. 2. Velocity sedimentation diagram of an
extract of blood vessels labeled with [³H]proline
for 40 min. Sedimentation was from right to left
on a 5-20% sucrose gradient containing 6M urea,
0.1M NaCl, 0.05M Tris-HCl pH 7.5 and 0.1% Triton-X100,
in a SW60 Beckman rotor at 56,000 rpm at 22° for
19 hrs. For better resolution smaller, 8 drop,
fractions were taken over peak A than over the rest
of the gradient which is analyzed in 16 drop fractions.
The fraction numbers shown as abscissae scale are in
8 drop fractions.

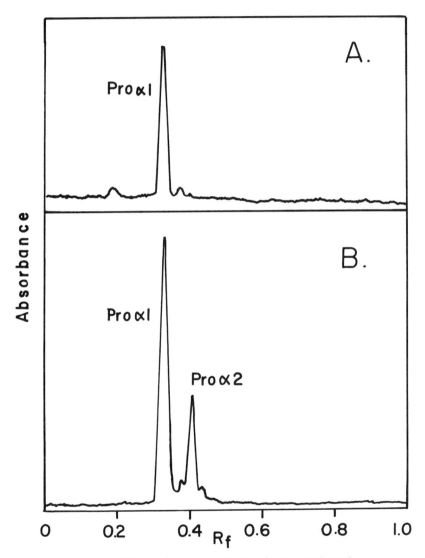

Fig. 3. Electrophoretogram of the reduced
components of peak A and peak B obtained by velocity
sedimentation (see Fig. 2).   Direction of migration
is from left to right on a sodium dodecyl sulfate,
5% acrylamide slab gel, fluorographic analysis is
shown.

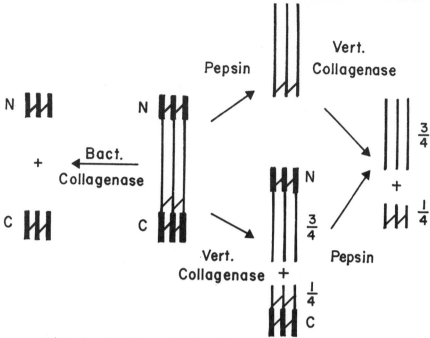

Fig. 4. Outline of the cleavage products produced when procollagen type III was cleaved with bacterial collagenase, vertebrate collagenase and pepsin.

are also S-S linked. Separately, Nowack et al. (16) have recently found that the amino propeptides of a procollagen III extracted from calf skin are disulfide-linked.

Our detailed analysis will be given elsewhere[2], but the higher sedimentation coefficient of type III procollagen relative to that of type I is well explained by a lower hydrodynamic resistance of the denatured molecules, because of a smaller radius of gyration caused by the interchain disulfide bridges at both ends of the molecule.

Now pulse-chase experiments[2] over the time span (90 min.) in which we observed processing of procollagens to collagen in chick embryo bone, tendon and cartilage, show partial conversion of the material of peak A to slightly slower sedimenting material. By analogy with the processing of type I procollagen to altered procollagen (3), we believe that some cleavage has occurred, but with the important restriction that from the continued high sedimentation coefficient we believe that this partly processed procollagen III molecule is still disulfide linked both at its amino and its carboxyl end. Although we have not extended these time studies,

as the viability of such organ cultures may be questionable
at longer incubation times, we have not been able to obtain
evidence for substantial amounts of a smaller soluble form
of type III procollagen. As type III collagen is otherwise
obtained from blood vessels by proteolytic treatment, with
pepsin, it could be that procollagen III is never processed
in vivo to a collagen comparable to collagen I or collagen II,
and that substantial residues of the "propeptides" of
type III remain attached and serve structural functions. If
this is so then basement membrane procollagen may not be
unique in retaining noncollagenous terminal peptide portions,
which are cleaved off in types I and II, and this could be a
feature of other collagens. One might expect that collagens
in which substantial noncollagenous portions are retained may
not show outstanding fiber·formation, but instead an intimate
interaction with other extracellular matrix compounds.

We gratefully acknowledge collaboration and discussion
with Dr. Robert Burgeson and financial support from USPHS
grants AM-13748 and CA-09056-ol, and we thank Dr. N. A.
Kefalides and the sponsors of this Symposium for the opportunity to discuss our work here.

REFERENCES

1. Meezan, E., Brendel, K., Hjelle, J. Thomas, and Carlson,
    E.C. *(This volume.)*
2. Fessler, J.H., Morris, N.P., Greenberg, G.M. and Fessler,
    L.I., *IN:* "Extracellular Matrix Influences on Gene
    Expression", (Slavkin, H.C. and Greulich, R.C., Editors),
    Academic Press, New York, p. 101, 1975.
3. Morris, N.P., Fessler, L.I., Weinstock, A. and Fessler,
    J.H., *J. Biol. Chem.* 250:5719, 1975.
4. Fessler, L.I., Morris, N.P. and Fessler, J.H., *Proc. Natl.*
    Acad. Sci., USA, 72:4905, 1975.
5. Goldberg, B. and Sherr, C.J., *Proc. Natl. Acad. Sci.,*
    USA, 70:361, 1973.
6. Fessler, J.H. and Fessler, L.I., *Fed. Proc.* 35 (and other
    observations in process of publication), 1976.
7. Laskey, R.A. and Mills, A.D., *Eur. J. Biochem.* 56:335,
    1975.
8. Bornstein, P., *Ann. Rev. Biochem.* 43:567, 1974.
9. Fessler, J.H., Doege, K.H. and Fessler, L.I., *Fed. Proc.*
    (in press), 1977.
10. Howard, B.V., Macarak, E.J., Gunson, D. and Kefalides,
    N.A., *Proc. Natl. Acad. Sci., USA,* 73:2361, 1976.
11. Clark, C.C. and Kefalides, N.A.*(This volume).*

12. Heathcote, J.G., Sear, C.H.J. and Grant, M.E.

13. Fessler, L.I., Burgeson, R.E., Morris, N.P. and Fessler, J.H., *Proc. Natl. Acad. Sci., USA,* 70:2993, 1973.

14. Fessler, L.I. and Fessler, J.H., *J. Biol. Chem.* 249: 7637, 1974.

15. Burgeson, R.E., Wyke, A.W. and Fessler, J.H., *Biochem. Biophys. Res. Commun.* 48:892, 1972; and Burgeson, R.E., *doctoral dissertation,* University of California, Los Angeles, 1973.

16. Nowack, H., Olsen, B.R. and Timpl, R., *Eur. J. Biochem. (in press), 1976.*

THE ENZYMIC BASIS FOR THE HYDROXYLATION

OF CARBON 3 IN PROLYL RESIDUES

OF CHICK PROTOCOLLAGEN

Marlene W. Karakashian, Charles C. Clark

and Nicholas A. Kefalides

University of Pennsylvania
Philadelphia, Pennsylvania

SUMMARY: Ammonium sulfate fractions of embryonic chick homogenates catalyze the formation of both 3- and 4-hydroxy[^{14}C]proline in Type I [^{14}C]proline-labeled protocollagen substrates. Both hydroxylase activities are dependent on the amount of ammonium sulfate protein added to an assay mixture and the time of incubation. The two hydroxylase activities are differentially inhibited by poly-L-proline and can be physically separated by affinity chromatography, suggesting that the hydroxylation of prolyl carbon 3 is probably catalyzed by a discrete and as yet undescribed prolyl hydroxylase.

Newly translated vertebrate procollagen α-chains undergo a series of chemical changes before they are released from cells (1-3). Among the processing steps is the hydroxylation of selected prolyl residues in the polypeptide chains (1-3). The hydroxylation of prolyl carbon 4 is extensive in all collagens and appears to be essential for the stability and normal secretion of the triple-helical procollagen molecules at physiological temperatures (4,5). A hydroxylation of prolyl carbon 3 also occurs during collagen biosynthesis (6),

383

but this affects very few prolyl residues in most collagens,
the exception being basement membrane (Type IV) collagen
where 10-15% of the total hydroxyproline present is 3-hydroxy-
proline (7). The functional significance of this second type
of prolyl hydroxylation remains unknown.

The enzyme responsible for the hydroxylation of prolyl
carbon 4 has been studied for some time (for reviews, see 3,
8) and has been purified by several laboratories (9-12).
This hydroxylase recognizes prolyl residues in the Y-position
of the repeating (X-Y-Gly) collagen triplet and requires O_2,
Fe^{2+}, α-ketoglutarate and a reducing agent such as ascorbate
in the reaction (8-12).

Much less is known about the basis for the formation of
3-hydroxyproline, but studies of [^{14}C]proline-incorporation
indicate that it is brought about by a reaction similar to
that producing 4-hydroxyproline (13,14). Sequence studies
show that 3-hydroxyproline occurs only in the X-position of
(X-Y-Gly) and is always followed by 4-hydroxyproline in the
Y-position (15,16). Berg and Prockop (11), unable to detect
hydroxylation of prolyl carbon 3 in experiments testing the
specificity of their highly purified prolyl hydroxylase, proposed that a separate enzyme might be responsible for the
carbon 3-hydroxylation. Although Nordwig and Pfab (17) reported some time ago that crude enzyme preparations from
earthworm body wall and embryonic calf skin catalyze the
formation of both 3- and 4-hydroxy[^{14}C]proline in [^{14}C]proline
labeled protocollagen and that the earthworm extract also
catalyzed the formation of large amounts of 3-hydroxyproline
relative to 4-hydroxyproline in (Pro-Pro-Gly)$_n$ substrates,
there have been no further reports until now of an enzymic
hydroxylation of prolyl carbon 3 in any substrate. The present study shows that ammonium sulfate fractions of embryonic
chick homogenates catalyze both types of prolyl hydroxylation
in Type I protocollagen substrates. Because the two hydroxylating activities are differentially inhibited by poly-L-proline and can be physically separated by affinity chromatography, we suggest that two discrete enzymes are indeed involved in collagen prolyl hydroxylation.

MATERIALS AND METHODS

Source of Hydroxylases

   Thirteen-day old chick embryos were homogenized in an equal volume of 0.01M KCl and the homogenates were fractionated with ammonium sulfate according to the procedure of Berg and Prockop (11). The material which precipitated between 30-65% saturation was dissolved in enzyme buffer (0.05M KCl, 0.1M NaCl, 0.1M glycine, 0.01M Tris-Hcl, pH 7.8 at 4°C), dialyzed against this buffer and centrifuged for 40 min. at 20,000 x g. The protein concentration of the fraction was estimated by absorbance at 230 nm and adjusted to 10 mg/ml before it was subdivided and frozen. Aliquots of such preparations were used directly for assays of prolyl hydroxylating activity and for affinity chromatography.

Type I [14C]proline-labeled Protocollagen Substrates

   Leg tendons were removed from 17-day old chick embryos and used to prepare   matrix-free tendon fibroblast suspensions according to established procedures (18,19). Approximately 10^7 cells/ml were incubated at 37°C in Krebs medium containing 10% fetal calf serum and 0.3mM α,α'-dipyridyl for 30 min., following which 100 μCi [14C]proline was added and the incubation continued for another 4 hrs. Cells were then collected by low speed centrifugation (10 min., 1200 x g, 22°C), resuspended in cold 0.5M acetic acid containing various protease inhibitors (20), gently homogenized and left stirring overnight. Homogenization and subsequent procedures were carried out at 4°C. Usually, rat tail tendon collagen (2 mg) in 0.5M acetic acid was added to a homogenate before it was neutralized with 5N NaOH and dialyzed against large volumes of protocollagen buffer (0.4M NaCl, 0.1M Tris-HCl, pH 7.5). After centrifugation (30 min., 20,000 x g), ammonium sulfate (176 mg/ml) was slowly added to the supernatant. The resulting precipitate was collected by centrifugation (30 min., 20,000 x g) and redissolved in protocollagen buffer containing 2M urea. After additional dialysis against buffer and a final centrifugation, a radioactive supernatant was obtained which contained no detectable hydroxy[14C]proline upon hydrolysis and chromatographic analysis. Aliquots of such Type I [14C]-proline-labelled protocollagen substrate preparations were used to assay both types of prolyl hydroxylase activity and stored frozen when not in use.

Hydroxylase Assay Procedure

Fifty-200 µl of a substrate preparation was preincubated at 37°C and then added to a reaction mixture (final volume 1.0 ml) containing 50 mM Tris-HCl, pH 7.8, 2 mM sodium ascorbate, 0.5mM α-ketoglutarate, 0.05mM $FeSO_4$, 0.1mM dithiothreitol, 0.06 mg catalase (Worthington), 2 mg bovine serum albumin, and water. 10-300 µl of a chick ammonium sulfate fraction was then added and the complete mixture incubated for a designated time, usually 60 min., at 37° in a shaking water bath. Reactions were stopped by the addition of an equal volume of 12 N HCl, following which the reaction tubes were sealed and hydrolyzed for 18 hrs at 107°. Upon cooling, the hydrolysates were evaporated to dryness in a rotary evaporator, dissolved in buffer [0.2N sodium (citrate), pH 2.95] and chromatographed on the long column of a JEOLCO amino acid analyzer at 54°. For purposes of comparing the hydroxyproline content of different hydrolysates, all results were routinely normalized to a base of 300,000 total recovered counts (cpm*), a figure which approximated the actual total number of counts recovered in each chromatographic analysis.

Affinity Chromatography and Gel Filtration

An affinity column was prepared according to the procedure of Tuderman, et al. (12), except that we used a commercially available cyanogen bromide-activated agarose (Sepharose 4B, Pharmacia) and poly-L-proline with an average molecular weight of 54,000 (Type IV, Sigma). Material bound to the affinity column was eluted with poly-L-proline with an average molecular weight of 9,000 (Type II, Sigma) which was dissolved in column buffer at a concentration of 3 mg/ml. Buffers and procedures used were those of Tuderman, et al, except that the appropriate effluent and eluate fractions were concentrated by precipitation with ammonium sulfate rather than ultrafiltration. Selected fractions obtained by gel filtration were assayed immediately for prolyl hydroxylating as was the concentrated effluent from the affinity column.

RESULTS AND DISCUSSION

The chromatographic identification of the hydroxy[^{14}C]-prolines and the basis for our measurements of the two types of prolyl hydroxylase activity are shown in Figure 1. As marked by the horizontal bars at the top of the figure, 3-hydroxyproline elutes several fractions earlier than 4-hydroxyproline and at a position identical to that of an unlabeled 3-hydroxyproline standard (not shown) co-chromatographed with a similar hydrolysate detected by a modified ninhydrin procedure (21). All hydroxy[^{14}C]proline-containing hydrolysates also have a small amount of radioactive material which elutes as an asymmetric peak after 4-hydroxyproline. We have observed that the amount of this material increases with the length of acid hydrolysis and that the amount of 4-hydroxyproline decreases with time under these conditions. Accordingly, we have tentatively identified this material as a 4-hydroxyproline derivative, possibly its cis isomer (22), and we have included it with 4-hydroxyproline in the measurements of prolyl hydroxylase activities.

The size of the hydroxyproline peaks obtained depends on the levels of prolyl hydroxylase activities in an extract (upper and middle panel, Fig. 1) or on the type and purity of the collagen being examined. For example, an analysis of Type I [^{14}C]proline-labeled procollagen secreted by tendon fibroblasts (lower panel, Fig. 1) reveals that approximately 3% of the total newly synthesized hydroxyproline is 3-hydroxyproline. Ogle, et al. (6) estimated that 0.26% of an acid hydrolysate of bovine Achilles tendon collagen is 3-hydroxyproline, an amount which would correspond to 1.8 residues/-1000 or 1.9% of the total hydroxyproline, if it is assumed that their hydrolysate had 94 residues of 4-hydroxyproline per thousand amino acid residues (23). Routine amino acid analyses of rat tail tendon collagen hydrolysates in our laboratory (24) show a similar 3-hydroxyproline content. The observed proportion of 3-hydroxy[^{14}C]proline in newly synthesized chick tendon procollagen is reproducible and twice as high as the best proportion we have been able to achieve to date by the hydroxylation of tendon protocollagen substrates using ammonium sulfate fractions having both prolyl hydroxylase activities.

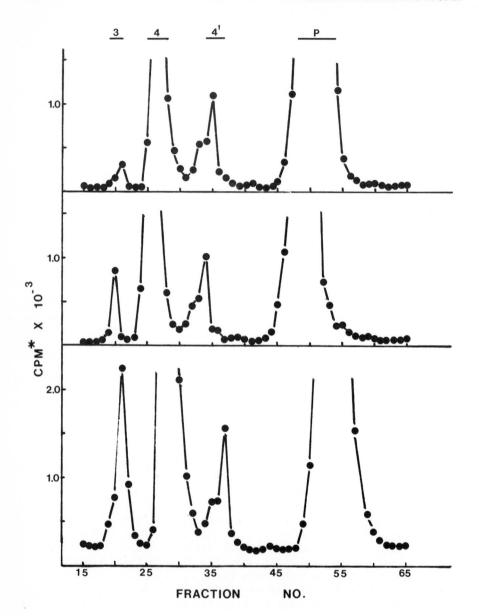

Fig. 1. Chromatographic analyses of hydroxy-
[^{14}C]prolines in hydrolysates of prolyl hydroxylase
reaction mixtures and newly synthesized [^{14}C]proline-
labeled procollagen. Upper panel: Type I embryonic
chick tendon [^{14}C]proline-labeled protocollagen
(50 µl) hydroxylated in a reaction mixture with
0.3 mg protein from a chick ammonium sulfate fraction.
The bars at the top of the panel indicate the respective chromatographic positions of 3-hydroxyproline
(3), 4-hydroxyproline (4), a 4-hydroxyproline derivative which arises during acid hydrolysis (4') and
proline (P). The results are expressed as normalized
cpm (cpm*) for purposeṣ of comparing different hydrolysates and represent the proportions of counts in a
total of 300,000 counts recovered after chromatography.
Proportion of counts as hydroxy[^{14}C]proline: 3-Hyp =
411 cpm*; 4-Hyp = 34,463 cpm*; 4'-Hyp = 2640 cpm*.
3-Hyp/Total Hyp x 100 = 1%. Middle panel: Type I
embryonic chick tendon [^{14}C]proline-labeled protocollagen (50 µl) hydroxylated in a reaction mixture
with 3.0 mg protein from the same chick ammonium
sulfate fraction. Proportion of counts as hydroxy-
[^{14}C]proline: 3-Hyp = 949 cpm*; 4-Hyp = 85,653 cpm*;
4'-Hyp = 2357 cpm*. 3-Hyp/Total Hyp x 100 = 1%.
Lower panel: Type I embryonic chick tendon [^{14}C]pro-
line-labeled procollagen isolated from the incubation
medium and partially purified by ammonium sulfate precipitation. Proportion of counts as hydroxy[^{14}C]-
proline: 3-Hyp = 3,534 cpm*; 4-Hyp = 114,754 cpm*;
4'-Hyp = 2738 cpm*. 3-Hyp/Total Hyp x 100 = 3%.

Using the hydroxylase assay conditions chosen, both
prolyl hydroxylase activities depend on the time of incubation
and the amount of ammonium sulfate fraction added to the reaction mixture (Table I). A more detailed examination of the
effect of adding varying amounts of an ammonium sulfate
fraction of known protein concentration (Figure 2) shows that
only at very low concentrations is the formation of each type
of hydroxyproline linearly correlated with the amount of
protein added. In each case, departure from linearity occurs
at a point which corresponds to approximately 60% of maximum
substrate hydroxylation, which we have defined as that amount
of hydroxylation catalyzed by an excess amount of the ammonium
sulfate fraction (300 µl or 3 mg protein added). Similar
observations were reported by Kivirikko and Prockop (25) in
their early studies of 4-hydroxyproline formation in protocollagen substrates using ammonium sulfate fractions with
prolyl carbon 4 hydroxylating activity. The similarities
between the results for both types of prolyl hydroxylation
strongly indicate that there is also an enzymic basis for
the prolyl carbon 3 hydroxylation and again raises the possibility that both hydroxylations are catalyzed by a single
enzyme, the well-known prolyl hydroxylase.

TABLE I

Effect of Incubation Time on
Prolyl Hydroxylase Activities

Extract	Hours	3-Hyp*	4-Hyp*
50 µl	2	639	51,827
	1	529	31,252
	1/2	209	16,137
20 µl**	2	556	70,011
	1	473	58,391
	1/2	326	42,565
4 µl**	2	290	36,075
	1	234	17,608
	1/2	168	14,741

  * *cpm per 300,000 total counts used in assay.*
 ** *same hydroxylase source and substrate used in assay.*

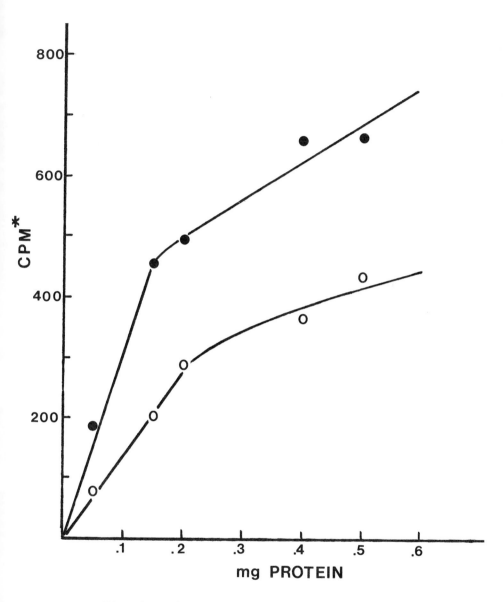

Fig. 2. Effect of protein concentration in chick
ammonium sulfate fraction on proline hydroxylation of
[^{14}C]proline-labeled protocollagen (50 µl) incubated
for 60 min. at 37°C. The hydroxyproline content of
each hydrolysate was estimated by chromatographic
analysis as illustrated in Fig. 1. cpm* = proportion
of counts per 300,000 total counts recovered after
chromatography. 3-Hyp (o——o); 4-Hyp (•——•, cpm* x
10^{-2}).

One indication that two enzymes may be involved in these reactions was provided by studies of the effect of poly-L-proline on prolyl hydroxylation (Table II). Poly-L-proline is a known competitive inhibitor of the prolyl carbon 4 hydroxylase (26), but it is only weakly inhibitory of the prolyl carbon 3 hydroxylating activity of our ammonium sulfate fractions. More compelling evidence for the existence of two discrete prolyl hydroxylases was obtained following affinity chromatography of an ammonium sulfate fraction containing both enzyme activities. As summarized in Table III, the effluent from the affinity column (which contained almost all of the protein from the starting extract) had a significantly altered ratio of 3- and 4-prolyl hydroxylating activities. Moreover, material eluted from the affinity column and separated from the eluting agent (low molecular weight poly-L-proline) by gel filtration had only carbon 4-hydroxylating activity.

TABLE II

Effect of Poly-L-Proline (Ave. MW 9000)
on Prolyl Hydroxylase Activities

mg. Protein Added	3-Hyp*	4-Hyp*	% 3-Hyp	% 4-Hyp
0.2	472	58,390	100	100
0.2 + 5 μg $(pro)_n$	403	24,332	85	42
0.1	249	35,211		
	234	42,048		
	242	38,630	100	100
0.1 + 5 μg $(pro)_n$	239	11,874		
	171	12,660		
	205	12,267	85	32

*
cpm per 300,000 total counts used in assay.

TABLE III

*Affinity Chromatography of Extract With 3- and
4-Prolyl Hydroxylating Activities*

Assayed Material	3-Hyp*	4-Hyp*	3-Hyp: 4-Hyp
Starting extract	588	62,124	1: 106
Column effluent, concentrated	373	11,850	1: 32
Eluted material after gel filtration, single fraction	0	46,047	0

*cpm per 300,000 total counts used in assay.

These results demonstrate that the hydroxylation of both
carbons 3 and 4 in prolyl residues of Type I tendon protocollagen is enzymic and suggest that the hydroxylation of
prolyl carbon 3 is probably catalyzed by a discrete and as
yet undescribed prolyl hydroxylase which differs in important
respects from the well-known prolyl hydroxylase. Work
related to the further purification and characterization of
this postulated second prolyl hydroxylase is in progress.

ACKNOWLEDGEMENTS

These studies were supported by NIH Grant AM-14526 and
HL-05061. We are pleased to acknowlege the technical
assistance of Marie Gleason and the helpful advice of
Dr. Peter Dehm. C.C.C. is a recipient of a NIH Research
Career Development Award 1 KO4 AM-00063.

REFERENCES

1. Bornstein, P., *Ann. Rev. Biochem.* 43:567-603, 1974.
2. Martin, G.R., Byers, P.H. and Piez, K.A., *Adv. Enzymol.* 42:167-191, 1975.
3. Prockop, D.J., Berg, R.A., Kivirikko, K.I. and Uitto, J., *In:* "Biochemistry of Collagen (Ramachandran, G.N. and Reddi, A.J., eds.) Plenum Press, New York, 1976.
4. Jimenez, S., Harsh, M. and Rosenbloom, J., *Biochem. Biophys. Res. Comm.* 52:106-114, 1973.
5. Berg, R.A. and Prockop, D.J., *Biochem. Biophys. Res. Comm.* 52:115-120, 1973.
6. Ogle, J.D., Arlinghaus, R.B. and Logan, M.A., *J. Biol. Chem.* 237:3667-3673, 1962.
7. Kefalides, N.A., *Biochem. Biophys. Res. Comm.* 47:1151-1158, 1972.
8. Cardinale, G.J. and Udenfriend, S., *Adv. Enzymol.* 41:245-300, 1974.
9. Halme, J., Kivirikko, K.I. and Simons, K., *Biochim. Biophys. Acta* 198:460-470, 1970.
10. Rhoads, R.E. and Udenfriend, S., *Arch. Biochem. Biophys.* 139:329-339, 1970.
11. Berg, R.A. and Prockop, D.J., *J. Biol. Chem.* 248:1175-1182, 1973.
12. Tuderman, L., Kuuitti, E.R. and Kivirikko, K.I., *Eur. J. Biochem.* 52:9-16, 1975.
13. Kaplan, A., Witkop, B. and Udenfriend, S., *J. Biol. Chem.* 239:2559-2561, 1964.
14. Fujimoto, D.F. and Adams, E., *Biochim. Biophys. Acta.* 107:232-246, 1965.
15. Gryder, R.M., Lamon, M. and Adams, E., *J. Biol. Chem.* 250:2470-2474, 1975.
16. Hulmes, D.J.S., Miller, A., Parry, D.A.D., Piez, K.A. and Woodhead-Galloway, J., *J. Mol. Biol.* 79:137-148, 1973.
17. Nordwig, A. and Pfab, F.K., *Biochim. Biophys. Acta.* 181:52-58, 1969.
18. Jimenez, S.A., Dehm, P., Olsen, B.R. and Prockop, D.J., *J. Biol. Chem.* 248:720-729, 1973.
19. Berg, R.A. and Prockop, D.J., *Biochem.* 12:3395-3401, 1973.
20. Fessler, L.I., Morris, N.P. and Fessler, J.H., *Proc. Nat. Acad. Sci., U.S.A.*, 72:4905-4909, 1975.
21. Rosen, H., *Arch. Biochem. Biophys.* 67:10-15, 1957.
22. Dziewiatkowski, D.D., Hascall, V.C. and Riolo, R.L., *Anal. Biochem.* 49:550-558, 1972.
23. Piez, K.A. and Likins, R.C., *In:* "Calcification in Biological Systems" (Sognnaes, R.F., ed.) A.A.A.S., Washington, 1960.

24.   Dehm, P., unpublished results.
25.   Kivirikko, K.I. and Prockop, D.J., *Arch. Biochem.
      Biophys.* 118:611-618, 1967.
26.   Prockop, D.J. and K.I. Kivirikko, *J. Biol. Chem.* 244:
      4838-4842, 1969.

C. Immunochemical Properties

ANTIGENIC PROPERTIES OF

BASEMENT MEMBRANE COLLAGENS

Diane E. Gunson, Bradley W. Arbogast

and Nicholas A. Kefalides

University of Pennsylvania
Philadelphia, Pennsylvania

SUMMARY: A radioimmunoassay has been developed for the detection of Type IV collagen using a specific antiserum raised against pepsin digested bovine anterior lens capsule (BALC). Specificity was determined by testing ^{125}I labelled Types I, II, III and IV bovine collagen in both a direct radioimmunoassay and a radioimmune inhibition assay; both showed that the antiserum had no reaction with Types I, II and III, but a marked reactivity with Type IV collagen. Full antigenic activity was retained after pepsin redigestion or heat denaturation of the antigen, while it was completely abolished by collagenase digestion. The anti-BALC sera reacted similarly with ^{14}C-labelled rat parietal yolk sac (Type IV) collagen, although unhydroxylated rat Type IV collagen, produced in the presence of $\alpha\alpha$-dipyridyl, was less reactive, unless it was enzymatically hydroxylated using prolyl hydroxylase. In this case the collagen became almost fully antigenic. Reduced and alkylated rat Type IV also had little reactivity with the antiserum. These results demonstrate that specific antisera to Type IV collagen can be produced without absorption with other types of collagen, and that there is some cross-reactivity between rat and bovine Type IV collagens. Antigenicity seems to depend, at least in

397

*part, on the hydroxylation of the proline residues and
on the integrity of the disulfide bonds, and residues
in the collagenase sensitive part of the molecule.*

INTRODUCTION

Type IV collagen is a major component of basement
membranes and has similar chemical and physical properties
to other collagen types (1). Antisera to the various collagens have been produced and used in a multiplicity of different tests for the purposes of studying collagen structure
and antigenic sites (2,3,4). Minute quantities of antigen
can be detected using the radioimmunoassay, so this test has
proved to be particularly useful in the study of biosynthetic
systems where only small amounts of antigen are available
(5,6). This paper summarizes recent progress in this
laboratory on basement membrane collagen immunology.

METHODS

Antisera

Rabbits were immunized with Type IV collagen from
bovine anterior lens capsule (4).

Antigens

Bovine Types I, II and III collagens and rat and bovine
Type IV collagens were prepared as described before (4,5).
Unhydroxylated rat Type IV collagen was prepared by the
incubation of rat parietal yolk sacs with $\alpha\alpha$-dipyridyl. Some
of this material was subsequently hydroxylated, either by
removal of the $\alpha\alpha$-dipyridyl, followed by further incubation,
or by the addition of the enzyme prolyl hydroxylase (6).
Reduction and alkylation, pepsin and collagenase digestion
were carried out as described previously (6).

Radioimmunoassay

This was carried out as described previously (4,5) using
either ^{125}I-labelled bovine or ^{14}C-labelled rat, Type IV
collagen. Goat antirabbit immunoglobulin was used to precipitate the immune complexes.

RESULTS AND DISCUSSION

Figure 1 shows the results of a radioimmunoassay using bovine, ^{125}I-labelled Types I, II, III and IV collagen. This demonstrates a marked reactivity of our antiserum with Type IV, but not with Types I, II and III collagen. Similarly, the radioimmune inhibition assay demonstrated that Type IV collagen was an effective inhibitor whereas Types I, II and III caused no inhibition of the reaction between the antiserum and ^{125}I-labelled Type IV collagen (4,5). Absorption of the antisera with Types I, II and III collagen caused no reduction in antibody titer whatsoever, so the serum was generally used without prior absorption (4). Collagenase digested Type IV collagen did not react with the antisera in either the direct radioimmunoassay or the radioimmune inhibition assay, while pepsin redigested material retained full activity. Heat denatured Type IV collagen also retained full activity and was, if anything, a slightly more effective inhibitor than native type IV collagen (4,5). These observations are somewhat conflicting as loss of reactivity after collagenase digestion suggests that the antigenic site(s) are in the triple-helix, whereas retention of reactivity after

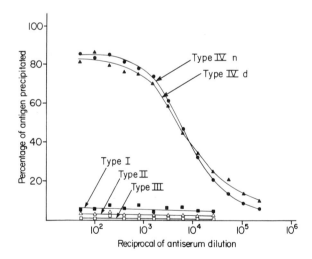

Fig. 1. Radioimmunoassay using rabbit antiserum to native Type IV collagen with ^{125}I-labelled Types I, II, III, and IV bovine collagen as antigens. (From Gunson and Kefalides, 1976)

denaturation suggests that the determinants are non-helical.
To resolve this dilemma, we would like to suggest that the
determinants lie at the junction of helical and non-helical
parts of the α chain in a region which is collagen sensitive
but pepsin resistant.

The antisera had much lower titers against rat Type IV
collagen than with bovine material (5), suggesting that these
are species specific and non-species specific antigenic
determinants. The effects of pepsin and collagenase on the
antigenicity suggested that both types of determinants were
similarly located. Reduction and alkylation lowered the antigenicity of rat collagen (5), indicating that the integrity
of disulfide bonds is of some importance for the antigenic
site or sites.

Further experiments using unhydroxylated rat Type IV
collagen showed that this material had very much lowered
antigenicity, suggesting that the hydroxylation of proline
and lysine may be important for full reactivity (5) (Figure 2).

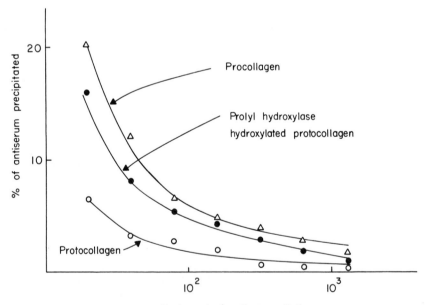

Fig. 2. *Radioimmunoassay using rabbit antiserum to
native bovine Type IV collagen with ^{14}C-labelled rat Type
IV procollagen, protocollagen and hydroxylated protocollagen. (From Arbogast, Gunson and Kefalides, 1976)*

A purified preparation of prolyl hydroxylase was used to
allow hydroxylation of the proline but not the lysine; the
product of this experiment had regained almost full antigenicity (6). Use of purified glucosyl-galactosyl-hydroxylysine
as an inhibitor in a radioimmune inhibition assay showed that
there was no inhibition with this substance, indicating that
it is not *per se* important for antigenicity (6). These results show that the antigenic determinants in rat Type IV
collagen appear to be highly dependent on the hydroxylation
of proline which is presumably situated predominantly in the
helical areas.
     The antisera have also been shown to react in a similar
fashion with bovine endothelial cell collagen (7) and rabbit
Descemet's membrane collagen as well as with rat and bovine
Type IV collagen. However, there was no reaction with rat
tendon procollagen. These results further demonstrate the
specificity of our antisera for basement membrane collagens
with a total lack of reactivity for the interstitial collagens.

ACKNOWLEDGEMENTS

     This work was supported by NIH Grants AM-14526 and
AM-20553.

REFERENCES

1.   Kefalides, N.A., *Biochemistry* 7:3103, 1968.
2.   Beil, W., Timpl, R. and Furthmayer, H., *Immunology* 24:13,
     1973.
3.   Hahn, E., Timpl, R. and Tuller, E.J., *Immunology* 28:561,
     1975.
4.   Gunson, D.E. and Kefalides, N.A., *Immunology* 31:563, 1976.
5.   Gunson, D.E., Arbogast, B. and Kefalides, N.A., *Immunolo-
     gy* 31:577, 1976.
6.   Arbogast, B.W., Gunson, D.E. and Kefalides, N.A., *J. Im-
     munol.* 117:2181, 1976.
7.   Howard, B.V., Macarak, E.J., Gunson, D.E. and Kefalides,
     N.A., *Proc. Nat. Acad. Sci., U.S.A.*, 73:2361, 1976.

THE SPECIFICITY OF THE ANTIGENIC DETERMINANTS

OF BOVINE RENAL TUBULAR BASEMENT MEMBRANE

Wijnholt Ferwerda

Vrije Universiteit
Amsterdam, the Netherlands

Cécile M. I. v. Loon*
and
Thea M. Feltkamp-Vroom*

Central Laboratory of the Netherlands
Red Cross Blood Transfusion Service
Amsterdam, the Netherlands

SUMMARY: Bovine renal TBM and its glycoprotein and collagen units were prepared in a pure state. Four rabbit
antisera were obtained, against a) the Gp units, b) the
Col. units, c) and d) particulate TBM i.m. and i.p. injected. By immunodiffusion three antibody populations
could be detected, two directed against the Gp component
and one against the Col component. Indirect immunofluorescence studies indicated that TBM and GBM of bovine
kidney reacted in the same degree with all sera. Although

Abbreviations used: TBM: tubular basement membrane;
GBM: glomerular basement membrane; Gp: glycoprotein; Col: collagen; i.m.: intramuscularly; i.p.: intraperitoneally; SDS:
sodium dodecylsulphate.

*
Present address: Department of Pathology, Slotervaart
Ziekenhuis, Louwesweg 6, Amsterdam, the Netherlands.

403

the reaction with human, Wistar rat and Brown Norway
rat kidney sections was less, an obvious TBM-specifi-
city was found in some cases. However, an interstitial
nephritis could not be induced with any of the four
antisera.

Antibodies directed against TBM and other epithelial
basement membranes have been prepared by a few investigators
(1-4). The occurrence of anti-TBM antibodies in man has
occasionally been found (5-7). Some investigators reported
the induction of interstitial nephritis in animals after the
injection of heterologous (8,9) or homologous (10) TBM,
accompanied with the occurrence of TBM-specific antibodies.
No information has been obtained about the TBM-specific
antigenic determinants.
    Recently, we have isolated bovine renal TBM (11) and
its glycoprotein and collagen components (12,13). Antisera
were raised in rabbits against these isolated components and
intact TBM. The specificity of the antisera was tested by
Ouchterlony diffusion tests and immunofluorescence studies.
The possibility of inducing an interstitial nephritis with
the various antibody populations has also been investigated.

METHODS

    Bovine renal TBM was prepared as described previously
(11,14). Bovine renal GBM was prepared by the method of
Spiro (15). The isolation and purification of the glycoprotein and collagen units was obtained by collagenase and
pepsin digestion, respectively (12,13). Polyacrylamide gel
electrophoresis was performed by the method of Davis (16).
All solutions contained 0.1% SDS.
    Four antisera were prepared in rabbits by repeated in-
jection of particulate TBM, intramuscularly (3 x 0.6 mg) or
intraperitoneally (6 x 4.8 mg), glycoprotein units (3 x 1.0
mg) and collagen units (2 x 0.2 mg, 1 x 0.5 mg). The antisera were heated for 1 hr at 56° and absorbed with bovine
erythrocytes before use. The immunoglobulin fractions were
obtained by precipitation with 33% ammonium sulphate. Double
diffusion analyses were performed by the method of Ouchterlony (17). To guarantee a perceptable precipitation in the
double diffusion experiments, antisera or globulin fractions
were used with a titer of at least 1:1600. Immunofluorescence studies were done as described by Roos et al. (18).
The titers of the antisera and globulin fractions were

recorded as the highest dilution still giving a detectable fluorescence.

Brown Norway rats were injected intravenously with 2 ml of globulin fractions of a high titer (see Table I) derived from the various antisera. The fixation of antibodies to GBM and TBM and the nephrotoxic effects were determined.

TABLE I

Titer of the globulin fractions of anti-Gp, anti-Col, anti-TBM (i.m.) and anti-TBM (i.p.) with respect to TBM and GBM of normal bovine, human, Wistar rat and Brown Norway rat kidney

Globulin Fraction	Bovine Kidney		Human Kidney		Wistar Rat Kidney		Brown Norway Rat Kidney	
	TBM	GBM	TBM	GBM	TBM	GBM	TBM	GBM
anti-Gp 1	10.000*	10.000	3.200	3.200	800	800		
2	25.600		1.600		800		3.200	
anti-Col 1	3.200	3.200	100	<20	50	<20		
2	6.400		800		800		1.600	
anti-TBM (i.m.) 1	1.600	800	800	800	400	200	3.200	
2	12.800	3.200	1.600	800	800	50	2.200	400
anti-TBM (i.p.)	12.800	3.200	1.600	400	400	50	800	20

*reciprocal values of the titer.

RESULTS

In understanding the results concerning the antigenic properties of TBM, it is important to mention the existence of two glycoprotein components, a high and a low molecular weight component. The two components can be visualized by polyacrylamide gel electrophoresis (Figure 1).

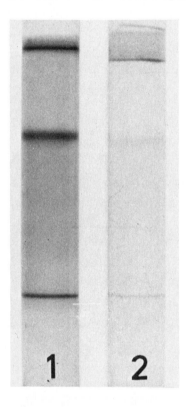

Fig. 1. Polyacrylamide gel electrophoresis pattern (7.5% gel, Tris-HCl buffer, pH 8.3, 0.1% SDS) of the glycoprotein fraction obtained from bovine renal TBM. Left: (1) stained with amido black; and right: (2) stained with periodic-acid Schiff stain.

The results obtained with the double diffusion technique are shown in Figure 2. The anti-Gp serum produced two precipitin lines with its immunogen, the glycoprotein fraction (Fig. 2A). Based on gel permeation chromatography (13) it could be pointed out that the heavy line of the most diffused component and the faint line of the least diffused component were brought about by the low and the high molecular weight glycoprotein component, respectively. Identical precipitin lines were obtained with the anti-TBM (i.m.) globulin fraction (not shown). It should be mentioned that with the unpurified anti-TBM (i.m.) serum only the precipitin line with the low molecular weight glycoprotein component could be visualized

(Figure 2A). The anti-TBM (i.p.) globulin fraction produced
with the Gp fraction only a precipitation line with the high
molecular weight Gp component (Figure 2C). The anti-Gp and
the anti-TBM (i.m.) globulin fractions were negative with the
Col fraction. The anti-TBM (i.p.) globulin fraction was not
tested with the Col fraction. The anti-Col globulin fraction
produced one precipitin line with its immunogen, the Col
fraction. This anti-Col serum was negative with the Gp
fraction (Figure 2B). Testing the four antisera with the
indirect immunofluorescence technique it appeared that all
antisera, when incubated with sections of normal bovine
kidney, produced a positive reaction with the basement membranes of tubuli, glomeruli, Bowman's capsule and with the
vascular wall.

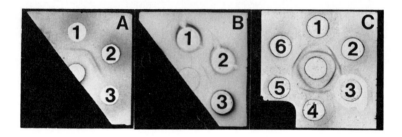

Fig. 2. Double diffusion test of the Gp and Col
fractions derived from bovine renal tubular basement
membrane against the various antisera.
A. Center well: Gp fraction; 1: anti-TBM (i.m.)
    serum; 2: anti-Gp serum; 3: anti-Col serum.
B. Center well: Col fraction; 1: anti-TBM (i.m.)
    globulin fraction; 2: anti-Col globulin fraction;
    3: anti-Gp serum.
C. Center well: collagenase-soluble material; 1, 3 and
    5: anti-TBM (i.p.) globulin fraction; 2,4 and 6:
    anti-Gp globulin fraction.

    In Figure 3, the result obtained with the anti-Gp serum
is shown. The unspecificity of the anti-Gp serum was further
evidenced by the fact that this antiserum produced two lines
of identity with the glycoprotein components derived from
both TBM and GBM (Figure 4). In this experiment collagenase
soluble material had been used, which was shown earlier to
produce the same precipitin lines as the Gp fraction (13).

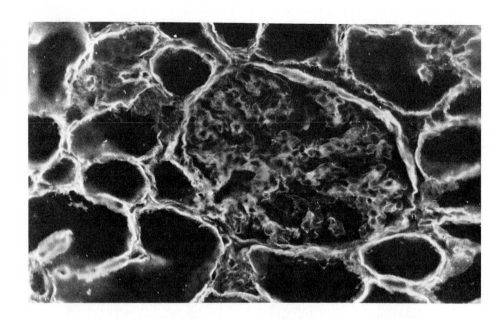

Fig. 3. Photomicrograph showing fixation of
anti-Gp antibodies on TBM, GBM, Bowman's capsule and
blood vessel wall in a normal bovine kidney section.
X 300.

Fig. 4.   Double diffusion
test of the Gp fractions derived
from bovine renal TBM and GBM.
Center well: anti-Gp serum; 1 and
4:   Gp fraction of GBM; 2 and 5:
collagenase-soluble material from
GBM; collagenase-soluble material
from TBM.

In order to obtain more information about the quantitative fixation of the antibodies and about the species-speci-
ficity, the titers of the globulin fractions were recorded
with respect to TBM and GBM of kidney sections of different

species (Table I). With bovine kidney, only anti-TBM (i.m.) and anti-TBM (i.p.) showed a slightly higher titer with respect to TBM. With sections of human, Wistar rat and Brown Norway rat kidneys the two anti-TBM as well as the anti-Col globulin fractions showed more or less higher affinity for TBM. Anti-Gp fixed in all cases in the same degree to TBM and GBM.

With all globulin fractions the titer measured with bovine kidney was the highest. Note the difference between Wistar and Brown Norway rats. The capacity of the four globulin populations to induce an interstitial nephritis was tested with Brown Norway rats. An interstitial nephritis could not be induced with any of the four sera.

Within half an hour after intravenous injection, rabbit immunoglobulins were detectable in a linear pattern along the GBM. After 4 to 5 weeks rat immunoglobulin was also observed together with Complement in a similar linear pattern. Signs of glomerulonephritis developed during the observation period of 9 weeks, as determined by light microscopy.

DISCUSSION

The existence of two antigens of glycoproteinic origin, a high and low molecular weight antigen, has also been reported by Denduchis and Kefalides (19) for sheep anterior lens capsule and by Kefalides for canine GBM (20). However, in these studies the antiserum was raised against GBM obtained after reduction and alkylation, while our glycoprotein material was in an unreduced state. When our glycoprotein material was reduced with β-mercaptoethanol, no precipitation occurred at all with the anti-Gp serum (unpublished results).

The occurrence of antigenic determinants common to various basement membranes (1,2,21-27), reticuline and basement membrane (1,21,24,28,29) and units derived from different basement membranes (19,20) has been reported earlier. At present, few examples are known of antisera reacting specifically with one particular basement membrane or group of basement membranes (1-10). It is questionable whether TBM-specific antisera can be obtained with units prepared with the aid of enzyme digestion or urea extraction, because the tertiary structure of the protein has most probably been disturbed by these methods.

Intactness of the membrane might be prerequisite for the occurrence of specific determinants. Although our results give some support to this idea, no conclusive evidence has been obtained.

Attempts were made to induce interstitial nephritis in
Brown Norway rats. Although, especially for the anti-TBM
globulin preparations the TBM-specificity was obvious,
neither perceptable deposition of immunoglobulins along TBM,
nor symptoms of an interstitial nephritis were found.   It
appeared that the bulk of the heterologous immunoglobulins
were fixed by the GBM, and induced a nephrotoxic serum
nephritis.

REFERENCES

1. Midgley, A.R. and Pierce, G.B., *Am. J. Pathol.* 43:929-
    943, 1963.
2. Pierce, G.B. and Nakane, P.K., *Lab. Invest.* 17:499-514,
    1967.
3. Pierce, G.B., *In:* "Chemistry and Molecular Biology of
    the Intercellular Matrix", (Balasz, E.A., Editor),
    Academic Press, New York, Vol. 1, pp. 471-506, 1970.
4. Steblay, R.W. and Rudofsky, U., *Science* 180:966-968,
    1973.
5. Klassen, J., Kano, K., Milgrom, F., Menno, A.B., Anthone,
    R., Sepulveda, M., Elwood, C.M. and Andres, E.A., *Int.
    Arch. Allergy* 45:675-689, 1973.
6. Border, W.A., Lehman, D.H., Egan, J.D., Sass, H.J.,
    Glode, J.E. and Wilson, C.B., *N. Eng. J. Med.* 291:381-
    384, 1974.
7. Morel-Maroger, L., Kourilsky, O., Mignon, F. and Richet,
    G., *Clin. Immunol. Immunopath.* 2:185-194, 1974.
8. Steblay, R.W. and Rudofsky, U., *J. Immunol.* 107:589-594,
    1971.
9. Lehman, D.H., Wilson, C.B. and Dixon, F.J., *Kidney Int'l.*
    5:187-195, 1974.
10. Sugisaki, T., Klassen, J., Milgrom, F., Andres, G.A. and
    McCluskey, R.T., *Lab. Invest.* 28:658-671, 1973.
11. Ferwerda, W., Meijer, J.F.M., Eijnden, D.H. v.d. and
    Dijk, W. v., *Hoppe Seyler's Z. Physiol. Chem.* 355:976-
    984, 1974.
12. Ferwerda, W., Smit, J.W. and Feltkamp-Vroom, T.M.,
    Biochem. Soc. Trans. 2:640-642, 1974.
13. Ferwerda, W., *Thesis,* Vrije Universiteit, Amsterdam, 1975.
14. Ferwerda, W. and Dijk, W.v., *Hoppe Seyler's Z. Physiol.
    Chem.* 356:1671-1678, 1975.
15. Spiro, R.G., *J. Biol. Chem.* 242:1915-1922, 1967.
16. Davis, B.J., *Ann. N.Y. Acad. Sci.,* 121:404-427, 1964.
17. Ouchterlony, O., *In:* "Diffusion-in-gel Methods for
    Immunological Analysis, Progr. Allergy", (Kallós, P., Ed.)
    S. Karger, Basel, Vol. 5, pp. 1-78, 1958.

18. Roos, C.M., Feltkamp-Vroom, T.M. and Helder, A.W., *J. Pathol.* 118:1-8, 1976.
19. Denduchis, B. and Kefalides, N.A., *Biochim. Biophys. Acta* 221:357-366, 1970.
20. Kefalides, N.A., *Connect. Tissue Res.* 1:3-13, 1972.
21. Cruickshank, B. and Hill, G.S., *J. Pathol. Bacteriol.* 66:283-289, 1953.
22. Goodman, M., Greenspon, S.A. and Krakower, C.A., *J. Immunol.* 75:96-104, 1955.
23. Roberts, D. St. C., *Brit. J. Ophthalm.* 41:338-347, 1957.
24. Steblay, R.W., *J. Immunol.* 88:434-442, 1961.
25. Steblay, R.W. and Lepper, M.H., *J. Immunol.* 87:636-646, 1961.
26. Pierce, G.B., Midgley, A.R. and Sri Ram, J., *J. Exp. Med.* 117:339-346, 1963.
27. Pierce, G.B., Beals, T.F., Sri Ram, J. and Midgley, A.R., *Am. J. Pathol.* 45:929-961, 1964.
28. Loewi, G., *Ann. Rheum. Dis.* 26:544-551, 1967.
29. Feltkamp-Vroom, T.M. and Balner, H., *Eur. J. Immunol.* 2:166-173, 1972.

CHEMICAL AND IMMUNOLOGICAL STUDIES ON

BASEMENT MEMBRANE COLLAGEN FROM A MURINE TUMOR

Rupert Timpl

Max-Planck-Institut fur Biochemie
Martinsried bei Munchen, Germany

Roslyn W. Orkin, Pamela Gehron Robey,
George R. Martin

National Institute of Dental Research
National Institutes of Health
Bethesda, Maryland

and Georg Wick

Institut fur Experimentelle Pathologie
University of Innsbruck
Innsbruck, Austria

ABSTRACT: A collagenous component was extraced with
0.1 M acetic acid from a murine tumor and purified on
DEAE cellulose in 2 M or 8 M urea. This protein was
larger than collagen γ components and had an amino
acid composition typical for basement membrane collagens.
After reduction two polypeptide chains with an apparent
m.w. of 140,000 could be identified in disc electro-
phoresis. Cleavage with CNBr produced a peptide pattern
different to that known for type I, II and III collagens
and indicated the presence of two disulfide-bonded
regions in the molecule. Digestion of the native colla-
gen with various proteases produced several fragments;

413

*some of them were smaller than α-chains. Prior
reduction of disulfides enhanced degradation. Rabbit
antisera against the basement membrane collagen showed
high titers for the immunizing antigen in agglutination
and radioimmune assays but negligible cross-reactions
with other collagens. Antibodies were purified by
immunoadsorption and used in indirect immunofluorescence studies. They stained basement membrane structures in sections of murine and human skin, kidney,
lung and blood vessels and reacted with the extracellular matrix of the murine tumor. The tumor could
also be stained by autoantibodies from patients with
Goodpasture syndrome.*

We have carried out studies on a collagenous protein
from a transplantable murine tumor (1). While the tumor
resembled a chondrosarcoma* at the histological level (2)
(Fig. 1), significant differences were noted at the ultrastructural level as well as in the biochemical properties
of the collagenous protein synthesized by the tumor (1).
Ultrastructurally an amorphous, fiberless matrix was
observed in the tumor resembling more closely basement
membrane than cartilage. When the tumor was grown in
lathyritic animals large amounts of a collagenous protein
could be extracted but this protein was not composed of
α chains as one would expect with cartilage collagen. Instead
larger components were obtained which resembled basement membrane collagens in composition (3). Now we have isolated the
native protein and characterized it by physical, chemical and
immunological methods.

Isolation and Chemical Characterization

The collagenous protein from the tumor was extracted with
0.5 M acetic acid and separated from other proteins in the
extract by precipitation with 10% NaCl followed by DEAE-cellulose chromatography (Timpl et al., in preparation). Amino
acid analyses showed that the protein contained 280 glycines/
1,000 amino acids. Other amino acids were present in proportions reported for authentic basement membrane collagens (3).

*We suggest that this tumor be designated the EHS-sar-
coma to acknowledge the contributions of Engelbreth-Holm and
R. L. Swarm. See ref. 2 for a discussion of the origins of
this tumor.*

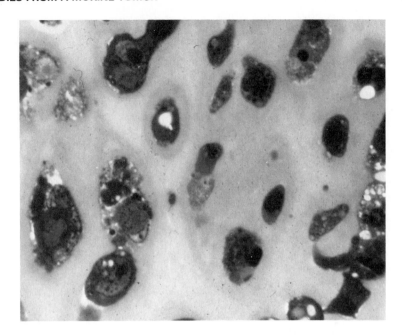

*Fig. 1. Appearance of the EHS-sarcoma showing
an abundant hyaline matrix separating pockets of
cells. The tumor was tentatively diagnosed as an
"undifferentiated" chondrosarcoma (2). The tissue
in which this tumor arose is not known.*

For example, very high levels of hydroxyproline and hydroxylysine are present while the levels of alanine, lysine and
arginine are low. The protein migrated on electrophoresis in
sodium dodecyl sulfate with a molecular weight of greater
than 300,000. After reduction of disulfide bonds under
denaturing conditions three chains with an apparent molecular
weight of 140-180,000 were obtained (1). In size the protein
resembles procollagen more than collagen.
    Physical studies (Bruckner et al., in preparation) showed
that the protein had a collagen-like helix. The circular
dichroism spectra qualitatively resembled that produced by
type I collagen. The spectrum of the protein changed at about
35° indicating that the helix of the protein was unstable
above that temperature. Reduction of disulfide bonds in the
protein under non-denaturing conditions did not alter its
stability.
    The studies outlined above indicated that we were dealing
with a collagenous protein. With other procollagens and
collagens, it is possible to determine the length of the helix,

since this domain is resistant to proteolytic digestion. Incubation with pepsin has often been used to solubilize basement membrane collagens (3) but a variety of molecular weight components are found in such preparations.

Following incubation of the purified tumor proteins with pepsin in 0.5 M acetic acid at 15°, we observed high molecular weight material. On reduction, some of the original chains in the preparation and two smaller components were found and estimated to be 55,000 and 70,000. Since it was possible that the disulfide bonds in the protein prevented digestion, we reduced and alkylated the protein under non-denaturing conditions prior to exposure to proteases. The reduced protein was more readily digested by pepsin but again gave rise to a 70,000 and 55,000 molecular weight species. A different result was obtained with trypsin where the reduced protein was converted in part to a component migrating somewhat more rapidly than the α2 chain of type I collagen (Fig. 2).

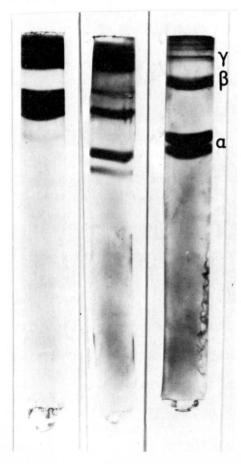

Fig. 2. Electrophoretic separation of collagenous components on polyacrylamide in sodium dodecyl sulfate. Migration is from top to bottom.

Left: The tumor collagen isolated by extraction with acetic acid was reduced and alkylated under non-denaturing conditions and an aliquot was electrophoresed.

Center: The reduced and alkylated tumor collagen was treated with trypsin (10% by weight) for 6 hrs at 15° at pH 8.0 and then electrophoresed. Note the loss of higher molecular weight components and the appearance of new lower molecular weight bands.

Right: Components of type I collagen used as standard α2 and α1 chains (94,500 M.W.) penetrate furthest.

Aliquots of the above preparations were dissolved in
dilute acetic acid and dialyzed against ATP to induce SLS
crystallite formation. Only amorphous deposits were observed
with the intact reduced and alkylated proteins and the non-
reduced protein digested with pepsin or trypsin. However,
when the reduced alkylated protein was treated with pepsin or
trypsin copious amounts of crystallites showing a distinct
cross-striation pattern were observed. Trypsin digestion
produced material forming crystallites about 3000 A in
length, while pepsin treatment produced shorter segments of
about 2000 A. The results suggest that the protein contains
a triple-helical domain and that non-helical disulfide
linked portions hinder proteolytic digestions. The band
pattern observed in the crystallites is distinct from that
obtained with other collagens (Fig. 3).

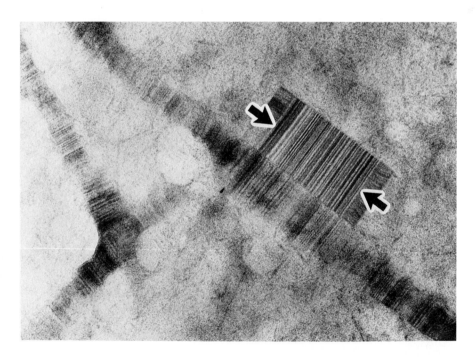

Fig. 3. An SLS-aggregate of basement membrane collagen.
The reduced and alkylated but native protein was incubated
with trypsin as indicated in the legend to Fig. 2. The enzyme
treated protein was dialyzed against 0.1% acetic acid and ATP
to produce SLS-crystallites.
An authentic type I collagen pattern (indicated by arrows)
has been placed on the picture adjacent to the basement mem-
brane collagen. Note the lack of correspondence of bands.

Immunofluorescent Studies

Rabbits were immunized with the DEAE-purified tumor
collagen and antiserum of high potency was obtained. This
antiserum was used to localize the protein in the tumor and
also to investigate the reaction of the antiserum with
authentic basement membranes in normal tissues from mouse and
man. Using the indirect immunofluorescent procedure, we
found that the matrix of the tumor reacted strongly with the
antiserum (Fig. 4). The same staining patterns were observed
with sera from patients with the Goodpasture syndrome known
to contain antibodies to basement membrane antigens. Similar
studies were done with sections of mouse and human skin,
lung and kidney and showed that the basement membranes in
these tissues reacted strongly with this antiserum.

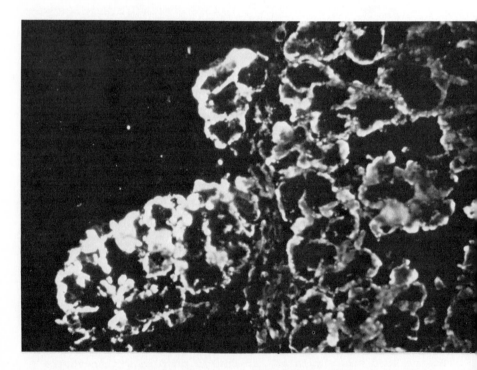

*Fig. 4. Localization of the basement membrane collagen
in the tumor. Antisera prepared to the basement membrane
collagen in rabbits was allowed to react with sections of the
tumor and then detected by the indirect fluorescence method
using fluorescein-conjugated goat anti-rabbit gamma globulin.*

SUMMARY

Our studies establish that the extracellular matrix of the EHS-sarcoma is a basement membrane-like structure. It contains a procollagen-like molecule with a large helical domain. Disulfide-crosslinks in this protein inhibit proteolysis while the reduction of disulfide bonds allows access of proteases to susceptible bonds. A helical segment of 3000 A can be isolated after tryptic-digestion while pepsin appears to produce cleavages in the helical domain. Since the physical studies indicate a typical collagen helix, it is unlikely that long non-helical stretches exist in the helical domain. As expected, immunofluorescent studies indicate that the tumor protein is predominantly located in the extracellular matrix of the tumor. We have found that the basement membranes in kidney, lung and skin from mouse and man react with the antisera. It may be that such basement membranes contain a similar molecule.

REFERENCES

1. Orkin, R.W., McGoodwin, E.B., Gehron, P., Martin, G.R., Valentine, T. and Swarm, R., *J. Exp. Med.* 145:204, 1977.
2. Swarm, R.L., *J. Natl. Cancer Inst.* 31:953, 1963.
3. Kefalides, N.A., *IN:* "International Review of Connective Tissue Research", (Hall, D.A. and Jackson, D.S., Editors) Academic Press, New York, Vol. VI, pp. 63-104, 1973.

III. Pathology of Basement Membranes
A. Alterations in Diabetes and Other Disease States

CHANGES OF

GLOMERULAR BASEMENT MEMBRANE IN DISEASE

Richard J. Glassock

University of California
School of Medicine
Los Angeles, California

*SUMMARY: This brief overview has attempted to illustrate
the spectrum of structural, chemical and functional
disturbances of the glomerular basement membrane in
selected human and experimental diseases. It is apparent
that many questions remain unanswered and that much needs
to be done to interrelate the parameters of structure,
chemistry and function into a meaningful and comprehensive picture of the response of basement membrane to
injury.*

The glomerular basement membrane (GBM) may be altered by
direct toxic injury (e.g., aminonucleoside of puromycin,
aminoglycoside antibiotics), immunologic injury (e.g., anti-
GBM antibody, immune complex deposition), congenital and
hereditary biochemical defects (e.g., diabetes mellitus,
Alport's syndrome), and by mechanisms which are largely unknown (e.g., "dense deposit" disease). This brief overview
of the changes in GBM in disease will focus on three areas:
mainly, abnormalities of structure; alterations of biochemistry; and pertubations of function. The abnormalities of
structure will be covered only superficially as others in
this symposium will address this area more directly. The
anatomical and biochemical changes of basement membrane which
accompany diabetes mellitus will be covered in depth in other
sessions of this symposium.

421

STRUCTURE

     Tables IA and B summarize some of the more important and
well characterized structural alterations of GBM and the
associated elements of the glomerular capillary wall in human
disease (1-34). Information pertaining to this area of
glomerular disease has been gleaned largely from detailed
light and electron microscopic analysis of renal biopsy
material. The spectrum of changes is quite extensive. Some
of these alterations will be covered elsewhere in the
symposium (e.g., diabetes), however, a few particularly
interesting disorders deserve special comment.
     Galle in 1962 first recognized the unique disorder of
GBM structure in an ultrastructural analysis of a form of
proliferative glomerulonephritis (12). In this study an extremely electron-dense "deposit" was noted within the substance of the GBM proper. The existence of this entity, often
referred to as "dense deposit disease" has been amply confirmed in succeeding years and is now generally regarded as one of
the variants of membranoproliferative glomerulonephritis
(MPGN) (15,16). It is a relatively uncommon disorder, chiefly
affecting children and young adults, and is most often clinically manifested by heavy proteinuria accompanied by recurring
bouts of hematuria (14-16). Some patients may present initially with features of the acute nephritic syndrome or
rapidly progressive glomerulonephritis (15,16). There may be
an association with partial lipodystrophy in some patients
(35). Striking abnormalities of the complement (C') system
are often present. These abnormalities consist of a persist-
ing reduction of the serum concentration of C3 ($Beta_1C/Beta_1A$
globulin) without an accompanying reduction of the early acting components of the classical pathway of C activation (Clq,
C2, C4) (36,37). A factor capable of cleaving native C3 may
be detected in the serum of such patients (C3 nephritic
factor, C3NeF) (36,37). This factor may be analagous to an
activated form of the initiating factor of the alternative
pathway of C' activation (38).
     By light microscopy, the lesions consist of variable
degrees of mesangial and endothelial cell proliferation with
an associated increase in mesangial matrix and conspicuous
thickening of the capillary wall (15,16,38). A striking
feature has been the strongly eosinophilic character of the
GBM by hematoxylin and eosin stains and the intense Periodic
acid-Schiff positivity (15,16,38). The GBM stains a blue-
green hue with Masson trichrome and has an increased affinity
for toluidine blue (16). The central aspect of the GBM
acquires a non-argyrophilic character (39). By immunofluores-
cense microscopy immunoglobulins, the early acting components

TABLE I

Alterations of Glomerular Capillary Wall
in Disease
Structural Changes

A. Glomerular Basement Membrane

Alteration	Disease - Prototype
Diffuse thickening	Diabetes mellitus (1-3)
Deposits of granular, fibrillar, striated or amorphous material	Immune complex (4-6) and anti-GBM (5,6) nephritis, amyloidosis (6), nail-patella (7) and other (8)
Focal discontinuities; "gaps"	RPGN*, acute GN (9)
Localized thinning rarefaction, splitting	Alport's syndrome (10,11)
Electron-dense transformation	MPGN** Type II (12-16)
Spike-like projections	Membranous nephropathy (17)
Wrinkling, accordianization	Nephrosclerosis, ischemia (18)
Polypoid change	Trisomy 21 and immune renal disease (19)

Rapidly progressive glomerulonephritis.
**Membranoproliferative glomerulonephritis.*

TABLE I

Alterations of Glomerular Capillary Wall
in Disease
Structural Changes

B. Cellular Constituents

1. *Epithelial cell*

Denudation; necrosis degeneration, detachment	Focal glomerular sclerosis (20,21) "Toxic" nephrosis (22,23)
Foot process effacement	Any proteinuric state (5), especially lipoid nephrosis (24)
"Myelin" figures	Fabry's disease (25,26)
Proliferation	RPGN (5,6)

2. *Endothelial cell*

Swelling and/or proliferation	Toxemia of pregnancy (27)
Intracytoplasmic inclusions (microtubular structures)	SLE and other disease (28,29)

3. *Mesangial cell*

Proliferation	Poststreptococcal GN, MPGN (5,6,30,31)
Circumferential interposition	MPGN (32)
Increased synthesis of matrix and/or basement membrane-like material	Diabetes (1-3), MPGN (5,6,30, 31)
Deposits of proteinaceous material	Immune complex nephritis (4-6) SLE (32,33), Berger's disease (34)

---

of complement, properdin and C3 proactivator (Properdin Factor
B) are not identified in the lesions (15,16). However, C3 is
found in a prominent linear fashion outlining the GBM (15,16).
Ultrastructural analysis provides the principal distinguishing
feature of this disorder; the electron-dense transformation of
the GBM, often referred to as a "deposit" (Figure 1). Similar
changes are observed in Bowman's capsule and tubular basement
membrane but not in any extrarenal site (14).

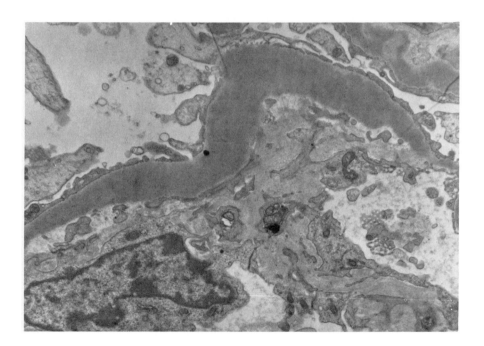

*Fig. 1. Electron photomicrograph of a peripheral glomeru-
lar capillary revealing electron "dense deposits" (X 85).
(Courtesy Arthur Cohen, M.D.)*

The biochemistry of the altered basement membrane in this disease has been extensively studied by Mahieu and Galle and will be summarized in a later section. The mechanisms by which the GBM is altered in this disease and its relationship to the complement abnormalities is entirely unknown. However, the observation that the structural abnormality recurs with great frequency in renal allotransplants is indicative that it is the result of a systemic disease process (14,40) in which prominent activation of the alternative pathway of C' activation is present (36,37,38). As emphasized by Galle and Mahieu this disorder is most properly viewed as a modification of the structural character of the GBM itself (14).

Another disease in which an abnormality of GBM structure is present is Alport's syndrome or hereditary nephritis and deafness (5,10,11,41,42). This disorder chiefly affects children and young adults (5,41,42). Males are more frequently affected than females (5,41,42). Recurrent gross or microscopic hematuria is the most common presenting feature (5,41,42). Detailed family studies have indicated an autosomal dominant inheritance with variable penetrance and expressivity (5,41). Progressive renal disease seems to be preferentially associated with the X chromosomes (5). The disease principally affects glomerular and tubular basement membrane (TBM) structure and function but in contrast to "dense deposit disease", other organs and systems may be affected. They include the organ of Corti, the cornea and the megakaryocyte (41,43). Morphologically, the GBM in this disease is irregularly thickened and thinned, frayed and split with marked distortion of the lamina densa. Lacunae and granulations within the GBM are also seen (10,11,43). The specificity of the ultrastructural abnormality for Alport's syndrome has been recently questioned (44). Lesions of a similar character are seen in the TBM but alterations of basement membrane structure outside the kidney have not been well studied (10,11,43). Immunoglobulins and complement are not found on immunofluorescent microscopy. The relationship of the defect in basement membrane structure of the kidney to the auditory, ocular and hematologic abnormalities is unknown. Recent observations on the immunochemistry of GBM in patients with Alport's will be discussed subsequently.

Experimentally induced alterations of GBM have also been extensively studied. Heterologous anti-GBM nephritis [nephrotoxic serum nephritis (NSN)] (45-49) and aminonucleoside of puromycin nephrosis [aminonucleoside nephrosis (AMN)] (22,23, 50-54) have been the most widely studied models. In acute NSN there is swelling of the visceral and parietal epithelial cell layer and the endothelial cell, alteration in the density of GBM and the development of a flocculant amorphous precipitate in the subendothelial area (46-49). Endothelial cells may appear to be partially detached from underlying GBM (49). Subsequently proliferation of endothelial and mesangial cells develops partially occluding the capillary lumina (45). In the complement-dependent variety of anti-GBM induced injury, polymorphonuclear leukocytes emigrate to sites of antibody deposition and induce further changes in GBM structure, perhaps via enzymatic processes (45,46). Complement independent, leukocyte independent alterations of GBM have also been induced by anti-GBM antibody (55).

In AMN the epithelial cells are distorted and replaced with a continuous sheet of cytoplasm (50-54). Areas of focal loss of epithelial cell cytoplasm may be seen (22,23). Slit pores are replaced by sites of closs approximation between adjacent epithelial cell membranes (22). These findings resemble those in human focal glomerular sclerosis (20,21).

In the autologous immune complex nephritis of rats [Heymann nephritis, (AICN)], the subepithelial deposits which characterize this disorder result in displacement of foot processes towards the urinary space. The slit diaphragm remains intact, but areas of epithelial cell detachment occur similar to that seen in AMN (56).

BIOCHEMISTRY

The biochemical alterations of GBM in human and experimental disease is summarized in Table II (57-75). The biochemical changes of basement membrane which occur in association with diabetes mellitus are discussed elsewhere in this symposium (3,57). Only fragmentary and incomplete data on the biochemistry of GBM is available concerning specific glomerulopathies. This area of investigation has been compromised by the lack of availability of tissue for study in the early stages of disease and the formidable problems encountered in obtaining purified material for study in patients with advanced disease. Not surprisingly then, the results of chemical analysis of such tissue has often been confusing and contradictory. In addition, many of the studies have been conducted in extremely heterogeneous populations of patients with "chronic glomerulonephritis" (CGN). Compared to normal, the GBM analyzed from patients with CGN (lobular and MPGN) has an increased content of proline, hydroxyproline and glycine and less lysine and hydroxylysine (58,59). This may be indicative of an increased collagen content known to be present by ultrastructural analysis of chronically diseased kidneys. The glycopeptides isolated

TABLE II

*Alterations of Glomerular Capillary Wall
in Disease*

Alteration	*Disease - Prototype*
Abnormalities of amino acid and/or carbohydrate composition	Diabetes mellitus (3), ? (57) Glomerulonephritis, human and experimental (58-60, 73-75)
Decreased sialoprotein	Glomerulonephritis, aging (61) Diabetes (57) Experimental GN (62-64)
Changes in lipid phosphorous and cholesterol	Human and experimental GN (46, 47, 65-67)
Increased affinity for serum proteins (albumin, lgG)	Diabetes (68,69)
Decreased content of antigen reactive with anti-GBM antibody	Alport's syndrome (70) MPGN (71)
Decreased cystine content	Diabetes (57), Dense deposit disease (14,57)
Disturbed ratio of disaccharide to heteropolysaccharide units	Diabetes (57), ? (3) Glomerulonephritis (58) Congenital nephrosis (72)

from glomerulonephritic basement membranes by collagenase
digestion, boiling or urea solubilization appear to have a
higher molecular weight than normal GBM. Although these
glycopeptides appear to have a relatively normal composition,
there may be a distortion in the ratio of the disaccharide to
heteropolysaccharide units in glomerulonephritic GBM (58,59).
On the basis of these data it has been suggested that in CGN
there is an accumulation of basement membrane material similar
in composition to that of the normal kidney either due to
increased synthesis or decreased degradation. There may be
an increase in the cholesterol and hexosamine content of CGN
basement membrane; however, it is thought that cholesterol is
not a constituent of normal GBM (65). Westberg and Michael
have reported that in chronic glomerulonephritis (particularly
MPGN) the collagen related amino acids; hydroxyproline,
hydroxylysine and glycine are decreased while tyrosine,
arginine and lysine content is increased (73). They also
noted a decrease in disaccharide content of diseased GBM (73).

Most studies have demonstrated a tendency for a decrease
in the sialic acid content of isolated GBM from CGN (58-60).
These observations correspond to those made in both human and
experimental proteinuric renal disease where a decrease in
histochemically detectable sialoprotein content (e.g.,
colloidal iron reaction) of glomeruli has been noted (60).
At the present time it is not clear if this reduction is due
to a decrease in the sialoprotein layer of the glomerular
epithelial cell or a decrease in the sialic acid residue
thought to be integral to the GBM.

Recently, Mahieu and Galle have reported on a careful
study of the composition of the abnormal GBM in "dense deposit
disease" and the results are summarized in Table III (14).
Plasma proteins could not be eluted from such GBM by acidic
buffers. The principal changes when compared to normal GBM
were increased content of sialic acid buffers. The principal
changes when compared to normal GBM were increased content of
sialic acid and reduced content of half-cystine. The staining
characteristics of these membranes are indicative of an
increase in content of 1,2 glycol or aminoalcohol groups and
a reduced number of aldehyde groups susceptible to oxidation.
The increased affinity for the heavy metals used (lead citrate,
uranyl acetate) in preparation for electron microscopy is
presently unexplained.

Recently, an important observation has been made regarding the immunochemistry of GBM in Alport's syndrome by McCoy
and colleagues (70). They found that GBM from such patients
is deficient in an antigen reactive with the anti-GBM antibody harvested from patients with Goodpasture's syndrome
(homologous anti-GBM antibody). It is believed that this

TABLE III

*Electron-dense Alterations of Glomerular
Basement Membrane
Biochemistry**

No plasma proteins elutable from isolated glomeruli

Decreased cystine content

Increased sialic acid

Normal disaccharide unit

Normal hydroxyproline

Normal glycine

Decreased affinity for silver after periodic acid
  oxidation

Increased (?) content of 1,2 glycol or amino-alcohol
  groups

Increased affinity for toluidine blue

---

*Data of Galle and Mahieu (14).

antibody is primarily reactive with the non-collagenous glycopeptide (heteropolysaccharide) constituent of GBM (78),
although some investigators have also demonstrated reactivity
of these antibodies with the disaccharide-peptide moiety of
GBM (79,80). The development of circulating and tissue bound
anti-GBM antibody in a few patients with Alport's syndrome
who received renal allotransplants containing a normal GBM
composition is further evidence favoring an inherited deficiency of one or more glycopeptide antigenic components of
GBM in this disease (70). The specificity of this immuno-
histochemical findings is open to question as Scheinman, Fish
and Michael have previously demonstrated a striking decrease
in heterologous anti-GBM and anti-collagen reactive antigen
sites in the GBM of patients with MPGN, some of whom might
have had "dense deposit disease" (71). In this study, no
decrease in anti-GBM or anti-collagen reactive antigens was
found in diabetes and membranous GN, but there was a striking

increase in anti-actomyosin reactive sites in the mesangium
of diabetic glomerulopathy. Only limited studies were conducted using a homologous anti-GBM antibody but in general
the results were similar to those found when heterologous
anti-GBM was used as the immunological probe. Finally, GBM
isolated from patients with congenital nephrotic syndrome has
revealed an increased content of hydroxylysine, hydroxyproline
and the disaccharide unit. No changes in sialic acid content
were noted (72).

The biochemistry of GBM has also been investigated in
some detail in several models of experimental glomerulonephritis. In the proliferative glomerulonephritis accompanying NSN there is a decrease in the phospholipid content of
glomeruli, a decrease in the sialic acid content and an increase in the turnover of hydroxyproline (46-48, 63). X-ray
diffraction studies have suggested an increased ordering
within the lattice framework of abnormal GBM (67). There is
a decrease in the electrophoretic mobility of isolated GBM in
this model (67). Fragments of GBM may be found in the urine
indicative of enzymatic degradation of GBM by proteases
present in polymorphonuclear leukocytes (48). The possible
relationship of these biochemical changes to function will be
commented upon subsequently.

Studies of AMN have revealed an alteration in the composition of the disaccharide unit and both an increase in the
incorporation of proline and hydroxylysine in the GBM (63,74,
75). These observations support an increased synthesis or
decreased degradation of GBM glycoproteins in this experimental model. The sialoprotein content of glomeruli is regularly
reduced in association with the development of proteinuria
and there is evidence of increased sialic acid turnover (62).
The urine may contain an increased quantity of GBM-like glycoproteins with an increase in the excretion rate of acidic and
neutral proteases (74). In AICN the reduction of sialoprotein
follows rather than precedes the development of proteinuria
(77).

FUNCTION

An abnormality of the molecular sieving properties of the glomerular capillary wall, manifested by proteinuria, is one of the hallmarks of glomerular disease (See also Table IV). As pointed out earlier in this symposium the complex organization of the glomerular capillary wall as a three-layered structure is responsible for the restriction of transglomerular passage of protein macromolecules (reviewed in 81). Thus it might be expected that abnormalities of one component independent of the other may lead to a recognizable abnormality of function.

TABLE IV

Alterations of Glomerular Capillary Wall in Disease Functional Changes in Experimental Models

Alteration	Disease Prototype
Localized increase in permeability to neutral macromolecules (ferritin, lanthanum)	NSN* (48) AICN** (90) AMN*** (76)
Increased permeability to cationic probes (enzymes)	AMN (91,92)
Decreased permeability to neutral macromolecules (dextran, PVP)	NSN (84) AMN (86)
Increased permeability to anionic macromolecules (dextran sulfate)	NSN (85) AMN (86)

* NSN = nephrotoxic serum nephritis.
** AICN = autologous immune complex nephritis.
*** AMN = aminonucleoside of puromycin nephrosis.

In order to examine the functional properties of the glomerular capillary wall in disease we have recently studied an experimental model of glomerulonephritis utilizing as molecular probes neutral and anionic dextrans (82-86). Polydisperse tritiated dextran was infused into a strain of rats possessing surface glomeruli (Munich-Wistar) in which a mild to moderate form of a proliferative glomerulonephritis had been induced by small amounts of heterologous anti-GBM antibody (i.e., NSN) (84). The fractional clearance of these neutral dextran molecules, ranging in effective molecular radii from 18 to 42 Å, were then compared to values for normal rats of the same strain. The intraglomerular pressures and flows which determine the rate of glomerular ultrafiltration (GFR) were also measured by micropuncture techniques in order to determine the influence of pertubations of hemodynamics upon macromolecular solute transfer. In NSN no restriction to the transglomerular passage of neutral molecules less than a radius of 17 Å was found. However, reduced transglomerular transport was found for molecules between 28 and 42 Å (Figure 2). These changes could not be accounted for by the resultant hemodynamic changes. Using a mathematical model of solute transport based on pore theory the average effective pore radius ($\bar{R}_o$) of the glomerular capillary wall was found to be similar in NSN and controls (approximately 50 Å) (Figure 3). However, the ratio of total pore area to pore length (S'/l) was reduced to one-third of normal in NSN, corresponding to a measured decline in the hydraulic permeability co-efficient based on the hemodynamic measurements. These observations indicated that proteinuria in this model was <u>not</u> due to an increase in number of pores or the pore radius.

Similar studies were then conducted using an anionic probe, dextran sulfate, in rats with similar degrees of NSN (85). In contrast to the <u>reduced</u> fractional clearance of neutral dextran, <u>enhanced</u> transglomerular passage of these anionic polymers was noted in NSN when compared to normal rats (Figure 4). The sialoprotein content of glomeruli was reduced by colloidal iron staining. Similar abnormalities of transglomerular transport of neutral and anionic dextran have been recently described in AMN (86). Taken together these studies indicate that the abnormal filtration of anionic serum proteins is at least in part due to a loss of the fixed negative charges characterizing the glomerular capillary wall. The nature and anatomic site of these negatively charged substances is largely unknown, but they may reside in the glomerular basement membrane proper or in association with the epithelial cell podocyte and slit pore membrane. The relationship of the reduced content of histochemically identifiable sialoprotein to the alterations in transglomerular transport of protein is uncertain. Protein overload proteinuria is associated with a reduction in sialoprotein (87). As noted previously, in AICN the decrease in sialoprotein follows rather than precedes the structural lesion of GBM and proteinuria (77).

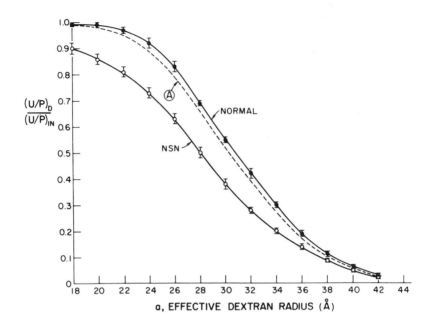

Fig. 2. *Comparison of fractional dextran clearance plotted as a function of effective dextran radius for NSN rats (O) and for a group of non-nephritic control rats (●). Values are expressed as ± 1 S.E. Curve A (----) is a plot of the effect upon dextran transport of changes solely in the hemodynamic determinants of GFR. (From J. Clin. Invest. 57:1272, 1976. Reprinted with permission of publisher).*

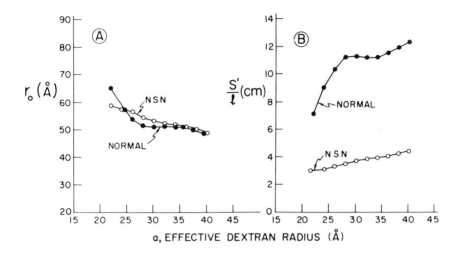

Fig. 3. (A) The relationship between r_O, the pore radius, and the effective dextran radius, a, for NSN rats and non-nephritic control rats. (B) The relationship between S'/l, ratio of the total pore area to pore length, and effective dextran radius, a, for NSN rats and non-nephritic control rats. (From J. Clin. Invest. 57:1272, 1976. Reprinted with permission of publisher).

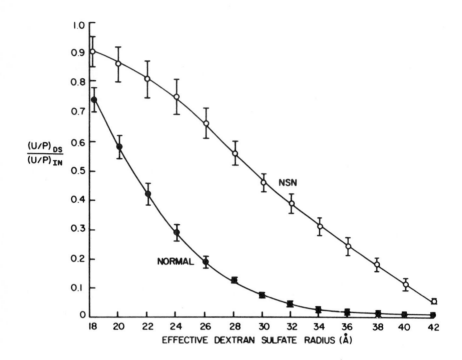

Fig. 4.   Comparison of fractional dextran
sulfate clearances plotted as a function of
effective dextran sulfate radius for NSN rats
(O) and for a group of non-nephritic control
rats (●).   Values are expressed as means ± 1 S.E.
(From J. Clin. Invest. 57:1287, 1976.   Reprinted
with permission of publisher).

Human proteinuric renal disease has also been studied using neutral molecular probes [e.g., dextran, polyvinylpyrollidone (PVP)] (88,89). Reduced transglomerular transport of neutral PVP was noted in a study of "minimal change disease" in children (89). Others have noted reduced clearance of lower molecular weight polymers and increased clearance of higher molecular weight polymers in proteinuric renal disease associated with overt structural lesions of the glomeruli (88). Localized and/or diffusely enhanced permeability to neutral and cationic ultrastructural probes (e.g., lanthanum, ferritin, enzymes) have been noted in experimental models of proteinuria (22,48,54,76,90,92-94).

ACKNOWLEDGEMENTS

Supported in part by NIH Grant No. AM-16565, The National Kidney Foundation and Kidney Foundation of Southern California.

REFERENCES

1. Farquhar, M.G., Hopper, J., Jr., Moon, H.D., *Am. J. Pathol.* 35:721-753, 1959.
2. Larsson, O., *Acta Med. Scand. Suppl.* 480:1-66, 1967.
3. Spiro, R.G., *Diabetologia* 12:1-14, 1976.
4. Dixon, F.J., Feldman, J.D. and Vasquez, J.J., *J. Exp. Med.* 113:899, 1961.
5. Glassock, R.J. and Bennett, C.M., *In:* "The Kidney", (B.M. Brenner and F.C. Rector, Jr., Eds.) W. B. Saunders Co., Philadelphia, Chapter 23, p. 941, 1976.
6. Heptinstall, R.H., *In:* "Pathology of the Kidney" (2nd edition), Boston, Little, Brown & Co., Pp. 737, 1974.
7. Ben-Bassat, M., Cohen, L. and Rosenfeld, J., *Arch. Pathol.* 92:350, 1971.
8. Bariety, J. and Callard, P., *Lab. Invest.* 32:636, 1975.
9. Stejskal, J., Pirani, C.L., Okada, M., Mandelanakis, N. and Pollak, V.E., *Lab. Invest.* 28:149, 1973.
10. Hinglais, N., Grunfeld, J.P. and Bois, E., *Lab. Invest.* 27:473, 1973.
11. Spear, G.S., *Clin. Nephrol.* 1:336, 1973.
12. Galle, P., "Mise en évidence an microscope électronique de'une lésion singuliere des membranes basales du rein et de la substance hyaline", Thesis, Paris, 1962.
13. Berger, J. and Galle, P., *Presse. Méd.* 71:2351, 1963.
14. Galle, P. and Mahieu, P., *Am. J. Med.* 58:749, 1975.

15. Habib, R., Gubler, M.C., Loirat, C., Ben Maiz, H. and Levy, M., *Kidney Int.* 7:204, 1975.
16. Vargas, R., Thomson, K.J., Wilson, D., Cameron, J.S., Turner, D.R., Gill, D., Chantler, C. and Ogg, C., *Clin. Nephrol.* 5:73, 1976.
17. Ehrenreich, T. and Churg, J., *In:* "Pathology Annual", (S.C. Sommers, Ed.) Appleton-Century-Crofts, New York, p. 145, 1968.
18. Heptinstall, R.H., *In:* "Pathology of the Kidney" (2nd Edition), Boston, Little, Brown & Co., p. 130, 1974.
19. Martin, S.A. and Kissane, J.M., *Arch. Pathol.* 99:249, 1975.
20. Grishman, E. and Churg, J., *Kidney Int.* 7:111, 1975.
21. Cohen, A.H., Mampaso, F. and Zamboni, L., "An ultrastructural study", submitted for publication, 1976.
22. Ryan, G.B. and Karnovsky, M.J., *Kidney Int.* 8:219, 1975.
23. Caulfield, J.J., Ried, J.J. and Farquhar, M.G., *Lab. Invest.* 34:43, 1976.
24. Folli, G., Pollak, V.E., Reid, R.T., Pirani, C.L. and Kark, R.M., *Ann. Intern. Med.* 49:775, 1958.
25. Hartley, M.W., Miller, R.E. and Lupton, C.H., *Lab. Invest.* 12:850, 1963.
26. Duncan, C., *Pathology* 2:9, 1970.
27. Spargo, B., McCartney, C.P. and Winemiller, R., *Arch. Pathol.* 68:593, 1959.
28. Györkey, F., *New Engl. J. Med.* 283:333, 1969.
29. Bariety, J., Ricker, D., Appay, M.D., Gossette, J. and Callard, P., *J. Clin. Pathol.* 26:21, 1973.
30. Mandalenakis, N., Mendoza, N., Pirani, C.L. and Pollak, V.E., *Medicine* 50:319, 1971.
31. Habib, R., Kleinknecht, C., Gubler, M.D. and Levy, M., *Clin. Nephrol.* 1:194, 1973.
32. Arakawa, M. and Kimmelstiel, P., *Lab. Invest.* 21:276, 1969.
32a. Koffler, D., Agnello, V., Carr, I. and Kunkel, H.G., *Am. J. Pathol.* 56:305, 1969.
33. Dujovne, I., Pollak, V.E., Pirani, C.L. and Dillard, M.G., *Kidney Int.* 2:33, 1972.
34. Berger, J. and Hinglais, N., *J. Urol. Nephrol.* (Paris) 74:694, 1968.
35. Sissons, J.G.P., West, R.J., Fallows, J., Williams, D.G., Boucher, B., Amos, N. and Peters, D.K., *N. Engl. J. Med.* 294:461, 1976.
36. Ooi, Y.M., Vallota, E.H. and West, C.D., *Kidney Int.* 9:46, 1976.
37. Gwynn Williams, D., Peters, D.K., Fallows, J., Petrie, A., Kourilsky, O., Morel-Maroger, L. and Cameron, S., *Clin. Exp. Immunol.* 18:391, 1974.

38. Schreiber, R.D., Gotze, O. and Muller-Eberhard, H.J., *J. Exp. Med.* 144:1062, 1976.
39. Anders, D. and Thoenes, W., *Virchows Arch. (Pathol. Anat.)* 369:87, 1975.
40. Turner, D.R., Cameron, J.S., Bewick, M., Sharpstone, P., Melcher, D., Ogg, C., Evans, D.J., Thafford, A.J.P. and Leibowitz, S., *Kidney Int.* 9:439, 1976.
41. Morris, R.C., Jr., McInnes, R.R., Epstein, C.J., Sebastian, A. and Scrwer, C.R., *In:* "The Kidney" (B.M. Brenner and F.C. Rector, Jr., Eds.), W. B. Saunders Co., Philadelphia, p. 1232, 1976.
42. Epstein, C.J., Sahud, M., Piel, C.F., Goodman, J.R., Bernfield, M.R., Kishner, J.H. and Albin, A., *Am. J. Med.* 52:299, 1972.
43. Rumpelt, H.J., Langer, K.H., Scharer, K., Straub, E. and Thoenes, W., *Virchows Arch. (Pathol. Anat.)* 364:225, 1974.
44. Hill, G.S., Jenis, E.H. and Goodloe, S., Jr., *Lab. Invest.* 31:516, 1974.
45. Unanue, E.R. and Dixon, F.J., *In:* "Advances in Immunology", (F.J. Dixon, Jr., and J. Humphrey, Eds.), Academic Press, New York, p. 1, 1967.
46. Gang, N.F. and Kalant, N., *Lab. Invest.* 22:531, 1970.
47. Gang, N.F., Mautner, W. and Kalant, N., *Lab. Invest.* 23:150, 1970.
48. Gang, N.F., Trachtenberg, E., Allerhand, J., Kalant, N. and Mautner, W., *Lab. Invest.* 23:436, 1970.
49. Blantz, R.C. and Wilson, C.B., *J. Clin. Invest.* 58:899, 1976.
50. Vernier, R.L., Papermaster, B.W. and Good, R.A., *J. Exp. Med.* 109:115, 1959.
51. Arakawa, M., *Lab. Invest.* 23:489, 1970.
52. Ryan, G.B., Rodewald, R. and Karnovsky, M.J., *Lab. Invest.* 33:461, 1975.
53. Ryan, G.B., Leventhal, M. and Karnovsky, M.J., *Lab. Invest.* 32:397, 1975.
54. Gang, N.F., Trachtenberg, E., Wheatley, P.J. and Mautner, W., *Proc. Soc. Exp. Biol. Med.* 140:449, 1972.
55. Simpson, I.J., Amos, N., Evans, D.J., Thompson, H.M. and Peters, D.K., *Clin. Exp. Immunol.* 19:499, 1975.
56. Schneeberger, E.E. and Grupe, W.E., *Lab. Invest.* 34:298, 1976.
57. Kefalides, N., *J. Clin. Invest.* 53:403, 1974.
58. Mahieu, P., Winand, R.J. and Nusgens, B., *In:* "Advances in Nephrology", (J. Hamburger, J. Crosnier and M.H. Maxwell, Eds.), Year Book Publishers, Chicago, p. 25, 1972.
59. Mahieu, P., *Kidney Int.* 1:115, 1972.

60. Blau, E.B. and Haas, J.E., *Lab. Invest.* 28:477, 1973.
61. DeBats, A., Gordon, A.H. and Rhodes, E.L., *Clin. Sci. Mol. Med.* 47:93, 1974.
62. Michael, A.F., Blaw, E. and Vernier, R.L., *Lab. Invest.* 23:649, 1970.
63. Lui, S. and Kalant, N., *Exp. Mol. Pathol.* 21:52, 1974.
64. Misra, R.P. and Kalant, N., *Nephron* 3:84, 1966.
65. Misra, R.P. and Berman, L.B., *Lab. Invest.* 18:131, 1968.
66. Misra, R.P. and Berman, L.B., *Am. J. Med.* 47:337, 1969.
67. Kalant, N., Misra, R.P., St. J. Manley, R. and Wilson, J., *Nephron* 3:167, 1966.
68. Westberg, N.G. and Michael, A.F., *Diabetes* 21:163, 1972.
69. Miller, K. and Michael, A.F., *Diabetes* 25:701, 1976.
70. McCoy, R.C., Johnson, H.K., Stone, W.J. and Wilson, C.B., *Lab. Invest.* 34:325, 1976.
71. Scheinman, J.I., Fish, A.J. and Michael, A.F., *J. Clin. Invest.* 54:1144, 1974.
72. Mahieu, P., Monnens, L. and van Haelst, V., *Clin. Nephrol.* 5:134, 1976.
73. Westberg, N.G. and Michael, A.F., *Acta Med. Scand.* 193:49, 1973.
74. Kefalides, N.A. and Forsell-Knott, L., *Biochim. Biophys. Acta* 203:62, 1970.
75. Blau, E.B. and Michael, A.F., *Proc. Soc. Exp. Biol. Med.* 141:164, 1972.
76. Gang, N.F. and Mautner, W., *Lab. Invest.* 27:310, 1972.
77. Couser, W.G., Stilmant, M. and Darby, C., *Lab. Invest.* 34:23, 1976.
78. Marquardt, H., Wilson, C.B. and Dixon, F.J., *Kidney Int.* 3:57, 1973.
79. Mahieu, P.M., Lambert, P.H. and Maghuin-Rogister, G.R., *Eur. J. Biochem.* 40:599, 1973.
80. McIntosh, R.M. and Griswold, W., *Arch. Pathol.* 92:329, 1971.
81. Karnovsky, M.J. and Ainsworth, S.K., *In:* "Advances in Nephrology", (J. Hamburger, J. Crosnier and M.H. Maxwell, Eds.), Year Book Medical Publishers, Chicago, p. 35, 1972.
82. Chang, R.L.S., Ueki, I.F., Troy, J.L., Deen, W.M., Robertson, C.R. and Brenner, B.M., *Biophys. J.* 15:887, 1975.
83. Chang, R.L.S., Deen, W.M., Robertson, C.R. and Brenner, B.M., *Kidney Int.* 8:212, 1975.
84. Chang, R.L.S., Deen, W.M., Robertson, C.R., Bennett, C.M., Glassock, R.J. and Brenner, B.M., *J. Clin. Invest.* 57:1272, 1976.

85. Bennett, C.M., Glassock, R.J., Chang, R.L.S., Deen, W.M., Robertson, C.R. and Brenner, B.M., *J. Clin. Invest.* 57:1287, 1976.

86. Baylis, C., Bohrer, M.P., Troy, J.L., Robertson, C.R. and Brenner, B.M., *Am. Soc. Nephrol,*, 9th Ann. Meeting, p. 68, 1976.

87. Roy, L.P., Vernier, R.L. and Michael, A.F., *Proc. Soc. Exp. Biol. Med.* 141:870, 1972.

88. Hardwicke, J., *In:* "Advances in Nephrology (Vol. 2), (J. Hamburger, J. Crosnier and M.H. Maxwell, Eds.), Year Book Medical Publishers, Chicago, p. 61, 1972.

89. Robson, A.M., Giangiacomo, J., Kienstra, R.A., Nagui, S.T. and Ingelfinger, T.R., *J. Clin. Invest.* 54:1190, 1974.

90. Schneeberger, E.E., Leber, P.D., Karnovsky, M.J. and McCluskey, R.T.,*J. Exp. Med.* 139:1283, 1974.

91. Venkatachalam, M.A., Karnovsky, MlJ. and Cotran, R.S., *J. Exp. Med.* 130:381,' 1969.

92. Venkatachalam, M.A., Cotran, R.S. and Karnovsky, M.J., *J. Exp. Med.* 132:1168, 1970.

93. Schneeberger, E.E., *Nephron* 13:7, 1974.

94. Farquhar, M.G. and Palade, G.E., *J. Exp. Med.* 114:699, 1961.

THE ORIGIN OF THE URINARY

GLUCOSYL-GALACTOSYL-HYDROXYLYSINE:

INTERSTITIAL COLLAGEN, GBM OR C1q?

Robert Askenasi

Hopital Brugmann
Universite Libre de Bruxelles
Bruxelles, Belgium

SUMMARY: Urinary glucosyl-galactosyl-hydroxylysine
originates mainly from interstitial collagen and there
is no doubt that skin collagen may be an important
source at least in pathologic states.   Less than 10%
of the glucosyl-galactosyl-hydroxylysine excreted in
urine could come from GBM and C1q.   Our data suggest a
more significant contribution of the latter.

INTRODUCTION

Glucosyl-galactosyl-hydroxylysine (glc-gal-hyl) and
galactosyl-hydroxylysine (gal-hyl) are two substances present
only in interstitial collagens and in basement membranes.
Their proportions in these proteins vary and it is well known
that the disaccharide unit predominates in basement membranes
(1,2,3), while the monosaccharide form is more abundant in
bond collagen than in skin collagen (4,5). Both hydroxylysyl
glycosides are present in urine (5,6) where they constitute
the most abundant form of hydroxylysine excreted in normal
subjects (6,7).

Since the amount of collagen in the interstitium is
much greater than the total amount of basement membrane (8),
it may be considered that both metabolites recovered from
urine are mainly derived from the former. It is generally
accepted that gal-hyl is predominantly excreted in bone
diseases and glc-gal-hyl in skin diseases (5,6). Although
it is known that gal-hyl, excreted in large amounts in
Paget's disease, originates from bone collagen, the origin of
glc-gal-hyl is less clear. Indeed, in Ehlers-Danlos Type VI,
due to a deficiency in hydroxylysine which affects mainly
skin collagen(9), the urinary excretion of hydroxylysine is
practically normal. Urinary glc-gal-hyl could consequently
have an origin other than skin collagen. Glomerular basement
membrane (GBM) particularly rich in glycosylated hydroxylysine (1,3), could directly yield significant amounts of
glc-gal-hyl in urine. On the other hand, the C1q subcomponent
of the complement system which has a composition similar to
that of a basement membrane (10) could be another source.
The present study was designed to verify these hypotheses.

MATERIALS AND METHODS

Hydroxylysyl glycosides were measured according to the
chromatographic method of Askenasi (11).
Determinations were performed in urine in the following
conditions:
- subcutaneous inflammatory granuloma were induced
  in young rats and the urinary excretion of
  hydroxylysyl glycoside was observed during the
  period of acute inflammation as described in (12);
- *in vitro* perfusions of isolated dog kidney allowed
  the measurement of hydroxylysyl glycosides independently of the important amounts of interstitial
  collagen present in the whole body as described
  in (13);
- the excretion of glc-gal-hyl and gal-hyl was
  determined after kidney transplantation in two
  anephric patients;
- the proportions of glycosylated and non-glycosy-
  lated hydroxylysine were calculated in the non-
  dialyzable fraction of the urine from two normal
  subjects and from two patients suffering from
  Paget's disease as described in (14).

RESULTS

The urinary excretion of glc-gal-hyl and gal-hyl during the first 15 days of acute inflammation is shown in Figure 1.

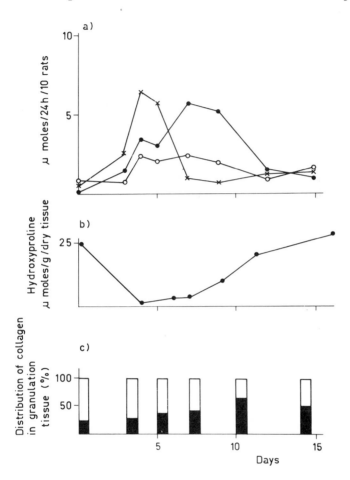

Fig. 1.   a) Urinary excretion of hydroxyproline (0), glc-gal-hyl (X) and gal-hyl (0) during acute inflammation.
b) Total collagen in granulation tissue measured by its content in hydroxyproline (according to Bailey, A.J., Bazin, S. and Delaunay, A., Biochimica et Biophysica Acta 328:383-390, 1973.
c) Soluble (open columns) and insoluble (solid columns) fractions of collagen in granulation tissue (according to Bailey, A.J., Bazin, S. and Delaunay, A., Biochimica et Biophysica Acta 328:383-390, 1973.

Normal rats excrete a little more gal-hyl than glc-gal-hyl. From the second to the sixth day, excretion of glc-gal-hyl is much higher than the excretion of gal-hyl. From the seventh to the fifteenth day, excretion of hydroxylysyl glycosides decreases, but urinary excretion of gal-hyl rises above the excretion of glc-gal-hyl. Urinary excretion of total hydroxyproline follows the pattern of gal-hyl until the fifth day but continues to rise when the hydroxylysine glycosides are returning to normal levels until the tenth day when it begins to decrease to near normal levels.

Figure 2 shows the urinary excretion of total hydroxylysyl glycosides (first column), gal-hyl (second column) and glc-gal-hyl (third column) during three periods of 30 minutes each by isolated dog kidneys. *In vitro* excretions are compared to excretions with the kidneys *in situ*. In period III, total hydroxylysyl glycoside excretion represents 10% of the amount excreted *in situ* while glc-gal-hyl excretion represents 7% of the glc-gal-hyl excreted *in situ*.

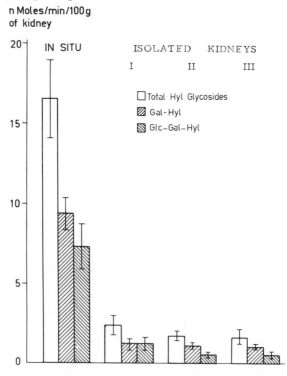

Fig. 2. *Urinary excretion of hydroxylysyl glucosides (first column) gal-hyl (second column) and glcgal-hyl (third column)* in situ *and* in vitro. *Height of column indicates the mean value.*

Figures 3 and 4 illustrate urinary excretion of hydroxylysyl glycosides after renal transplantation in two human
subjects. Their excretion parallels creatinine excretion.
In Figure 3, on day 2 after transplantation and before
rejection, the excretion of glc-gal-hyl is going down while
the excretion of gal-hyl follows the excretion of creatinine.
The same pattern is observed with the second patient but the
excretion of glc-gal-hyl remains low during the whole period
of rejection.

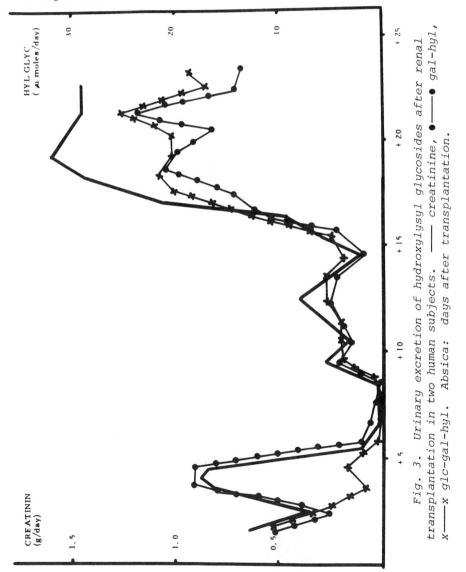

Fig. 3. Urinary excretion of hydroxylysyl glycosides after renal
transplantation in two human subjects. ——— creatinine, ●——● gal-hyl,
x——x glc-gal-hyl. Absica: days after transplantation.

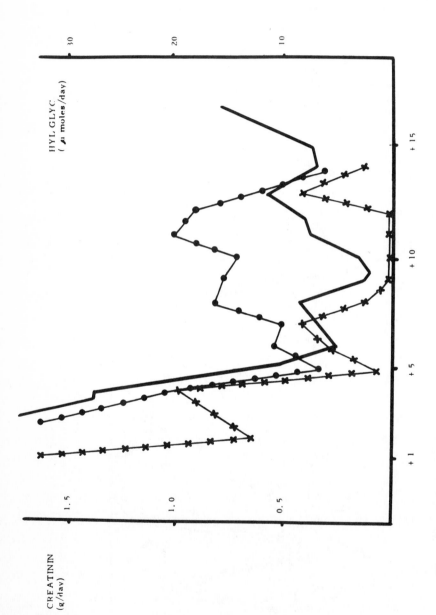

Fig. 4. Urinary excretion of hydroxylysyl glycosides after renal transplantation in two human subjects. —— creatinine, •——• gal-hyl, x——x glc-gal-hyl. Absica: days after transplantation.

Table I gives the distribution of hydroxylysyl glycosides in the non-dialyzable fraction of urine. The degree of glycosylation of this non-dialyzable hydroxylysine is comparable in retentates coming from normal subjects and from urine from subjects with Paget's disease. However, the glc-gal-hyl/gal-hyl ratio is higher in the urine from normal subjects.

TABLE I

DISTRIBUTION OF HYDROXYLYSYL GLYCOSIDES
IN NON-DIALYZABLE HYDROXYLYSINE

(μ moles/24 h)

	Total Hydroxylysine		glc-gal-hyl		gal-hyl		Unglycosylated Hydroxylysine	
Control	2.6	100%	1.7	65%	0.2	8%	0.7	27%
Control	1.75	100%	1.16	66%	0.15	9%	0.44	25%
Paget's disease	9.5	100%	3.2	34%	2.8	29%	3.5	37%
Paget's disease	15.4	100%	9.0	58%	3.0	19%	3.4	23%

It is noteworthy that the excretion of glc-gal-hyl is considerably increased during the first week of an acute skin inflammation. During this period, leucocytes, and macrophages release many degradatvie enzymes and the locally pre-existing collagen is degraded. This degradation process affects also the newly formed collagen, not yet fully polymerized, in the granuloma resulting from inflammation. There seems to be a correlation between this collagen degradation in inflammed tissue and increased excretion of glc-gal-hyl. Furthermore, the decreased excretion of hydroxylysyl glycosides (and of hydroxyproline) corresponds to the accumulation of the newly synthesized collagen of much lower solubility than the collagen of normal subcutaneous tissue.

In vitro perfusion of the isolated kidney allows the measurement of hydroxylysyl glycosides independently of the important amount of interstitial collagen present in the whole body. In this experimental system the only other source of glc-gal-hyl and gal-hyl is the blood necessary to perfuse the organ. This blood contains only traces of free hydroxylysine glycosides and the glc-gal-hyl contained in Clq. The total amount of Clq in a dog of 20 kg can be estimated as follows: Clq in 100 ml of plasma: 0.3% of total protein (10) $\simeq$ 20 mg plasma volume in a dog of 20 kg = 800 ml. Clq contained in a whole dog of 20 kg $\simeq$ 160 mg.

The amount contained in the experimental system of isolated kidney: 250 ml plasma contains: $\simeq$ 50 mg Clq.

The excretion of glc-gal-hyl by the kidney *in situ* results from the degradation of: Interstitial collagen + GBM (60 mg/100 g kidney) + Clq (160 mg) = 100%.

The excretion of glc-gal-hyl by the isolated kidney results from the degradation of: GBM (60 mg/100 g kidney) + Clq (50 mg) = 7%.

It is obvious that the main source of glc-gal-hyl is the interstitial collagen.

There are several arguments to believe that Clq contributes more than GBM to the 7% of glc-gal-hyl not coming from the body collagen. Indeed, the turnover of GBM seems very slow (15,16) and glc-gal-hyl is excreted less than gal-hyl during an allograft rejection which is in agreement with the fact that the complement generally decreases at this moment.

The non-dialyzable fraction of urine contains large polypeptides related to collagen synthesis (17) and supposedly arising from bone collagen. Although bone collagen is characterized by high amounts of gal-hyl, these large fragments are rich in glc-gal-hyl. It can be assumed that a small amount of glc-gal-hyl could come from degradation of polypeptides synthesized by osteoblasts but not incorporated in bone collagen because of their abnormal carbohydrate composition.

REFERENCES

1. Kefalides, N.A., *In:* "Chemistry and Molecular Biology of the Intercellular Matrix", (Balazs, E. A., ed.) Academic Press, London, p. 535, 1970.
2. Kefalides, N.A. and Forsell-Knott, L., *Biochim. Biophys. Acta* 203:62-66, 1970.
3. Spiro, R.G., *J. Biol. Chem.* 244:602-612, 1969.
4. Pinnell, R.S., Fox, R. and Krane, S., *Biochim. Biophys. Acta* 229:119-122, 1971.
5. Segrest, J.P. and Cunningham, L.W., *J. Clin. Invest.* 49: 1497-1509, 1970.
6. Askenasi, R., *J. Lab. Clin. Med.* 83:673-679, 1974.
7. Askenasi, R., *Clin. Chim. Acta* 59:87-92, 1975.
8. Man, M. and Adams, E., *Biochem. Biophys. Res. Comm.* 66: 9-15, 1975.
9. Pinnell, S.R., Krane, S.O., Kenzora, J.E. and Glimcher, M.J., *New Eng. J. Med.* 286:1013-1020, 1972.
10. Calcott, M.A. and Müller-Eberhard, H.J., *Biochemistry* 11:3443-3450, 1972.
11. Askenasi, R., *Biochim. Biophys. Acta* 304:375-383, 1973.
12. Askenasi, R., LeLous, M., Bazin, S. and Rao, V.H., *Clin. Sci. Mol. Med.* ·50:195-197, 1976.
13. Askenasi, R., VanHerweghem, J.L. and Ducobu, J., *Bio- medicine* 22:233-236, 1975.
14. Askeansi, R., Rao, V.H. and Devos, A., *Europ. J. Clin. Invest.* 6:361-363, 1976.
15. Daha, M.R., Spijckerman, M.G. and deGraeff, J., *Europ. Soc. for Clin. Invest.,* (abst. 54), p. 61, 1973.
16. Blau, E. and Michael, A., *J. Lab. Clin. Med.* 77:97-109, 1971.
17. Krane, M.S., Munoz, A.J. and Harris, E., *J. Clin. Invest.* 49:716-729, 1970.

BIOLOGICAL IMPLICATIONS OF CROSS-REACTIONS

BETWEEN GROUP A STREPTOCOCCI AND RENAL TISSUES

John B. Zabriskie, Vincent Fischetti,

Ivo van de Rijn,** Howard Fillit, and

Herman Villarreal, Jr.*

The Rockefeller University
New York, New York

ABSTRACT: A number of studies have now demonstrated
that renal antigens (GBM) share cross-reactions with the
antigens in the Group A streptococcus (STM). These
cross-reactions may be summarized as follows: lines of
common identify have been noted in extracts prepared
from GBM antigens and STM fractions. Amino acid analysis
of these fractions are suggestive of certain biochemical
similarities between the two structures. Immunization
of a number of animal species with STM results in a progressive glomerulonephritis.
Evidence has accumulated with respect to cellular

This research has been supported by Grants No. PHS 5
P01 AM-16944-03 from the Hektoen Institute and HRC-173 from
the New York State Health Research Council.

*Dr. H. Villarreal, Jr. is a fellow of the National
Kidney Foundation.

**Dr. I. van de Rijn is a recipient of a Senior Investi-
gatorship from the New York Heart Association.

453

immunity indicating that cellular reactivity to STM
or GBM antigens may play a role in progressive renal
disease, homograft rejection and the detection of renal
allograft crises.

All these studies suggest that the common denominator
in these observed serological and biological cross-reac-
tions between glomerular and bacterial membrane antigens
may well be a close structural similarity between STM and
mammalian transplantation antigens at the level of "dif-
ferentiation" antigens. The implication of these cross-
reactions in certain disease states will be discussed.

Students of the organism, the streptococcus, have long
known that this particular bacterium has the ability to mimic
host tissues. Starting with the outside capsule which is
composed of hyaluronic acid identical to that found in human
tissues (1), investigators have uncovered a host of serological cross-reactions between cellular structures of the streptococcus and tissue antigens (2). The majority of these
reactions appear to be related to antigens in the plasma
membrane of this organism.

For the purposes of the present discussion, attention
will be primarily centered on the cross-reactions between
these streptococcal membranes and renal antigens. Serologi-
cally at least two investigators (3,4) have demonstrated lines
of common identify in extracts prepared from renal glomerular
membrane antigens and streptococcal membrane fractions. Amino
acid analysis of these fractions has suggested (but does not
prove) that certain biochemical homologies exist between the
two fractions. Biologically, immunization of a number of
animal species has resulted in a progressive glomerulonephritis in dogs, with deposition of gamma globulin in the glomerular membrane region of the recipient host (6,7).

While the evidence is still fragmentary, the common
denominator in these observed serological and biological
cross-reactions between glomerular and bacterial membrane
antigens may well be a close structural similarity between
streptococcal membrane antigens and mammalian histocompatibility antigens. Support for this hypothesis is derived from
at least two lines of investigation. First, prior immunization of guinea pigs with Group A streptococcal membrane
antigens results in an accelerated rejection of implanted
skin grafts, and antisera to these membranes will evoke a
rapid rejection of these grafts when injected around the site
of the skin graft (8,9). The specificity of the reaction in

these studies was attested to by the fact that membranes prepared from other streptococcal groups did not evoke the
accelerated skin graft rejection when injected into these
animals. Among other gram-positive and gram-negative organisms tested, only *Staphylococcus aureus* and *S. albus* shared
this antigenic determinant with Group A streptococci. Second,
prior sensitization of dogs with Group A streptococci resulted
in an accelerated rejection of implanted renal allografts in
these animals (10). Finally, a number of investigators have
now shown a close biochemical similarity between mammalian
histocompatibility antigens, structural glycoproteins, and
streptococcal membrane fractions (9).

Given these series of observations, is there evidence to
suggest that these cross-reactions may play a role in the
initiation or continuation of renal disease? The clinical
evidence for this implication is controversial. Most large-
scale follow-up studies have failed to show an increased
incidence of chronic nephritis in patients following an
attack of acute poststreptococcal glomerulonephritis (11).
Other studies have shown on-going nephritis in a significant
number of patients. Earle and Jennings (12) studied a group
of 35 patients with clinical and biopsy evidence of chronic
progressive glomerulonephritis. Two-thirds of these patients
had a history of acute nephritis at the onset of their
disease. One-third of them had serum antibodies to Type 12
streptococci, in contrast with a control population of normals
and patients with other renal disease, among whom only 10 to
12% were positive for these antibodies. These type-specific
antibodies are persistent and protective against further
infection by the same nephritogenic type.

In one study (13), 20 out of 28 exacerbations in chronic
glomerulonephritis patients were associated with well-documen-
ted Group A streptococcal infections. Only eight nonstrepto-
coccal infections out of a total of 282 documented infections
in 68 patients resulted in exacerbations. No M-typing of the
streptococci was performed. If type-specific antibodies are
protective and persistent, we must assume that the repeated
infections in these patients were due to streptococci with
different M-types. Thus, if we wanted to implicate a streptococcal antigen in the pathogenesis of these exacerbations, it
would have to be an antigen common to all M-types. The most
likely candidate would be a streptococcal membrane antigen.

Our initial efforts in this direction began with the use
of the direct capillary migration technique of peripheral
blood leucocytes obtained from patients with progressive
glomerulonephritis. Using membranes and cell walls obtained
from both T12 and T5 Group A streptococci, we demonstrated

that these patients exhibited an increased reactivity to these
streptococcal structures (Table I). In contrast, patients who
had chronic pyelonephritis, lipoid nephrosis or chronic immune
complex diseases such as Lupus erythematosus were not reactive.
Thus renal tissue damage *per se* was not responsible for the
observed reactivity. Very similar results were obtained when
another technique of *in vitro* cellular reactivity to a given
antigen; namely, ^{14}C thymidine incorporation of sensitized
lymphocytes in the presence of streptococcal antigens was
used (14).

TABLE I

A Comparison of the Degree of Migration
Inhibition of Leucocytes Obtained from Progressive
Glomerulonephritic and Non-glomerulonephritic Patients

| Disease States | No. of Patients | Degree of Inhibition to Streptococcal Antigens | | |
		Type 12 Membranes	Type 12 Cell Walls	Type 5 Membranes
Progressive Glomer- ulonephritis	19	34 (±2)	27. (±3)	22.2 (±2)
Non-glomerular renal disease	15	8.1 (±1)	8.1 (±2)	5.1 (±1)
Normal subjects	10	8.8 (±2.6)	10.9 (±2.3)	---

Table II demonstrates that other investigators also noted
increased cellular reactivity to both streptococcal and renal
glomerular membrane antigens. Of interest was the finding by
both groups that the reactivity was primarily in the group of
proliferative chronic nephritis. The fact that this reactivity was primarily cellular was the observation by Macanovic
et al. (15) that many of these patients did not exhibit humor-
al antibodies to these antigens. Experimental support for the
hypothesis that cells alone may be involved in the pathogenesis of chronic forms of glomerulonephritis is provided by the
work of Heyman et al. (16) and Hess and Niff (17) who independently were able to transfer Heyman-type autoimmune nephritis by cells alone.

TABLE II

*The Cellular Response to Streptococcal and Renal
Antigens in Patients with
Progressive Glomerulonephritis*

Subjects	Antigens			
	GBM*	SCM[1]**	GBM	SCM[2]
Normals	0/21	0/21	0/21	0/21
Minimal Change	0/9	0/19	0/12	0/12
Proliferative glomerulo-nephritis	16/16	18/23	7/17	11/17
Membranous	0/8	0/8	0/6	0/6
Goodpasture's	2/4	3/6	6/6	0/6
Pyelonephritis			0/11	0/11

[1] *M. Macanovic et al., Lancet, p. 207, 1972.*

[2] *M. Dardenne et al., Lancet, p. 126, 1972.*

*Glomerular Basement Membrane.
**Streptococcal Cell Membrane

   To further assess the biological and structural properties of streptococcal protoplast membranes and renal glomerular antigens, we have pursued two other lines of investigation. The first avenue of approach was based on the hypothesis that the existing antigenic similarities between both
membrane structures might result in the use of either membrane
as an antigen to detect cellular sensitization to renal tissues in kidney transplant recipients. With this view in mind,
leukocytes obtained from a number of patients undergoing renal
transplantation were studied both before and after renal
transplantation. The pathological diagnosis in these patients
ranged from patients with chronic glomerulonephritis to poly-
cystic disease of the kidneys, chronic pyelonephritis, and
medullary cystic disease. The technique for the isolation of
peripheral blood leukocytes from these patients, the measurement of the degree of capillary migration, and per cent inhibition of these cells in the presence of sensitizing antigen
has been described previously (14). The details will not be
repeated here. The streptococcal protoplast membrane antigens
used in these studies were obtained from a type-5 strepto-
cocal strain; methods for the preparation of these antigens

have also been described in detail (18). In preliminary
experiments a number of different antigen concentrations were
used (100, 10, 1 µg/ml), but the best results were obtained
by using a concentration of 10 µg/ml of streptococcal membranes. Accordingly, these concentrations were used throughout all experiments.

When the leukocyte migration inhibition patterns of
patients who had undergone renal transplantation within the
past 1-2 year were examined, the observed values fell into
two general categories (see Table III). Using the degree of
creatinine clearance (CC) as a measure of renal function,
patients who had < 50 ml/min., and laboratory and clinical
evidence of rejection exhibited increased cellular reactivity
to streptococcal membranes. The average degree of inhibition
in these patients was 25%. In contrast, patients who demonstrated laboratory and clinical evidence of good renal function and no rejection (CC > 50 ml/min.) had only minimal
signs of altered cellular reactivity to these membranes (9%)
($p < 0.01$). The specificity of the reaction was strengthened
by the fact that cellular reactivity to streptococcal cell
walls was essentially the same in both groups of patients.

The second main area of investigation has centered
around the concept that sensitization to either streptoccal
membrane antigens or renal glomerular antigens would result
in accelerated organ rejection. As mentioned previously,
this concept was based on a body of data that suggests common
structural similarities between streptococcal membrane antigens and mammalian histocompatibility antigens. To test this
hypothesis, 22 mongrel dogs were injected subcutaneously s.c.

TABLE III

*A Comparison of the
Migration Inhibition Values and Creatinine
Clearance in Patients with Renal Transplantation*

Subjects	Creatinine Clearance/ml/minute	Percent Migration Inhibition
8	< 50 cc	25%
7	> 50 cc	9%

with streptococcal protoplast membranes (6-8 mg/kg) in incomplete Freund's adjuvant.   Renal tissue antigens were
isolated as described (10).

     Table IV summarizes the results of these experiments.
Eight animals rejected renal allografts in an accelerated
manner (5 days or less) and also showed definite evidence of
inhibition of leukocyte migration to both renal and streptococcal antigens.   In contrast, 12 animals rejected their
implanted kidneys in a first-set manner (6 days or longer)
and showed no evidence of altered cellular reactivity to
either antigen.   In most cases, the degree of leukocyte
migration of cells obtained from these animals actually
exceeded the control values obtained in the absence of
antigen (stimulation).

TABLE IV

A Comparison of Leukocyte Migration Inhibition
Values and the Type of Renal Allograft
Rejection in Dogs

Animals	Migration Inhibition	Type of Rejection
8/10	Positive	Accelerated
12/13	Negative	First Set

     How might we link these findings to an overall scheme
for the pathogenesis of some forms of chronic progressive
glomerulonephritis?   In Figure 1, we have outlined a possible
pathway.   In a susceptible individual, sensitization to streptococcal membrane components through repeated streptococcal
infections may lead to autosensitization to glomerular basement components.   Streptococcal membrane antigens are sero-
logically cross-reactive with glomerular basement membrane
components (4).   That there may also be cellular cross-
reactivity is suggested in the work of Dardenne, et al. (19)
and Macanovic, et al. (15); patients with chronic progressive
glomerulonephritis exhibiting cellular sensitization to
streptococcal antigens also show cellular sensitization to
glomerular basement membrane.   Once sensitization to autologous products is achieved, the process can be self-perpetuat-
ing, resulting in progressive deterioration of renal function.

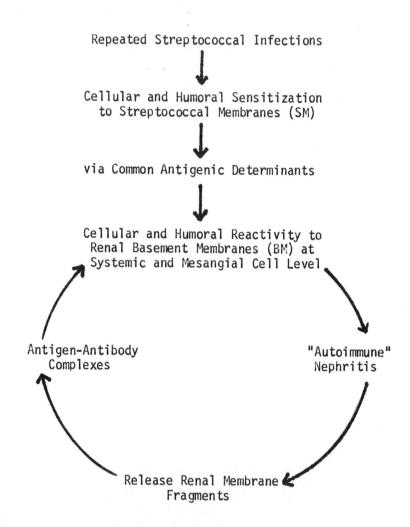

Fig. 1. *Possible mechanisms of renal damage in chronic nephritis.*

Intercurrent streptococcal infection could then enhance the already developing process. This scheme depends on the hypothesis that the cross-reactivity between streptococcal components and glomerular basement membrane is more than an incidental finding in chronic progressive glomerulonephritis. This point, as well as other questions raised by the scheme, is open to experimental and clinical investigation.

The relationship of the streptococcus to renal disease seems to be an intimate one. The exact nature of this relationship has given rise to much speculation. In this paper we have presented only some of the possible mechanisms by which the streptococcus may interact immunologically with its host and suggested some approaches for further investigation.

REFERENCES

1. Meyer, K., *Physiol. Rev.* 27:335, 1947.
2. Zabriskie, J.B., *Adv. in Immunology* 7:147, 1967.
3. Markowitz, A.S. and Lange, C.F., Jr., *Immun.* 92:565, 1964.
4. Holm, S.E., *Acta Path. Microbiol. Scand.* 70:79, 1967.
5. Markowitz, A.S., *Transplant. Proc.* 1:985, 1969.
6. Rappaport, F.I., Markowitz, A.S., McCluskey, R.T. and Hanaoka, T., *Surgery* 66:34, 1969.
7. Rappaport, F.T., Markowitz, A.S., McCluskey, R.T., Hanaoka, T. and Shimada, T., *Transplant. Proc.* 1:981, 1969.
8. Chase, R.M., Jr., and Rappaport, F.T., *J. Exp. Med.* 122:721, 1965.
9. Rappaport, F.T., Chase, R.M., Jr., and Solowey, A.C., *Ann. N.Y. Acad. Sci.* 129:102, 1966.
10. Zabriskie, J.B., Read, S.E., Ellis, R.J., Markowitz, A.S., Rappaport, F.T., *Transp. Proc. IV.* 2:259, 1972.
11. Zabriskie, J.B., Utermohlen, V., Read, S.E. and Fischetti, V.A., *Kidney Int.* 3:100, 1973.
12. Earle, D.P. and Jennings, R.B., *In:* "CIBA Symposium on Renal Biopsy" (Wolstenholme, G.E.W. and Cameron, M.P., Eds.) London Churchill, p. 156, 1961.
13. Seegal, D., Lyttle, J.D., Loeb, E.N., Jost, E.L. and Davis, G., *J. Clin. Invest.* 19:569, 1940.
14. Zabriskie, J.B., Lewshenia, R., Moller, G., Wehle, B. and Falk, R.E., *Science* 168:1105, 1970.
15. Macanovic, M., Evans, D.J. and Peters, D.K., *Lancet* 2:207, 1972.

16.   Heymann, W., Hunter, J.L.P., Hackel, D.B. and Cuppage,
      F., *Proc. Soc. Exp. Biol. Med.* 3:568, 1962.
17.   Hess, E.V., Ashworth, C.T. and Ziff, M., *J. Exp. Med.*
      115:421, 1961.
18.   Zabriskie, J.B. and Freimer, E.H., *J. Exp. Med.* 124:661,
      1966.
19.   Dardenne, M., Zabriskie, J.B. and Bach, J.F., *Lancet*
      1:126, 1972.

STUDIES ON DIABETIC NEPHROPATHY

Alfred F. Michael, Jon I. Scheinman,

Michael W. Steffes, Alfred J. Fish, David M. Brown

and S. Michael Mauer

Department of Pediatrics
University of Minnesota Medical School
Minneapolis, Minnesota

SUMMARY: Although the pathogenesis of diabetic nephropathy is unknown, morphologic, immunopathologic and
biochemical studies have demonstrated certain unique
features of this disease. No evidence for an immune
mechanism in the genesis of the mesangial or basement
membrane abnormalities could be demonstrated. A consistent finding in glomerular and tubular basement
membrane and Bowman's capsule was the demonstration of
striking immunofluorescence for IgG and albumin - findings not present in other diseases. Similar findings
can be demonstrated in normal kidneys 2-4 years after
transplantation into a diabetic environment. Whether
this is secondary to decreased crosslinking as reflected
by the decreased cystine content in GBM or is a consequence of increased vascular permeability and protein
entrappment in basement membranes with secondary thickening is unknown. These studies indicate that the epithelial renal extracellular membranes in diabetic nephropathy are uniquely abnormal. How these basement membrane
changes relate to the widened mesangium containing acto-
mysin is not clear. However, experimental and human
studies support the concept that changes in the kidney
are a consequence of the diabetic state.

INTRODUCTION

     The development of significant renal disease leading to
kidney failure is a well-recognized complication of diabetes
mellitus and is often associated with other manifestations of
microangiopathy such as blindness and complications related
to large vessel disease. Although there is an extensive
literature concerned with the morphology and clinical conse-
quences of diabetic microangiopathy, the pathogenesis of this
lesion(s) has not been defined. Basement membrane thickening
- the hallmark of small vessel disease - has been considered
a consequence of the diabetic state by some investigators and
a separately inherited abnormality by others. The morphologic
changes in vascular basement membranes have been studied most
extensively in muscle although similar changes have been
observed in a variety of other tissues. It has been assumed
that diabetic nephropathy developes as a consequence of
microvascular disease and thickening of the glomerular basement membrane (GBM).

MORPHOLOGY OF DIABETIC NEPHROPATHY

     The morphologic changes in the renal glomeruli in
diabetes mellitus consist of diffuse thickening of the GBM
and, less commonly, the presence of characteristic intercapillary mesangial nodules - the Kimmelstiel-Wilson lesion.
An increase in the size of the mesangium has been suggested
as an early and important feature of the developing glomerulopathy by Kimmelstiel and co-workers (1,2). The pathogenesis
of these lesions has not been defined although the thickened
GBM is thought to reflect a process that induces similar
changes in capillary basement membranes elsewhere in the body.
Other pathological findings - equally not understood - include the presence of exudative glomerular lesions - capsular
drops, arteriolar hyalinosis, and the so-called fibrin caps.
     The amount of information regarding the evolution of the
glomerular disease in juvenile diabetes mellitus is limited.
Østerby (3), however has demonstrated in a relatively small
number of young patients that the GBM is of normal thickness
at the onset of diabetes; is statistically thickened after
3 1/2 - 5 years of disease; and, by paired analysis in the
five patients studied sequentially, shows increased thickening after 1 1/2 - 2 1/2 years. Widening of the mesangium as
an early lesion was not observed by Østerby at the onset of
diabetes and the relative mesangial size remained normal

during the first five years; however on paired analysis the mesangial basement membrane-like material expressed as a percent of mesangial area was increased after two years of the disease. Recent studies by Østerby, Gunderson and Kroustrup (4) have demonstrated that the glomerular volume and the glomerular capillary area are significantly increased in early diabetes at a time when the thickness of the GBM and mesangium is normal.

FUNCTIONAL CHANGES

Pnysiologic changes in the kidney have been demonstrated in patients with juvenile-onset diabetes: (A) increase in kidney size (5) as well as the diameter and mass of individual glomeruli (6); (B) increase in glomerular filtration rate and filtration fraction unassociated with alteration in fractional dextran clearance (7-9); and (C) increase in urinary albumin excretion induced by exercise (9). Although the precise mechanism of these changes has not been elucidated there is evidence that these abnormalities may be corrected in part by good metabolic control of the diabetic state (9-11). An increase in microvascular permeability in the extremity has also been demonstrated using techniques that permit quantitative measurements of capillary filtration and diffusion as well as whole body transcapillary escape of albumin (10-12). These renal and vascular changes have been considered to reflect a functional and structural microangiopathy. In addition it has been suggested that the high GFR in diabetes might be related to increased filtration pressure and an enlarged capillary area. These functional abnormalities in diabetes mellitus could be caused by dysfunction of the glomerular mesangial contractile system as noted below.

DIFFUSE GLOMERULAR DISEASE

During the last 15 years a considerable body of evidence has supported the concept that immunological mechanisms play a decisive role in the development of most forms of diffuse glomerular disease or glomerulonephritis. These evidences which will not re reviewed in detail have been derived from (A) studies on experimentally-induced forms of immune mediated glomerulonephritis which are strikingly similar to human disease; (B) immunopathologic studies demonstrating immune reactants (Ig and complement components) within or adjacent to the GBM, tubular basement membrane (TBM), or mesangium;

(C) studies showing decreased serum levels of components of the classical and alternative complement pathways or the presence of circulating split components (e.g., C3d, Ba, Bb); (D) the presence of antibody in serum and eluted from kidney that has specificity for certain host antigens (e.g., GBM, TBM, nDNA, etc.) or foreign antigens (e.g., Australia antigen, bacterial antigens, etc.); (E) the demonstration of circulating immune complexes in certain diseases such as lupus erythematosus; and (F) the association of deficiency of the second component of complement (C2), immune complex disease, and the HLA complex as well as the recent demonstration of a B-cell alloantigen with membranoproliferative glomerulonephritis (13).

Three general types of immune-mediated renal disease have been identified: (A) anti-basement membrane disease in which antibody fixes to renal basement membranes; (B) immune-complex disease in which circulating antigen-antibody complexes localize along renal basement membranes; and (C) in-situ complex disease in which an antigen is sequestered in the glomerulus (e.g., in the mesangium or at an epithelial site) and circulating antibody then combines with the antiben to form local complexes. Only the first two mechanisms have been clearly shown in human disease.

STUDIES OF RENAL BASEMENT MEMBRANES IN DIABETES

Immunopathological Observations

In view of the importance of immunologic mechanisms in the pathogenesis of most forms of human glomerulonephritis, an intriguing hypothesis is that the glomerular lesions may be related to an autoimmune mechanism or one involving immune complexes. Immunofluorescent studies in diabetic nephropathy have been reported from a number of laboratories with varying interpretations (evidence reviewed in reference 14). Detailed immunopathologic studies from our laboratory have not supported an immune mechanism as an acceptable explanation for the genesis of the GBM and mesangial changes, but those observations bear directly on the pathogenesis of the disease (14, 15).

Immunofluorescent studies of diabetic kidneys have demonstrated intense staining for IgG and albumin in all renal extracellular basement membranes: GBM, TBM, and Bowman's capsule (Figure 1). When compared with normal tissue and tissue from other types of kidney disease, these findings - especially the extensive immunofluorescent change in TBM - are specific for diabetic nephropathy and are not seen in

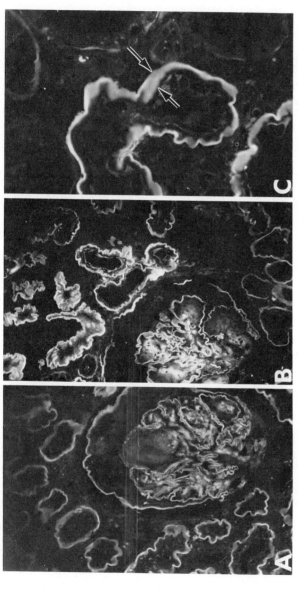

Fig. 1. Immunofluorescent micrograph showing brilliant staining for albumin (A) and IgG (B) along the TBM, GBM, and Bowman's capsule of a diabetic kidney. Weak auto-fluorescence of the mesangium is present. A higher magnification of a tubule (C) shows deposition of albumin predominantly along the outer aspect of the TBM (outer arrow); the inner aspect of the TBM is not stained (inner arrow) (Magnification A – X191, B – X174, C – X437). (Reprinted from reference 15)

other diseases (Figure 2). Other proteins such as B2 microglobulin, IgM, C4, C1q and Tamm-Horsfall protein are not detected in the basement membrane suggesting that size may not
be as important a determinant as extracellular concentration
of the protein. In the TBM, IgG and albumin are present along
the outer and middle aspects of the TBM. When diabetic kidney
sections are stained with heterologous antibody to basement
membrane, there is bright fluorescence along the inner aspect
of the TBM and a weaker reaction along its outer margin; the
middle portion of the TBM remains unstained and is very wide
when compared with normal TBM. The nature of this non-
reactive expanded region is unknown (Figure 3).

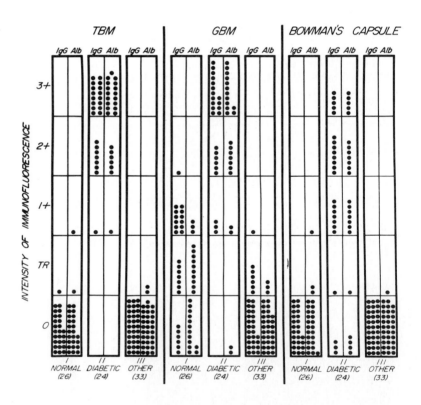

Fig. 2. Correlation of the intensity of immunofluorescence for IgG and albumin in TBM, GBM, and Bowman's capsule in kidneys from normal subjects, patients
with advanced diabetic nephropathy, and those with other
renal diseases. (Reprinted from reference 15)

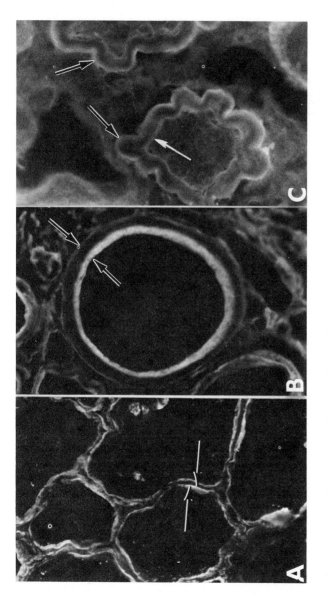

Fig. 3. A. Kidney section stained with rabbit antibody to human BM from a normal control. Note the thin delicate TBM; the total thickness of the membrane is enclosed by the two arrows (X 485). B. In a similar stained section from a patient with diabetic nephropathy there is intense staining primarily of the inner but also the outer aspect of the TBM. The two arrows enclose the entire thickness of the tubular basement membrane (X 534). C. Tissue sections from a diabetic patient stained sequentially with antibody to human BM (FITC) and anti-albumin (TMR). The black arrow represents the location of albumin predominantly in the outer aspect of the TBM, with some diffusion toward the center of the membrane, which is unstained. The BM antiserum stained both inner and outer membranes with accentuation of the inner aspect of the membrane as depicted by the white arrow (X 305). (Reprinted from reference 15)

469

Evidence not supporting an immune mechanism for the immunofluorescent findings noted above include: the absence of detectable anti-BM antibody in sera or kidney eluates; the inability to detect insulin, or insulin-antibody in the membranes; the failure to demonstrate *in vitro* complement fixation; and the presence of the non-immune protein, albumin. These studies indicate however that all renal extracellular membranes are uniquely abnormal in diabetes mellitus since similar findings are rarely noted in other diseases. We have also shown that these immunofluorescent findings are present relatively early (within 4 years) after the onset of the disease. Further, following transplantation of a normal kidney into a diabetic patient, linear fluorescence of membranes for IgG and albumin develop after a period of 2 years.

Although the mechanism of these changes in renal extracellular membranes is unknown it is possible that these findings denote a unique biochemical abnormality of basement membranes resulting in an increased affinity for serum proteins; or that the changes reflect an increase in vascular permeability leading to imbibition by membranes with resultant thickening; or a combination of both of these mechanisms. This concept may bear directly on the suggestion of Parving et al. (10,11) and Ditzel (16) that increased extravasation of proteins and a functional microangiopathy might precede and lead to the development of structural changes in diabetic vascular disease.

Ig and complement components but rarely fibrin may be observed in the diabetic insudative and exudative glomerular and vascular lesions - particularly the so-called fibrin caps - which are located along the endothelial aspect of the GBM. In contradistinction to the basement membrane changes noted above, it is possible but by no means proven that immune mechanisms may play a role in the pathogenesis of these lesions. Somewhat similar immunopathological findings may be observed in hyalinizing glomeruli in disease states other than diabetes (17).

Biochemical Studies on Diabetic GBM

As described in other parts of this symposium, numerous studies have been carried out describing the biochemical characteristics of the normal GBM. A limited number of studies have described the composition of diabetic GBM (Table I). Beisswenger and Spiro (18,19) demonstrated an increase in hydroxylysine and glucosylgalactose dissacharide linked to hydroxylysine in diabetic GBM suggesting an over-production of these subunits. However, our own studies (20,21) failed

TABLE I

Composition of Diabetic GBM as Described in Four Different Studies

	Composition of Certain Components in Diabetic GBM[+]					
	Hydroxyproline	Hydroxylysine	GGH*	Cystine	Sialic Acid
Beisswenger and Spiro (18,19)	Increase	Increase	Increase	Normal	–
Westberg and Michael (20,21)	Normal	Normal	Normal	Decrease	Decrease
Kefalides (22)	Normal	Normal	Normal	Decrease	Decrease
Sato et al. (23)	Increase	Normal	–	Decrease	Decrease

*Glucosylgalactosylhydroxylysine

[+]Other changes include: increase in glycine and decrease in tyrosine, valine, lysine (18,19);
increase in lysine, histidine, phenylalanine, glucose and decrease in
serine, proline (20,21);
increase in isoleucine, tyrosine (22);
decrease in glucose, mannose (23).

to confirm these observations but demonstrated a highly
significant decrease in half-cystine content. Similar findings have been shown for cystine and sialic acid content of
GBM by Kefalides (22) and Sato et al. (23). However, caution
must be expressed in the interpretation of these data because
of methodological differences in isolation of GBM and also
because soluble components are lost during ultrasonic disruption of GBM. Nevertheless, these findings may indicate
alteration in crosslinking of proteins in the GBM.

THE GLOMERULAR MESANGIUM

The glomerular mesangium consists of cells surrounded by
a matrix located in the intercapillary region of the glomerulus (Figure 4) (24-27). This mesangial zone is separated from
the capillary lumen by endothelial cells on its inner surface
and externally by the GBM. It extends into the axial region
of the glomerulus and lies in close contact with the juxtaglomerular regions. The nature of the mesangial matrix is
unknown. Although it is PAS positive and lies in close
approximation to the GBM, there is no evidence that it is
biochemically or immunologically similar to GBM. The mesangium is altered in many human renal diseases; is a frequent site
for deposition of immune complexes in various forms of glomerulonephritis; and has been shown to be expanded in diabetes
mellitus.

A number of substances have been found to localize in
the mesangium and the reticuloendothelial system following
administration to experimental animals, viz., carbon, thorotrast, antigen-antibody complexes, aggregated IgG and aggregated albumin (27a). Our understanding regarding the uptake
and processing of macromolecules by the mesangium is relatively incomplete. If radiolabelled aggregated human IgG is
administered to rats, glomeruli isolated, and the content of
aggregated IgG quantitated, it is possible to study the
kinetics of mesangial uptake and disappearance of the macromolecules (28,29). A number of factors are known to influence
localization of protein macromolecules in the mesangium:
(A) The size would appear to be a critical determinant since
monomeric 7S IgG does not localize appreciably whereas polydisperse aggregated IgG (>7S) and antigen-antibody complexes
(>11S) do localize; (B) The blood level is also an important
determinant. Earlier studies using carbon suggested that
glomerular uptake was increased when the reticuloendothelial
system became saturated (30,31). Recent studies by Haakenstad,
Striker, and Mannik (32,33) demonstrated that circulating
reduced and alkylated antigen-antibody complexes are not

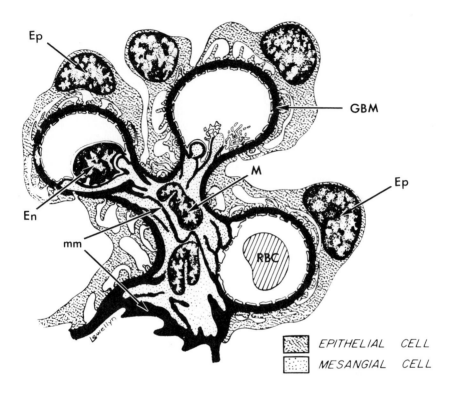

Fig. 4. Artist schematic drawing of a glomerular lobule with three capillary loops. Note the
central location of the mesangial (M) cells surrounded
by mesangial matrix which are separated from the capillary lumen by endothelial (E) cells and from the epithelial (Ep) cell foot processes by the glomerular
basement membrane (GBM). In diabetes there is expansion of the mesangium and a uniform increase in thickness of the GBM. (Reprinted in part from reference 27)

metabolized normally by the hepatic reticuloendothelial system and the resulting higher blood levels lead to increased mesangial uptake. Similarly, recently completed studies using endotoxin and hydrocortisone to modify the reticuloendothelial system suggest that mesangial uptake appears to be a critical function of the concentration in the blood (34). (C) A unique relationship between mesangial function and injury to the glomerular capillary with increased permeability to protein has also been demonstrated. When rats are given the aminonucleoside of puromycin or antibody to GBM which cause alteration in the glomerular capillary wall - a striking increase in mesangial uptake of aggregated IgG occurs (Figure 5) (28,29). However, disposal or disappearance of aggregates from the mesangium remains normal. (D) The role of physiologic factors on mesangial kinetics have been incompletely evaluated. Recent studies have suggested that uptake may be

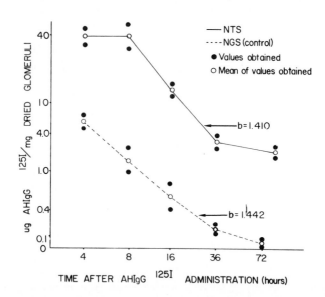

Fig. 5. *Kinetic analysis of mesangial sequestration of* ^{125}I *aggregated human IgG (AHIgG) after administration of aggregates to control rats given normal goat serum (NGS) and nephritic rats given antisera reacting with the GBM in the capillary wall (NTS). Note the significant increase in glomerular mesangial uptake but the normal disappearance of radioactivity in rats with glomerular capillary injury. (Reprinted from reference 29)*

independent of glomerular filtration rate but dependent on renal plasma flow and possibly on intraglomerular pressure (34a). (E) The precise mechanism of mesangial uptake and processing of macromolecules has not been defined. Although mesangial cells have been shown to engulf ferritin, carbon particles, and thorotrast, it is not at all clear that this is the major pathway for mesangial macromolecular processing. The fact that aggregated proteins and complexes have been found mainly in mesangial channels between cells experimentally - a site where immune deposits are found in human glomerulonephritis - suggests that extracellular transport through the mesangial matrix system may be the principle route of transport through the glomerulus. Egress from the mesangium could be through the matrix with regurgitation back into the glomerular capillary lumen and into the circulation; through the mesangium into the axial and lacis region as demonstrated for carbon (35); into the juxtaglomerular region with penetration into the distal tubule or into the renal lymphatics; and finally by way of mesangial cell digestion.

Evidences for abnormalities of the mesangium in diabetes include: (A) The presence of mesangial expansion with increase in matrix early in the course of human diabetes, and the presence of Kimmelstiel-Wilson nodules in this site; (B) The demonstration in experimental diabetes (streptozotocin or alloxan) of prominent widening of the mesangium and the early appearance of mesangial IgG, IgM and C3 (36). The reason for these mesangial changes are unknown. Preliminary studies suggest that the mesangial clearing mechanism may be impaired in these animals (36). Of interest is the fact that this mesangial deposition of immune proteins is not seen in human diabetes; further the linear basement membrane fluorescense so characteristic of human diabetes is not seen in the experimental model (Table II). (C) The demonstration by immunofluorescent studies in human diabetes that the widened mesangium contains significant amounts of the smooth muscle protein - actomyosin (37) and fibroblast-surface antigen (38) (Figure 6). Although these proteins are present in the normal kidney, they are more extensively present in diabetic kidneys in man and experimentally in diabetic rats.†

It is possible that the presence of smooth muscle in mesangial cells normally plays an important contractile function - possibly by controlling glomerular perfusion. The presence of increased actomyosin in diabetes is particularly provocative in view of the known physiologic changes seen early in the course of juvenile diabetes mellitus - such as increased glomerular filtration rate.

†*Unpublished data.*

TABLE II

Morphologic and Immunopathologic Changes in the Basement Membrane
and Mesangium in Experimental and Human Diabetes Mellitus†

	Basement Membrane		Mesangium			Exudative or Insudative Lesions
	Thickening	Presence of IgG and Albumin	Expansion	Presence of IgG and C3	Increased Actomyosin	
Experimental Diabetes	+	−	+	+	+ˣ	+
Human Diabetes	+	+	+*	−	+	+

†The table indicates the major abnormality observed: + denotes an increase
  − indicates no difference from normal.

*Kimmelsteil-Wilson nodules may be present.

ˣScheinman, J., Steffes, M. and Mauer, S.M., Unpublished Data.

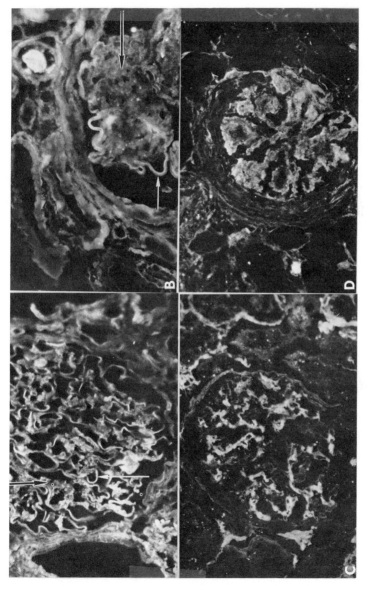

Fig. 6. Diabetes Mellitus. Immunofluorescent staining with anti-GBM (A and B) and anti-actomyosin (C and D) antisera. Anti-GBM staining in early (A) and late (B) diabetic nephropathy demonstrates thickening of basement membranes (white arrows) and unstained expanded mesangium (black arrows). Anti-actomyosin staining of early (C) and late (D) diabetic nephropathy demonstrates expansion of positively stained mesangium. Magnification X 800. (Reprinted from reference 37)

477

DIABETIC NEPHROPATHY AND TRANSPLANTATION

As noted above, experimental studies carried out by Maure et al. (reviewed in reference 36) in animals with diabetes mellitus (alloxan or streptozotocin) have demonstrated mesangial expansion and deposition of IgG, IgM and C3. These abnormalities, which become more severe as the duration of the diabetes increases, are reversed by transplantation of the diabetic kidney into a normal rat or by correction of the diabetes by islet transplantation from a syngeneic donor (36,39). These studies clearly demonstrate that the glomerular abnormalities observed in diabetic glomeruli are secondary to the diabetic state. At issue is the relationship of these experimental changes - primarily in the mesangium, to the morphologic and immunopathologic alterations observed in human diabetic kidneys where basement membrane thickening and immunofluorescence abnormalities predominate. This difference is unresolved at present.

In human diabetes mellitus recurrence of disease in the transplanted kidney is suggested by several observations: (A) The demonstration of arteriolar hyalinosis - in kidneys transplanted into diabetic patients - with a frequency higher than that seen in kidneys transplanted into nondiabetic patients (40). (B) The presence of increased renal basement membrane fluorescnece for IgG and albumin (41) (Figure 7) and increased mesangial actomyosin (38) within 4 years after grafting a kidney into a diabetic recipient.

These findings indicate that the renal abnormalities described above are a consequence of the diabetic state. This offers further support to the hypothesis that the pathogenesis of diabetic nephropathy is related more to the diabetic melieu than to an unrelated or linked disorder.

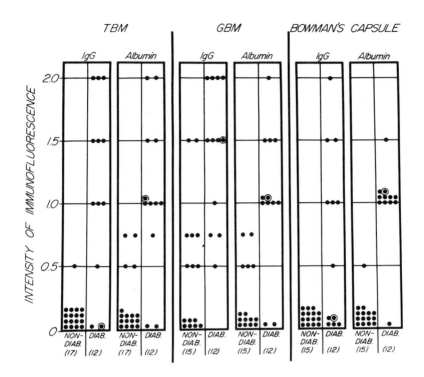

Fig. 7. Intensity of immunofluorescence for
IgG and albumin in the TBM, GBM, and Bowman's capsule
of kidneys transplanted into diabetic and nondiabetic
patients. The numbers in parenthesis are totals of
patients studied. ⊙ represents the patient with post-
transplantation sterioid-induced diabetes. (Reprinted
from reference 39)

REFERENCES

1. Iidaka, K., McCoy, J. and Kimmelstiel, P., *Lab. Invest.*
    19:573, 1968.
2. Kawanao, K., Arakawa, M., McCoy, J., Porch, J. and
    Kimmelstiel, P., *Lab. Invest.* 21:269, 1969.
3. Østerby, R., *Acta Medica Sandinavica Suppl.* 574, 1975.
4. Osterby, R., Gunderson, H.G. and Kroustrup, J., *(This
    volume)*, 1976.
5. Mogensen, C.E. and Anderson, M.J.F., *Diabetes* 22:706,
    1973.

6. Butcher, D., Kikkawa, R., Klein, L. and Miller, M., *J. Lab. Clin. Med.*, 89:544, 1977.
7. Ditzel, J. and Junker, K., *Brit. Med. J.*, II:13, 1972.
8. Mogensen, C.E., *Acta Med. Scand.*, 194:559, 1973.
9. Mogensen, C.E., *Diabetes* 25:872, 1976.
10. Parving, H.H., et al., *Diabetologia* 12:161, 1976.
11. Parving, H.H., et al., *Diabetes* 25:884, 1976.
12. Alpert, J.S., Coffman, J.D., Balodimos, M., Koncz, L. and Soeldner, J., *New Eng. J. Med.* 286:454, 1972.
13. Friend, P., Yunis, E., Noreen, H. and Michael, A.F., *Lancet* i:562, 1977.
14. Westberg, N.G. and Michael, A.F., *Diabetes* 21:163, 1972.
15. Miller, K. and Michael, A.F., *Diabetes* 25:701, 1976.
16. Ditzel, J., *Diabetes* 17:388, 1968.
17. Velosa, J., Miller, K. and Michael, A.F., *Am. J. Path.* 84:149, 1976.
18. Beisswenger, P. and Spiro, R., *Science* 168:596, 1970.
19. Beisswenger, P. and Spiro, R., *Diabetes* 22:180, 1973.
20. Westberg, N.G. and Michael, A.F., *Biochemistry* 9:3837, 1970.
21. Westberg, N.G. and Michael, A.F., *Acta Medica Scand.* 194:39, 1973.
22. Kefalides, N.A., *J. Clin. Invest.* 51:403, 1974.
23. Sato, T., Munakata, H., Yoshinaga, K. and Yosizawa, Z., *Clin. Chim. Acta* 61:145, 1975.
24. Vernier, R.L., Mauer, S.M., Fish, A.J. and Michael, A.F., *In:* "Advances in Nephrology", Year Book, Chicago, 1971.
25. Latta, H., Maunsbach, A. and Madden, S.C., *J. Ultrastruct. Res.* 4:455, 1960.
26. Farquhar, M. and Palade, G., *J. Cell Biol.* 13:55, 1962.
27. Vernier, R.L., Resnick, J. and Mauer, S.M., *Kidney Int.* 7:224, 1975.
27a. Michael, A.F., Fish, A.J. and Good, R.A., *Lab. Invest.* 17:14, 1967.
28. Mauer, S.M., Fish, A.J., Blau, E. and Michael, A.F., *J. Clin. Invest.* 51:1092, 1972.
29. Mauer, S.M., Fish, A.J., Day, N. and Michael, A.F., *J. Clin. Invest.* 53:431, 1974.
30. Benacerraf, B., McCluskey, R. and Patras, D., *Am. J. Path.* 35:75, 1959.
31. Benacerraf, B., Sebestyn, M. and Cooper, N.S., *J. Immunol.* 82:131, 1959.
32. Haakenstad, A.O., Striker, G.E. and Mannik, M., *Lab. Invest.* 35:293, 1976.
33. Haakenstad, A.O. and Mannik, M., *Lab. Invest.* 35:283, 1976.
34. Shvil, Y., Mauer, S. and Michael, A.F., *Clin. Res.* 23:237, 1975.

35. Elema, J., Hoyer, J. and Vernier, R.L., *Kidney Int*. 9: 395, 1976.
36. Mauer, S.M., Steffes, M.W., Michael, A.F. and Brown, D.M., *Diabetes* 25:850, 1976.
37. Scheinman, J.I., Fish, A.J. and Michael, A.F., *J. Clin. Invest*. 54:1144, 1974.
38. Scheinman, J.I., Fish, A.J., Matas, A.J. and Michael, A.F. The immunohistopathology of glomerular antigens. II. The glomerular basement membrane, actomyosin and fibroblast surface antigens in normal, diseased and transplanted human kidney. Submitted for publication, 1977.
39. Mauer, S.M., Steffes, M.W., Sutherland, D.E.R., Najarian, J.S., Michael, A.F. and Brown, D.M., *Diabetes* 24, 280, 1975.
40. Mauer, S.M., et al., *New Eng. J. Med*. 295:916, 1976.
41. Mauer, S.M., Miller, K., Goetz, F., Barbosa, J., Simmons, R.L., Najarian, J.S. and Michael, A.F., *Diabetes* 25:709, 1976.

EFFECTS OF AGING AND DIABETES

ON BASAL LAMINA THICKNESS OF SIX CELL TYPES

Rudolf Vracko

Veterans Administration Hospital
and
University of Washington School of Medicine
Seattle, Washington

ABSTRACT: Basal lamina (BL) investment was measured
in six different anatomic sites from seven age-matched
male diabetics (DM) and controls, 8 to 69 years old.
The sites were selected to represent the three ways
with which different cell types interact with their BL
during process of cell replenishment: (a) new cell
generations do not deposit a new BL layer (skin and
probably alveolar epithelium and capillaries); (b)
new BL layer is deposited and old one is removed
(muscle fibers); and (c) new BL layer is formed and
the old one is retained (muscle capillaries and renal
tubules). In all anatomic sites the average width
of BL investment was measured. In addition, in three
sites the number of BL layers that comprise the BL
investment were counted and the theoretical width of
individual BL layers determined by dividing measured
width by number of layers. The objective was to test
whether or not (a) BL accumulates in DM and in process
of aging preferentially in those structures in which
BL layers are deposited normally in course of tissue
repair; (b) accumulation of BL in DM occurs in same
structure as it does with normal aging; and (c) the
thickness of a single BL layer changes with aging or DM.

*The data reveal that (a) BL accumulates primarily in
tissues in which BL layering is a normal byproduct of
cell replenishment; (b) the same tissues are involved
in DM as are with aging; (c) in all tissues in which
the amount of BL increases with aging more BL accumulates in DM; (d) increased deposits of BL in DM and
with aging are caused by collection of multiple layers
of BL (except alveolar epithelial BL); and (e) the
thickness of a single BL layer remains essentially
unchanged in aging and in DM (except alveolar epithelial BL).*

INTRODUCTION

   Increased deposits of basal lamina (BL) that develop in
patients with diabetes mellitus (DM) 2,3,4,7,9,15,20,21,23,
29,32,33,34) and, to a lesser degree during aging (1,3,20,32)
and in several other conditions (5,6,8,11,12,14,17,18,19,27,
30) have a peculiar anatomic distribution. It appears that
BL's of some cell types become much thicker than those of
others and that in some structures excessive deposits are not
distributed uniformly but instead collect focally and
regionally. For example, the BL width of skeletal muscle
capillaries increases substantially with aging (3,15,32,33)
and in DM (2,9,15,20,21,23,32,33,34) while the BL of skeletal
muscle fibers does not (9,21). Furthermore, in individual
capillaries of skeletal muscle BL deposits are spotty and
patchy (2,9,21,23,34) and the accumulation is, on the average,
much more extensive in capillary beds of distal parts of
limbs than in more proximal sites (23,24,32).
   To obtain additional information about the distribution
of BL accumulation and possibly an insight into the nature of
the process the following questions were asked: (a) Is the
pattern of increased BL deposits similar in diabetics to what
it is in older people without DM? (b) Are these deposits
related to ecto-, meso- or endodermal cell origin of BL? and
(c) Is the pattern of BL deposition in aging and in diabetics
similar to that in experimental animals after episodes of
cell death and cell replenishment? The last question is concerned with observations which indicate that during cell
replenishment some cell types deposit a new layer of BL causing BL to accumulate like growth rings on cross-sections of
tree trunks while other cell types do not (25,27,38). Of the
six sites sampled in this study, skeletal muscle capillaries
(25,27) and renal tubules (5,30) deposit new BL layers and
retain the old ones, while of the remaining four tissues,

alveolar endothelium and epithelium (26) and epidermis (10, 16) do not. Young skeletal muscle cells also deposit a new layer but the old layer is removed so that when healing is completed only one layer remains (25,27,30).

METHODS

The thickness of BL investment of six different cell types was measured in tissues obtained during autopsies from seven male controls (mean 46.4 years ± 19.5 S.D.) and seven age-matched male diabetics (mean 45.0 years ± 20.4 S.D.). Skin was taken from mesial aspect of the left upper arm for measuring epidermal BL. Muscle capillary BL and BL of muscle fibers were measured in skeletal muscle from mid-portion of left anterior tibial muscle. The width of tubular BL was determined in samples from left kidney cortex and alveolar epithelial BL and alveolar endothelial BL were measured in tissues from upper lobes of lungs.

Fixation was with 2.0% osmium tetroxide in 0.1 M S-collidine buffer at pH 7.48. The width of BL was measured along the course of each BL at equidistant 3.0 cm intervals on 10 electronmicrographs (magnified 10,500 x) per specimen except skin where only five pictures were used. In circular structures as in capillaries or renal tubules the measuring sites were never closer than 3.0 cm. When the 3.0 cm mark transected pericytes only the BL peripheral to pericytes was measured. In the lungs the joined epithelial-endothelial BL was not measured.

Since increased amounts of BL were frequently composed of two or more layers, as shown in Fig. 1, in structures in which BL layering was sufficiently distinct, as in muscle fibers, muscle capillaries and lung capillaries the number of BL layers that made up the BL investment were counted. By dividing the mean measured width of BL by the mean number of layers for each subject an approximate width of a single BL layer for the subject and anatomic sites was calculated. Although BL of renal tubules was also layered, accurate counts were not possible because of focal fusion and blending of layers.

Standard methods were used for statistical calculations (22).

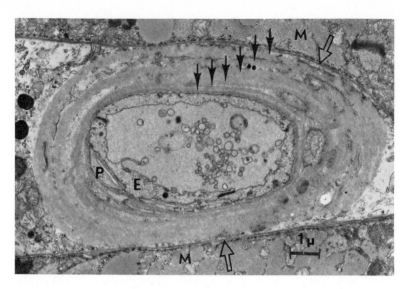

Fig. 1. Cross-section of a capillary from the
68-year-old diabetic. BL investment is increased
apparently by apposition of multiple BL layers (full
arrows). The innermost layer which is closely applied
to the endothelium is narrowest and densest in appearance. The more peripheral layers are wider and of
looser texture. Some are fused. In the spaces between
the layers is cell debris (suggesting that cells have
died) and collagen fibers. The presence of a single
layer of BL between endothelial cell (E) and pericyte
(P) suggest that BL does not accumulate either by
continuous or intermittent production of BL material
by either epithelial cell or pericytes. The bubbly
material in the capillary lumen is a post mortem
change. The two skeletal muscle fibers (M) have a
single layer of BL (open arrows).

RESULTS

The data are listed in Table I. They show that the
amount of BL is affected by at least three factors: type of
cells, presence of DM and by the subject's age. The effects
of the first two are shown in Figs. 2 and 3. Three degrees
of responses seem to occur: (a) epidermis has what appears
to be a "normal" amount of BL in controls and diabetics;
(b) in controls muscle capillaries and renal tubules have
three to four times as much BL as epidermis and in diabetics
the BL investment of muscle capillaries is 5-6 times and that

TABLE I

Width of basal lamina investment in six different sites of seven control and seven diabetic subjects: mean ± SD(n)

Controls (Age)	Epidermis	Muscle fibers	Lung epithelium	Lung capillaries	Muscle capillaries	Renal tubules
9	83.1±49.3 (49)	75.2±52.6 (162)	85.7±56.9 (104)	159.2±91.5 (79)	217.3±190.1 (50)	300.1±136.6 (154)
40	107.9±46.9 (48)	77.6±65.9 (186)	94.2±37.8 (136)	117.0±96.9 (90)	349.2±183.4 (72)	262.3±140.4 (146)
43	118.9±68.4 (67)	119.5±65.9 (144)	96.0±38.3 (128)	164.3±85.5 (96)	412.0±146.0 (65)	372.7±228.3 (131)
49	93.5±46.8 (43)	106.0±47.7 (178)	80.4±34.0 (138)	109.8±44.5 (89)	542.9±295.9 (66)	304.2±135.1 (127)
54	113.1±57.1 (45)	135.2±99.3 (181)	102.5±41.8 (173)	136.4±62.0 (95)	231.4±159.3 (59)	not measured
62	108.0±53.4 (46)	122.5±65.1 (121)	102.1±41.1 (162)	165.4±120.3 (102)	452.2±266.9 (49)	488.2±303.4 (140)
69	84.4±37.1 (35)	112.0±50.0 (143)	121.2±56.8 (129)	161.6±118.0 (97)	601.7±376.9 (51)	659.1±455.3 (100)
All controls	102.8±54.8 (333)	105.8±69.8 (1115)	97.6±45.3 (970)	145.3±94.9 (648)	400.9±271.5 (412)	383.7±277.3 (798)
Diabetics (Age)						
8	90.4±50.0 (59)	78.3±41.2 (163)	92.6±31.6 (89)	141.7±84.3 (77)	177.5±122.4 (51)	484.4±187.3 (149)
33	98.5±59.7 (64)	96.8±43.7 (154)	184.6±82.1 (127)	155.7±87.4 (97)	422.6±312.2 (53)	1296.0±539.0 (91)
41	78.0±36.4 (49)	115.0±67.2 (150)	154.5±78.8 (153)	215.1±130.8 (105)	521.1±305.4 (63)	1493.1±814.6 (84)
47	96.4±58.1 (43)	205.6±128.8 (140)	160.8±93.2 (123)	226.2±146.8 (107)	500.2±267.1 (62)	1885.2±1030.3 (86)
55	113.1±57.1 (45)	117.5±63.0 (148)	137.1±68.2 (162)	209.0±126.2 (104)	571.2±357.3 (79)	1253.1±707.8 (72)
53	120.6±53.8 (41)	134.1±75.2 (113)	215.0±109.7 (121)	191.9±107.6 (112)	731.9±483.3 (64)	672.0±338.2 (120)
68	99.5±65.7 (64)	116.8±93.2 (142)	166.4±124.8 (136)	140.5±90.7 (119)	753.4±538.4 (69)	1718.4±674.2 (111)
All diabetics	97.3±55.4 (365)	121.6±85.7 (1010)	160.2±94.8 (911)	184.0±118.1 (721)	542.5±408.6 (441)	1177.1±807.0 (713)

of renal tubules 12 times that of epidermal BL; and (c) of
the remaining three cell types muscle fibers and lung epithelium also have "normal" amounts in controls but have slightly
higher values when DM is present while lung capillaries in
both the controls and diabetics have moderately elevated
values.

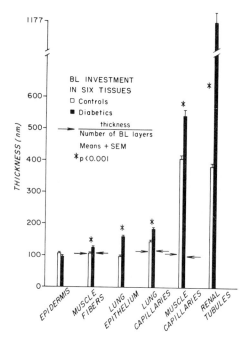

Fig. 2. Mean (± SEM) for all measurements of BL thickness in control subjects (open bars) and diabetic subjects (solid bars) for each of the six types of tissues. The differences between diabetics and controls are statistically significant (p < 0.001) in all sites except epidermis where p > 0.1. Arrows indicate a calculated average width for a single BL layer arrived at by dividing the mean BL width of each subject by the corresponding mean for the number of BL layers.

The ranges of measurements are much wider for muscle
capillaries and renal tubules than for other sites as seen in
histograms in Fig. 3. In every site except skin diabetics
have significantly greater amounts of BL. It is of interest
that each population of measurements includes values in the
low (presumably normal) ranges and that the differences among
population means are not due to shifting of all measurements
but rather due to different proportions and different degrees
of higher values. This relationship is demonstrated by a
positive correlation of subject means for each anatomic site
with the corresponding standard deviation (R = 0.998, DF = 10,
p < 0.001) indicating that BL amassment is not a uniform process but that in each BL type some normally thick parts of BL
remain while in other parts the BL investment is thickened to
various degrees.

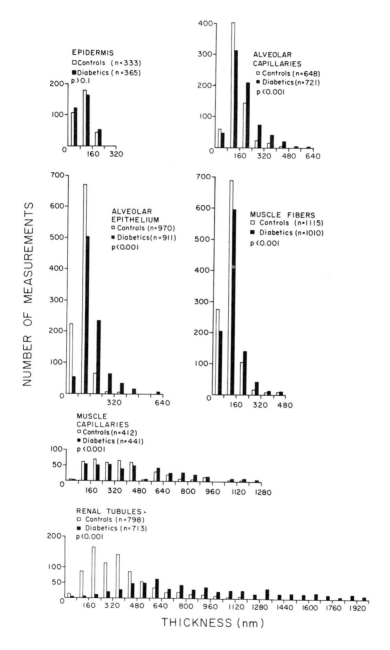

Fig. 3. Histograms showing distributions and ranges of measurements of BL thickness in each of the six tissues of controls (open bars) and diabetics (solid bars).

Of further interest is the calculated average width of
single BL layer arrived at by dividing each subject's mean
width of BL with the mean number of layers. In the three
sites in which such calculations were made the results
(arrows in Fig. 2) are remarkably alike and statistically
indistinguishable when tested with a student t-test from the
measured BL width in epidermis of controls and diabetics as
well as in muscle fibers and lung epithelium of controls
($p > 0.1$).

The effects of age on the amount of BL are demonstrated
in Figure 4. It is apparent that (a) in controls and
diabetics no changes occur in BL of epidermis and alveolar
capillaries; (b) alveolar epithelium of controls also
remains unchanged; (c) small increments seem to occur in BL
of muscle fibers of controls and diabetics as well as in
alveolar epithelium of diabetics; and (d) marked changes
occur with advancing age in muscle capillaries and renal
tubules of controls and diabetics. Particularly impressive
is the increase in slope and intercept on the Y axis for the
BL of renal tubules in diabetics suggesting that BL accumulation in renal tubules may be an early change of DM.
(The scale for renal tubules on the Y axis is reduced as
compared with scales for other cell types.)

DISCUSSION

The anatomic distribution of BL accumulation is not
related to ecto-, meso- or endodermal origin of cells. The
BL deposits in diabetics and controls are increased pre-
dominately in the two sites in which BL reduplication occurs
normally during cell replenishment, i.e., in muscle capillaries and renal tubules. In the remaining four sites where BL
accumulation is not a normal byproduct of cell turnover, BL
does not accumulate significantly with either aging or DM.
There also seems to be no difference in the structure of the
BL in controls and diabetics. In both, the accumulations
appear to be due to apposition of multiple BL layers, each
layer measuring on the average 100 nm in thickness.

The differences between the two types of subjects are
quantitative, diabetics having more BL than controls. This
suggests that presence of DM amplifies or advances in chrono-
logic time the normal effects of aging. Some recent tests
which appear to be measures of biologic age, as for example
in vitro growth potential of fibroblasts (31) and *in vivo*
"aging" of collagen (13) seem to support this concept. Both
tests indicate that tissues from diabetic donors behave not

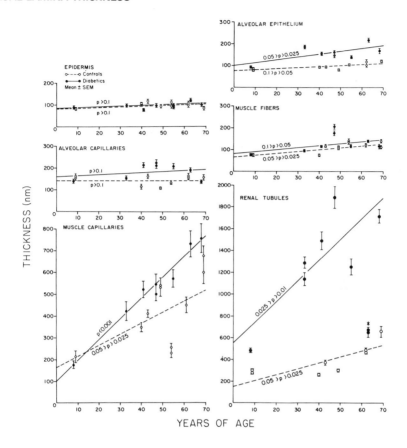

Fig. 4. Regression lines indicating the effects of age on BL thickness in each of the six tissues of controls (discontinuous lines) and diabetics (continuous lines). Each point represents the mean BL width (± SEM) of a subject. Where two points are marked per subject's age they represent independent duplicate determinations to show reproducibility of sampling and of measuring techniques. Only one of these points, selected by the flip of a coin, is listed in Table I and was used for calculation of regression lines. The point for renal tubules of the 62-year-old diabetic was considered an exceptional value and not used for calculation of regression lines. The "p" values listed with each regression line are an indication of whether the slopes are significantly different from "0" or not. Of the six pairs of regression lines only those for renal tubules are significantly different from each other in slope and in intercept on the Y axis (p < 0.05). It should be noted that to accomodate the data, the scale on the Y axis for renal tubules is smaller than that for the other five sites.

like those from chronologically older, nondiabetic subjects.
    If BL accumulation is related to rates of cell turnover,
as some data indicate (28,29,31), the findings reported in
this study support the idea that excessive accumulation of
BL in DM is related to an acceleration of the rates with which
cells are dying and are being replenished. Whether metabolic
aberrations of the diabetic state speed up the cell turnover
rates, or diabetic individuals have an inherent cellular
defect that renders these cells excessively vulnerable to
injury cannot be resolved at the present time.
    An alternative possibility that explains BL accumulation
but not other findings (13,31) is a decreased capacity to
resorb old BL layers. Because of similarities in the patterns
of BL accumulation and differences in its degree a failure to
resorb old BL layers would also be a normal aging phenomenon
that is advanced in time or severity in the presence of DM.

REFERENCES

1. Ashworth, G.T., Erdmann, R.R. and Arnold, N.J., *Am. J.
    Path*. 36:165, 1960.
2. Bencosme, S.A., West, R.O., Kerr, J.W., Wilson, D.L.,
    Am. J. Med. 40:67-77, 1966.
3. Bloodworth, Jr., J.M.B., Engerman, R.L., Camerini-Davalos,
    R.A., Powers, K.L., *IN:* "Early Diabetes", (Camerini-
    Davalos, R.A. and Cole, H.S., Editors), Academic Press,
    New York, pp. 279-295, 1970.
4. Churg, J. and Dachs, S., *Path. Anal*. 1:148-171, 1966.
5. Cuppage, F.F., Neagoy, D.R. and Tate, A., *Lab. Invest*.
    17:660-674, 1967.
6. Danowski, T.S., Khurana, R.C., Gonzalez, A.R and Fisher,
    E.R., *Am. J. Med*. 51:757-766, 1971.
7. Danowski, T.S., Fisher, E.R., Khurana, R.C., Nolan, S.
    and Stephan, T., *Metabolism* 21:1125-1132, 1972.
8. DiScalla, V.A., Salomon, M., Grishman, E. and Churg, J.,
    Arch. Path. 84:474-485, 1967.
9. Fuchs, U., *Frank.Zeitsch.fur Path*. 73:318-327, 1964.
10. Giacometti, L. and Parakkal, P.F., *Nature* 223:514-515,
    1969.
11. Gonzales-Angulo, A., Fraga, A. and Mintz, G., *Am. J. Med*.
    45:873-879, 1968.
12. Green, C.R., Ham, K.N. and Tamge, J.D., *Arch. Path*. 98:
    156-160, 1974.
13. Hamlin, C.R., Kohn, R.R. and Luschin, J.H., *Diabetes* 24:
    902-904, 1975.
14. Kalderon, A.E., Bogaars, H.A. and Diamond, I., *Am. J.
    Med*. 55:485-491, 1973.

15. Kilo, C., Vogler, N. and Williamson, J.R., *Diabetes* 21: 881-905, 1972.
16. Krawczyk, W.S., *J. Cell Biol.* 49:247-263, 1971.
17. Madroza, A., Suzuki, Y. and Churg, J., *Am. J. Path.* 61: 37-56, 1970.
18. McFadden, P.M. and Berenson, G.S., *Circulation* 45:808-814, 1972.
19. Norton, W.L., *Lab. Invest.* 22:301-308, 1970.
20. Pardo, V., Perez-Stable, E., Alzamora, D.B. and Cleveland, W.W., *Am. J. Path.* 68:67-80, 1972.
21. Siperstein, M.D., Unger, R.H. and Madison, L.L., *J. Clin. Invest.* 47:1973-1999, 1968.
22. Snedecor, G.W. and Cochran, W.G., *Statistical Methods,* Sixth Ed., Iowa State Univ. Press, Ames, Iowa, 1967.
23. Vracko, R., *Circulation* 41:271-283, 1970.
24. Vracko, R., *Circulation* 41:285-297, 1970.
25. Vracko, R. and Benditt, E.P., *J. Cell Biol.* 47:281-285, 1970.
26. Vracko, R., *Virchows Arch. (Path. Anat.)* 355:264-274, 1972.
27. Vracko, R. and Benditt, E.P., *J. Cell Biol.* 55:406-419, 1972.
28. Vracko, R. and Benditt, E.P., *Am. J. Path.* 75:204-224, 1974.
29. Vracko, R., *Diabetes* 23:94-104, 1974.
30. Vracko, R., *Am. J. Path.* 77:314-338, 1974.
31. Vracko, R. and Benditt, E.P., *Fed. Proc.* 34:68-70, 1975.
32. Williamson, J.R., Vogler, N.J. and Kilo, C., *Am. J. Path.* 63:359-370, 1971.
33. Williamson, J.R., Vogler, N.F. and Kilo, C., *Med. Clin. N. Am.* 55:847-860, 1971.
34. Yodaiken, R.E., Seftel, H.C., Kew, M.C., Lillenstein, M. and Ipp, E., *Diabetes* 18:164-175, 1969.

Supported by Veterans Administration Research funds and in part by NIH Grant HL-03174-20.

INCREASE IN BASEMENT MEMBRANE MATERIAL

IN ACUTE DIABETIC GLOMERULAR HYPERTROPHY

Ruth Østerby,

H. J. G. Gunderson and J. P. Kroustrup

The University Institute of Pathology
and
The Second University Clinic of Internal Medicine
Aarhus, Denmark

ABSTRACT: The renal hyperfunction which is present in early diabetes (increased GFR) prompted stereological studies of the glomerular structures. Determination of the size-distribution of glomeruli from a kidney biopsy material showed that the volume of the glomerular tuft is increased in early diabetes. Measurements of the surface area of the capillary walls showed furthermore that the total surface area of the glomerular capillaries is increased by 80%. This morphological change may well account for the increased filtration rate in early diabetes.

Earlier studies of the initial phases in the development of basement membrane (BM) thickening - the characteristic feature of diabetic microangiopathy - have shown that the thickness is normal at the onset of disease. The increased capillary surface area in conjunction with a normal BM thickness imply that a considerable increase in the absolute amount of BM material is present at the time of diagnosis. This finding is taken to indicate that the synthesis of BM material has been increased in the weeks of metabolic aberrations preceeding the diagnosis. That this change is the earliest

495

manifestation of the diabetic microangiopathy is
advanced as a working hypothesis. These rapid
changes in BM metabolism in the acute diabetic state
are presently studied further in animal models.

It has been well known for many years that one consequence of the metabolic disorder, diabetes mellitus, is
widespread alterations of the vascular walls, constituting
the diabetic angiopathy (3). It has also been known for
some years that the morphologic lesion of the diabetic
vessels appears as increased thickness of the basement membrane. When the diabetic disease presents clinically -
acutely as it usually does in young patients - the basement
membrane thickness is normal (11). But over the years of
diabetic life thickening of the basement membrane in vessel
walls progresses steadily, and this may finally lead to
death due to organ damage in the kidney, the heart or the
brain. The increasing thickening of the basement membrane
with increasing duration of diabetes has been firmly established for vessels at many different sites.

We have recently observed that also in acute diabetes -
i.e., after only a few weeks' duration of symptoms - changes
are present in the renal corpuscles which indicate that
already after such a short time period of metabolic derangement there is a demonstrable change in basement membrane
metabolism in the diabetic organism. The term "acute
diabetic glomerular hypertrophy" is used for this abnormality,
since it shows as an increase in the size of the glomerular
tuft. As it will be discussed later, the glomerular hypertrophy develops in all probability as a consequence of some
metabolic factor or factors. It will be considered here
especially with regard to the implications that the findings
have on the concepts of basement membrane turn-over.

One obvious question to raise is whether these early,
acute events are in any way linked to those developing in
long-term diabetes.

The stage of kidney hyperfunction covering usually the
first one or two decades of diabetes has been repeatedly
described (1,4,9). Glomerular filtration rate is elevated
shortly after the acute clinical onset of juvenile diabetes
and remains high for many years.

In the lack of any hallmarks for the mechanisms involved
in the increased GFR all the theoretically possible explanations have been considered, namely an increased intravascular
pressure, a qualitative change in the filtering apparatus
and finally, a change in the area available for filtration.

Clinical studies of the quality of the filter showed that in diabetics with increased GFR, the profile of graded dextran clearances was normal. The pressure gradient over the glomerular capillary wall has not been determined in patients.

The third theoretically possible explanation - increase in filtering surface - could be checked by measurements of the glomerular structures, and we, therefore, set out to do so.

The first part of this study concerned the determination of glomerular size. Kidney biopsies obtained from young diabetics at their first admission to the hospital with a recently diagnosed diabetes were available for the study. Part of the biopsy was embedded in paraffin for light microscopy. The remaining part was embedded for electron microscopy.

The glomerular profiles appearing on a section of kidney tissue show a wide range of sizes. The variation has two different sources. One is that one particular section is located at varying distances from the equatorial plane of the different glomeruli. The other is the variation in actual tuft size. It is possible, however, statistically to eliminate the variation due to section level and thus calculate the true distribution of glomerular sizes. We have applied the method of Saltikov (6), which from the distribution of cross-sectional tuft areas, leads to distributions of glomerular volume.

Figure 1 shows the calculated glomerular volumes in millions of cubic microns ($M\mu^3$). In the recently diagnosed diabetics the glomerular volumes were about twice those in non-diabetics (12). With point-counting methods the volume of the capillary lumina expressed as percentage of tuft volume was determined. This relative volume multiplied by the mean glomerular volume in turn leads to the estimate of total capillary lumen in each patient. Luminal volumes were also nearly doubled in the diabetics.

Increase in luminal volume could be due either to an unfolding of the capillary wall or to growth of the surface. This question was clarified by measuring the capillary wall area at low magnification-electron microscopy.

Within the tuft three different interfaces were defined and delineated as illustrated in Figure 2. One was the interface between capillary lumen and mesangial region, the second was that between mesangial region and urinary space, and the third was the surface of the peripheral basement membrane. A line grid as shown was superimposed upon the electron micrographs and the number of intersections

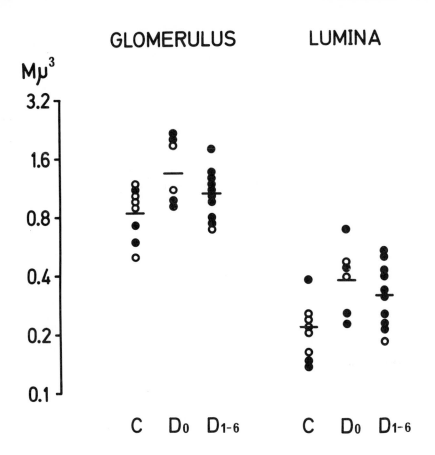

Fig. 1. *Volumes of the glomerular tuft and of the capillary lumen (log scale, in millions of μ^3 per 1.73 m^2 body surface) C: controls; D_O: newly diagnosed diabetics; D_{1-6}: patients with 1-6 years' duration of diabetes.*

between the test-lines and each of the three interfaces was counted. From the number of these intersections the densities of the respective surfaces are calculated by standard stereo-logical methods (2). Multiplying the surface density by the mean glomerular volume - as calculated from the paraffin embedded part of the biopsy, the absolute surface area per mean glomerulus was obtained.

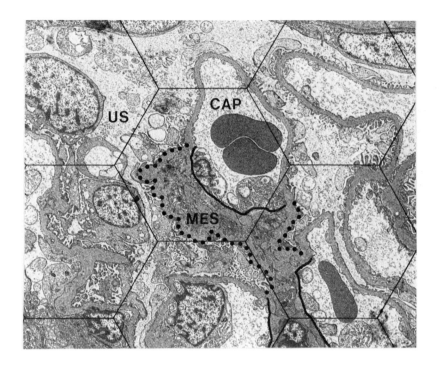

Fig. 2. Part of a glomerular tuft with the grid employed for the determination of surface densities of three interfaces between: 1) capillary lumen and urinary space; 2) mesangial region and urinary space (indicated for one mesangial region with a dotted line); and 3) mesangial region and capillary lumen (full drawn line). The size length of the hexagon corresponds to 7.4 μ.

These measurements showed that the absolute area of the peripheral glomerular capillary wall was increased in the recently diagnosed diabetic patients (Figure 3). The other two interfaces, i.e., the surfaces of the mesangial regions, did not show any statistically significant increase (2). There was an 80% increase in the capillary wall area and it may well account for the increased glomerular filtration rate. This positive morphologic finding makes other speculative assumptions on possible mechanisms of the hyperfunction unlikely.

The observed morphological changes have some noteworthy implications in terms of basement membrane metabolism. The increase in the absolute area of the peripheral basement membrane together with the fact that the thickness of the same basement membrane is unchanged signify that glomeruli in early diabetics contain increased quantities of basement membrane.

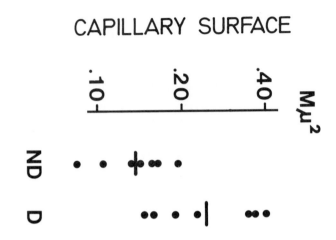

CAPILLARY SURFACE

Fig. 3. The absolute surface area per mean glomerulus (log scale, millions of μ^2 per 1.73 m^2 body surface) of the peripheral capillary wall in controls (ND) and newly diagnosed diabetics (D).

Considering the mechanisms that may be responsible for this accumulation we now switch over to the theoretical plane. Obviously nothing can be concluded regarding the steps which are affected in the normal balance between synthesis and break-down. However, taking into account some experimental data and a number of rough calculations one may come to the suggestion that increase in synthesis has occurred. A number of facts support the assumption that the enlargement of the renal corpuscles is brought about by the metabolic derangement: An increase in kidney size has been demonstrated by X-ray studies in similar groups of diabetics (5). Kidney size is again reduced to normal upon strict insulin treatment. It is furthermore possible to provoke similar changes in the nephron in rats made diabetic with streptozotocin. It has been shown in these animals that the kidney weight increases in parallel with the blood glucose level (8). Biochemical studies of such enlarged kidneys have excluded mere water retention as the cause. A few days after the onset of diabetes in rats, the enlarged kidneys

contain increased amounts of protein, and have an increase
in RNA to DNA ratio (7). The total kidney size is not necessarily an expression of glomerular size, since the glomerular
volume is only about 2 per cent of the total kidney volume.
However, increase in glomerular size has also been shown to
be present in diabetic rats after a few weeks.

Considering what the mechanism that may lead to this -
for the present, hypothetical - increase in basement membrane
synthesis, one moves further into the theoretical plane.

For the moment our working hypothesis is that enlargement of the glomerular capillary wall, which most likely
reflects increased synthesis of the peripheral basement
membrane, is triggered by the metabolic aberrations characterizing diabetes mellitus; this enlargement constitutes the
very earliest phase in the development of diabetic microangiopathy. After a short time, growth of the capillary
surface is stopped by mechanisms counteracting the glomerular
hyperfunction. The persistently high rate of synthesis then
results in an increasing thickness of the basement membrane,
eventually leading to the glomerulo-sclerosis of long-
standing diabetes. This unitary hypothesis implies that
information about the diabetic microangiopathy can be obtained from experiments of much shorter duration than previously
thought necessary.

REFERENCES

1. Ditzel, J. and Schwartz, M., *Diabetes* 16:264-267, 1967.
2. Kroustrup, J.P., Gundersen, H.J.G. and Østerby, R.,
   "Glomerular size and structure in diabetes mellitus,
   III. Early enlargement of the capillary surface."
   Submitted for publication.
3. Lundbaek, K., *In:* "Long-term Diabetes" Munksgaard,
   Copenhagen 1953.
4. Mogensen, C.E., *Dan. Med. Bull.* 19, Suppl. 3, 1972.
5. Mogensen, C.E. and Andersen, M.J.F., *Diabetologia* 11:
   221-224, 1975.
6. Saltikov, S.A., *In:* Stereology" (H. Elias, Ed.)
   Springer Verlag, Berlin, pp. 163-173, 1967.
7. Seyer-Hansen, K., *Clin. Sci. Mol. Med.* 51:221, 1976.
8. Seyer-Hansen, K., "Renal hypertrophy in experimental
   diabetes: Relation to severity of diabetes. Submitted
   for publication.
9. Stalder, G., Schmid, R. and Wolff, M.V., *Dtsch. med.
   Wschr.* 85:346-350, 1960.

10. Williamson, J.R. and Kilo, C., "Current status of capillary basement membrane in diabetes mellitus. *Diabetes* (in press).

11. Østerby, R., *Acta med. Scand.*, Suppl. 574, 1975.

12. Østerby, R. and Gundersen, H.J.G., *Diabetologia* 11:225-229, 1975.

CORRELATION BETWEEN BASEMENT MEMBRANE ANOMALIES

AND THE INCORPORATION PATTERN OF ^{14}C-PROLINE

AND ^{3}H-GLUCOSAMINE IN DIABETIC CONNECTIVE TISSUES[†]

L. Robert, P. Kern*, F. Regnault*,

H. Bouissou**, G. Lagrue***, M. Miskulin

and A. M. Robert

Laboratoire de Biochimie du Tissu Conjonctif
Institut de Recherche
sur les Maladies Vasculaires
C. H. U. Henri Mondor
Creteil, France

SUMMARY: A biopsy technique was used for the exploration
of the regulation of matrix macromolecule biosynthesis
in diabetes. Conjunctiva, skin and gingiva of human
diabetics and non-diabetics, skin and conjunctiva of KK
mice and conjunctiva of streptozotocin treated rats were
investigated. The incorporation of ^{3}H-proline, ^{14}C-
lysine and ^{14}C-glucosamine was determined (four hours,
37°C, in vitro) in sequential extracts obtained by one

  * Centre de Rech. sur les Maladies de la Rétine,
    CHU Cochin, 75014 Paris, France.
 ** Service d'Anatomie Pathologie, CHU Rangueil,
    31000 Toulouse, France.
*** Service de Néphrologie, Inst. Rech. Mal. Vasculaires,
    CHU Henri-Mondor, 94000 Créteil, France.

  † Supported by CNRS (ER No. 53), DGRST (74-7-0590, 74-7-
    0591, 74-7-0599) and INSERM (ATP 7) of France.

*of several "chemical dissection" methods devised for
this purpose. The specific activity and the distribution of total label was determined and compared to
control tissues.*

*Several parameters of the incorporation pattern
were found to be significantly different between
diabetic and non-diabetic controls. These biochemical
modifications were parallel to the morphological modifications such as increased basement membrane width, the
presence of cross striated collagen fibrils in the
basement membrane, the lysis of subpapillary elastic
fibrils and degenerative alterations of dermal collagen
bundles and fibroblasts. These results suggest that
diabetic connective tissue is characterized by a
(qualitative and/or quantitative) disorder of the
regulation of the biosynthesis of the macromolecules
of the intercellular matrix. The described biopsy
technique is a useful one for the early diagnosis of
these disorders.*

Increasing thickness of capillary basement membranes is
one of the important stigmata of diabetes at the level of the
intercellular matrix (1,2,3). It appears to be attributable
to a disorder of the biosynthesis of basement membrane
collagen and/or of the associated (structural) glycoproteins;
the nature of the disorder is still controversial (4,5,6).
The question, therefore, arises whether this anomaly of matrix
macromolecule biosynthesis is strictly localized to the cells
involved in basement membrane synthesis (capillary endothelial
cells) or whether it also concerns other cells involved in
matrix macromolecule biosynthesis. In order to answer this
question, a connective tissue biopsy technique was developed
and used to explore matrix macromolecule biosynthesis in
diabetic patients, normal controls as well as in some animal
models of diabetes such as KK-mice and streptozotocin induced diabetes in rats. Normal controls were Swiss mice and
untreated Wistar rats of the same age and sex. The results
obtained by this biopsy technique were compared to morphological findings such as the estimation of the width of capillary
basement membranes and the quality of the dermal subpapillary
connective tissue. A short review of these experiments will
be given here, the details were described elsewhere (6,7,8,9,
10,11,12).

MATERIALS AND METHODS

The diabetic patients and non-diabetic controls were selected in the Ophthalmology Department of the Cochin Hospital Medical School of Paris, the Purpan Hospital of Toulouse and at the Nephrology Department of the Henri-Mondor Hospital at Créteil. The glucose tolerance test, serum insulin levels and standard medical and biological data were recorded. The biopsies were carried out at the level of the conjunctiva (6), skin (11) or gingiva (10). The freshly excised tissue (approx. 50 to 500 mg fresh weight) was placed in a vial containing 5 ml of Krebs-Ringer phosphate or MEM or other appropriate culture medium, 5 μCi of ^{3}H-proline (25 Ci/mM) and 5 μCi ^{14}C-glucosamine (120 mCi/mM) or 5 μCi ^{14}C-lysine (200 mCi/mM) (CEA, Saclay) and placed for four hours in a shaking incubator. After centrifugation and over-night washing in frequent changes of cold medium with 0.5% proline and glucosamine or lysine, the tissues were sequen-tially extracted with the following solvent: a) 1M $CaCl_2$-tris-citrate (CTC buffer) pH 7.5. The precipitate which occurs on dialysis and contains soluble collagen was separately analyzed (CSC-fraction, see Table I); b) collagen-ase or trichloracetic acid 2.7%, 30 min. at 90°C; c) 8M urea 0.1M mercaptoethanol and the final residue was "solubilized" with elastase or with 1M KOH in 80% v/v aqueous ethanol (6, 11). The order of the extractions was varied in some experi-ments (for details see (3,6,10,11)). Each extract was ex-haustively dialyzed and the radioactivity and protein contents determined. Results are expressed as dpm/mg prot. and as percent of total radioactivity incorporated in a given fraction. The streptozotocin treatment of rats and the biological parameters of the spontaneously diabetic KK-mice were described (7,12,13). The measurement of basement mem-brane thickness was carried out on electron micrograms of conjunctival capillaries essentially as recommended by Siperstein (1,2,6,7). Light microscopic and electron micro-scopic evaluation of elastin and collagen morphology of the subpapillary layer of skin was done as described (14,15,16).

The statistical evaluation of the results obtained on 260 human conjunctival biopsies, on about 20 skin biopsies and on about 60 gingival biopsies were carried out by Professor Valensi (Faculté de Droit, Paris) on an IBM 360 computer.

RESULTS

Morphological Observations

The increase of capillary basement membrane thickness in
diabetics could be confirmed and extended. Figure 1 shows
that this increase is progressive with age, follows a very
similar shape to the one of the control (non-diabetic) curve
but starts at a higher level. It also could be observed in
about half of the diabetic population that fine, cross
striated fibrils appear within the normally amorphous basement
membrane (3,6) (Figure 2). This observation suggested the
assumption that a different collagen from the one normally
present (Type IV) may occur in these basement membranes. The
morphology of the fibrils would be compatible with either
Type I or III collagens. Identical observations were done
on the thickened glomerular basement membranes of KK-mice (7).

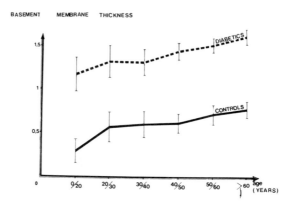

Fig. 1.   Basement membrane thickness (μ) of
conjunctival capillaries of normal and diabetic
subjects as a function of age (for details see (3,
6). The differences between normal and diabetics
are statistically significant (P <0.01).

*Fig. 2. Diabetic conjunctival capillary.
Cross striated fibrils (arrow) appear in the
thickened basement membrane (BM).*

   The microscopical and electron microscopical studies of
human dermal biopsies revealed the following modifications:
the fine subpapillary elastic fibrils are disappearing
(elastolysis) at an earlier age in most diabetics than in
non-diabetic controls (14,15). The collagen bundles also
show signs of disaggregation and the fibroblasts are filled
with electron-dense deposits and have a dilated endoplasmic
reticulum (11). Similar modifications appear at a significantly later age and on a milder level in the control
population. The morphological changes suggest therefore a
precoceous aging of connective tissues in diabetics.

Biochemical Results

   The comparison of the two kinds of parameters chosen:
specific radioactivity and the distribution of total label,
both determined in the sequential extracts obtained by
several "chemical dissection" procedures on the three different human tissues (conjunctiva, skin, gingiva) and on the
skin and conjunctiva of KK-mice and the conjunctiva of
streptozotocin treated rats all showed significant deviations
as compared to the same tissues taken from untreated or non-
diabetic controls. Table I shows the modifications observed
for conjunctival biopsies in humans, in KK-mice and in
streptozotocin treated rats. In all three species studied
the incorporation of proline in the TCA, expressed as percent

L. ROBERT *et al.*

TABLE I

Incorporation of ^3H-proline and ^{14}C-glucosamine in diabetic and control conjunctival biopsies. The extraction of biopsies was carried out as described in methods: CSC (soluble collagen fraction) obtained from the 1M $CaCl_2$-tris-citrate extract by dialysis; TCA, trichloracetic acid extract containing polymeric collagen; urea extract, containing mainly the structural glycoprotein components. Those parameters which exhibited statistically significant variations calculated from 260 human and 50 mouse and rat biopsies are underlined (P <0.01). Standard deviation for most parameters was ± 10%. For more details see reference 3, 6, 7 and 12.

SA = specific radioactivity of fraction, dpm/mg/prot. 10^3.
% TR = percent of total radioactivity in a given fraction.

Species	Extract	Normal N or Diab. D	proline		glucosamine	
			SA	% TR	SA	% TR
Human	CSC	N	450	8	32	11
		D	350	3	26	6
	TCA	N	60	6	8	12
		D	70	13	2	14
	Urea	N	650	50	22	48
		D	950	65	10	52
Mice (°) (KK and Swiss)	CSC	N	12	12		
		D	10	4		
	TCA	N	0.4	4		
		D	1.5	9		
	Urea	N	6	36		
		D	15	55		
Rats Wistar Strepto-zotocin	CSC	N	500	12	36	12
		D	400	8	20	16
	TCA	N	80	6	15	12
		D	100	11	3	13
	Urea	N	900	46	80	44
		D	1700	61	20	40

(°) For these experiments ^{14}C-proline (200 mCi/mM) was used.

of total label incorporated, increased and the specific
activity of glucosamine in all extracts decreased compared
to the controls. These differences were statistically significant. Essentially similar results were obtained with human
skin and gingival biopsies (10,11). In diabetic human dermis,
the specific activity of ^{14}C-lysine increased strongly in the
urea extract and decreased in the collagenase and elastase
extracts (11). In diabetic gingiva the results were very
similar to those found for the conjunctival biopsies (10).
The deviation from normal may be more conspicuous in different extracts according to the tissue investigated and the
extraction procedure used. In skin biopsies the specific
activity of ^{14}C-lysine of the urea extract showed the strongest deviation from normal (11). In streptozotocin treated
rats and in KK-mice conjunctival biopsies, the specific
activity of proline was much higher in the urea extract than
in the normal controls.

These results suggest the existence of a more general
disturbance of the regulation of matrix macromolecule biosynthesis in diabetics (6,7,8,9,10).

DISCUSSION

The above results suggest that besides the endocrinologi-
cal disturbances, diabetes appears also to be characterized
by a disorder of the regulation of the biosynthesis of intercellular matrix macromolecules. This disorder can be detected
by the described connective tissue biopsy technique. The
results obtained on several tissues and several species, in
human and experimental diabetes, suggest a more widespread
disorder of matrix macromolecule biosynthesis than formerly
assumed and not limited to the basement membrane only. A
satisfactory correlation was found between the biochemical
parameters and the morphological findings such as the width
of the capillary basement membranes of conjunctiva, the
presence of fine cross striated (collagen different of Type
IV) fibers in the basement membranes, the early disappearance
of subpapillary elastic fibrils of skin accompanied by
altered morphology of the collagen bundles and of the fibroblasts. The lipid content of the skin (sterols) does also
increase (14,15,16). Most of these results may be interpreted
as an accelerated aging of connective tissues in diabetics as
compared to non-diabetics. These changes may also be related
to the alterations of blood vessels seen in diabetics and
incriminated in the earlier and more severe outcome of
arteriosclerotic changes observed in these patients (17,18).

It can therefore be concluded that connective tissue involvement is a basic feature of diabetes which can be considered
as a disease resulting partially at least from the anomalie(s)
of the regulation (qualitative and/or quantitative) of the
biosynthesis of the macromolecules of the intercellular matrix.

REFERENCES

1. Siperstein, M.D., Unger, R.N. and Madison, L.D., *J. Clin.
    Invest.* 47:1968, 1973.
2. Regnault, F. and Bregeat, P., *In:* "6th European Confer-
    ence on Microcirculation, Aalborg, 1970", (Karger,
    Basel) p. 149-156, 1971.
3. Regnault, F. and Kern, P., *Bibl. Anat.* 11:459-467,
    (Karger, Basel), 1973.
4. Beisswenger, P.S. and Spiro, R.G., *Diabetes* 22:180, 1973.
5. Kefalides, N.A., *J. Clin. Invest.* 53:403, 1974.
6. Kern, P., Regnault, F. and Robert, L., *Biomedicine* 24:
    32, 1976.
7. Kern, P., Laurent, M. and Regnault, F., *Europ. J. Clin.
    Biol. Res.* 17:882, 1972.
8. Robert, L., *Presse Méd.* 79:2277, 1971.
9. Robert, L. and Robert B., *In:* "Journées Ann. de
    Diabétologie de l'Hôtel-Dieu", (Flammarion, Paris)
    p. 203-216, 1970.
10. Robert, L., Rivault, A., Moczar, M., Miskulin, M. and
    Robert, A.M., *In:* "Journées de Diabétologie de l'Hôtel-
    Dieu", (Flammarion, Paris) p. 35-44, 1976.
11. Moczar, M., Allard, R., Ouzilou, J., Robert, L.,
    Bouissou, H., Julian, M. and Pieraggi, M.T., *Pathol.
    Biol.* 24:329, 1976.
12. Duhault, J., Lebon, F., Regnault, F. and Kern, P.,
    Biorheology 11:167, 1974.
13. Kern, P., Picard, J., Caron, M. and Veissière, D.,
    Biochim. Biophys. Acta 389:289, 1975.
14. Bouissou, H., Pieraggi, M.T., Julian, M. and Douste
    Blazy, L., *In:* "Frontiers of Matrix Biology" (Karger,
    Basel) Vol. 1, p. 190-211, 1973.
15. Douste Blazy, L., Bouissou, H., Pieraggi, M.T., Julian,
    M. and Charlet, J.P., *In:* "Frontiers of Matrix Biology"
    (Karger, Basel) Vol. 2, p. 76-88, 1976.
16. Bouissou, H., Pieraggi, M.T., Julian, M. and Douste
    Blazy, L., *In:* "Frontiers of Matrix Biology" (Karger,
    Basel) Vol. 3, p. 242-255, 1976.
17. Malathy, K. and Kurup, P.A., *Diabetes* 21,1162, 1972.
18. Telner, A. and Kalant, N., *Atherosclerosis* 20:81, 1974.

ELEVATED KIDNEY GLUCOSYLTRANSFERASE ACTIVITY:

A BIOCHEMICAL DEFECT IN

INHERITED DIABETIC MICROANGIOPATHY

R. A. Camerini-Davalos, A. S. Reddi,

C. A. Velasco, W. Oppermann,

and T. Ehrenreich

New York Medical College
New York, New York

ABSTRACT: KK mice with genetically transmitted diabetes develop impaired tolerance to oral glucose at 100 days of age (chemical diabetes). Light and electron microscopy of the kidney revealed significant increase in mesangial matrix by 60 days (prediabetes) with nodular and exudative lesions in older animals.

Determination of glucosyltransferase activity in the renal cortex of KK mice showed that this enzyme was significantly elevated between the 25th and 55th days of life when compared to that of age-matched Swiss albino (SA) mice. Later on, no difference in enzyme activity was found between both strains of mice. In addition, the glucosyltransferase activity in the 40-day-old F_1 hybrid (♂ KK X ♀ SA) mice was intermediary to that of parental strains, suggesting that this enzyme responds to changes in the genetic make-up of the animal.

Morphometric analysis of glomeruli showed that KK mice have significantly decreased number of epithelial cells and increased number of mesangial cells at 40 days

of age, when compared to those from age-matched SA mice.
However, the total number of cells remained the same in
both groups of mice. At 70 days of age, the number of
epithelial cells was still decreased but the number of
mesangial cells was equal. The total glomerular cell
count, however, was significantly decreased in KK mice.
These studies indicate that increase in the kidney
glucosyltransferase activity caused by hyperplasia of
mesangial cells precedes the increase in mesangial
matrix, which appears to be a biochemical defect in
inherited diabetic microangiopathy.

In recent years, the nature of the renal glomerular basement membrane has been the subject of numerous investigations
both in health and disease.   In diabetes mellitus, the renal
glomerulus undergoes biochemical and morphologic changes
characterized primarily by thickening of the basement membrane
and accumulation of basement membrane-like material in the
mesangial region.   These changes can ultimately lead to an
impairment in the filtration process of the glomerulus with
the presence of large quantities of protein in the urine.
Although the pathogenesis is poorly understood, the extent of
these changes in the renal glomerular basement membrane is
related to the duration of the disease.
Compositional studies have shown that the renal glomerular basement membrane (GBM) is glycoprotein in nature and
contains substantial amounts of glycine, hydroxyproline,
hydroxylysine, glucose and galactose (Kefalides and Winzler,
1966; Spiro, 1967a; Kefalides, 1968, 1973).   Two types of
carbohydrate units are present in the GBM (Spiro, 1967b).   One
unit is a heteropolysaccharide, while the other is a disaccharide unit consisting of glucose and galactose linked to the
hydroxylysine residues in the peptide chain.   Recent studies
have shown that the assembly of the disaccharide units is
mediated by two highly specific glycosyltransferases, namely
galactosyl- and glucosyl- transferases (Spiro and Spiro,
1971a, 1971b, 1971c).
In a subsequent study, Spiro and Spiro (1971d) reported
significantly elevated levels of the kidney glucosyltransferase activity in alloxan hyperglycemic rats, suggesting that
enhanced basement membrane synthesis occurs in diabetes.   The
observation that the glucosyltransferase activity of a number
of other tissues of the diabetic rat was not elevated, in
contrast to the kidney, further substantiates the involvement,
both morphologically and clinically, of kidney basement membrane in diabetes (Spiro and Spiro, 1971d).   Grant et al.
(1976) also observed an increase in kidney glucosyltransferase

activity in streptozotocin diabetic rats. It is apparent that
an understanding of the metabolism of the kidney basement
membrane in normal as well as in diabetic states provides a
basis for the rational therapy and possible prevention of
basement membrane abnormalities in diabetes.

Human diabetic-like microangiopathy (glomerulosclerosis,
retinopathy and peripheral gangrene due to small vessel disease) was described by us (Oppermann et al., 1968; Camerini-
Davalos et al., 1970; Oppermann et al., 1973; Ehrenreich
et al., 1973) and other investigators (Wehner et al., 1972;
Kern et al., 1972; Duhault et al., 1973) in the Japanese KK
mice. These mice are a strain of animals with genetically
transmitted spontaneous diabetes. Studies with light microscopy revealed that the diabetic glomerulosclerosis in KK mice
involves an increase in the mesangial matrix as early as two
months of age. About 60% KK mice demonstrate glomerulosclerosis by two months of age (Oppermann et al., 1973).
In some KK, nodular and exudative-like lesions are also
present.

Electron microscopy revealed an increase in mesangial
matrix which in the early phases and in young animals did
not result in widening of the mesangial space (Figure 1).

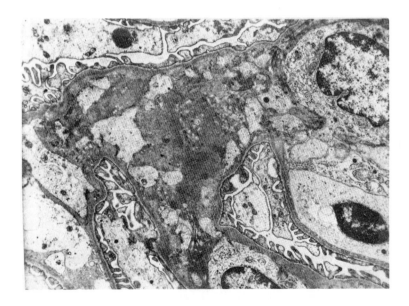

*Fig. 1. Electron micrograph illustrating slight
increase in mesangial matrix without widening of the
mesangial space. KK mouse 3 months of age. X 7800.*

With increase in mesangium, the mesangial space may be markedly widened (Figure 2). Usually, the mesangial accumulation
is accompanied by cells, although in a few instances there are
noted small, dense accumulations of mesangial matrix without
mesangial cells. As the mesangium proliferates, it grows into
the capillary wall and tends to become inserted between the
basement membrane and the endothelium. In advanced cases the
capillary lumen may be partly occluded by the mass of mesangium. Collagen fibrils are found in the mesangial matrix in
some instances. Electron-dense deposits are also seen in the
mesangial matrix, beneath the capillary endothelium, in the
basement membrane, and occasionally partly occluding the
capillary lumen. The latter appear to correspond to the
exudative lesions seen in light microscopy. The deposits
have an electron density generally less than that of the
basement membrane.

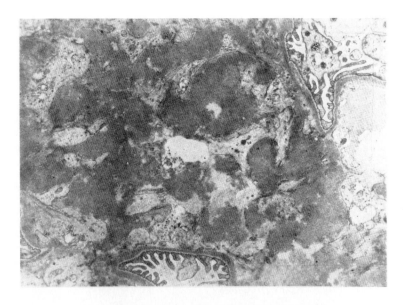

*Fig. 2.   Marked increase in mesangial matrix
with widening of the mesangial space.   KK mouse
14 months of age.   X 7800.*

Mean basement membrane thickening (BMT) of KK and control
mice Swiss albino I, calculated according to the methods of
Williamson and Siperstein, are summarized in Table I.
Although the mean basement membrane width values are higher
in KK than in control mice according to both methods, these

differences are not significant. The standard deviation for
mean BMT of KK mice is higher than that of control mice,
indicating that there is a greater variation in individual
values for KK mice.

TABLE I

Evidence of Segmental and Focal Basement Membrane
Thickening (BMT) in Control Swiss Albino I and KK Mice
(1 day to 21 months of age)

	Swiss Albino I (A) (n=16)	KK (A) (n=19)
Mean BMT		
Williamson method	1810 ± 610*	2310 ± 1660
Siperstein method	2390 ± 820	2960 ± 1980
Segmental variation in BMT		
Williamson method (SD$_1$)	442 ± 300	507 ± 556
Siperstein method (SD$_2$)	533 ± 316	609 ± 489
Focal variation in BMT		
Siperstein method (SD$_3$)	680 ± 538	803 ± 591

*
Mean ± S.D.

SD$_1$ and SD$_2$ reflect the segmental variation in mean
basement membrane thickening per capillary loop are averaged
in order to obtain the focal variation (SD$_3$) in the basement
membrane width around the circumference of the loop. The
difference for SD$_3$ between KK and control mice is greater
than those in segmental variation but also not statistically
significant (Table I). This observation indicates that
variation in basement membrane width around the circumference
of glomerular capillaries is greater than variation in mean
basement membrane width from vessel to vessel.

Oral glucose tolerance tests in KK mice, after an 18
hour fast, showed that these mice develop impaired tolerance
at 100 days of age (chemical diabetes). Before this age,
they are genetic prediabetics. Since the KK mice demonstrate
an increase in mesangial matrix by two months of age
(Oppermann et al., 1973), it was of interest to investigate
biochemical changes that lead to the development of glomerulosclerosis.

A significant increase in total protein synthesis by the
renal cortical tissue was observed in KK mice at 20 and 40
days of age, when compared to that of age-matched normal
control Swiss albino mice (Reddi et al., 1975). At 100 days
of age, no difference in protein synthesis was found between

KK and control mice.   Similarly, an increase in kidney ribo-
somal protein synthesis was observed in 40-day-old KK mice.
Determination of prolyl hydroxylase in the renal cortex of
KK mice showed an increase at 40 and normal activity at 100
days of age (Reddi et al., 1976, 1977).

   In addition, the glucosyltransferase activity was determined in KK and Swiss albino mice at different days of age
(Velasco et al., 1974).   As shown in Figure 3, the enzyme
activity gradually decreased with increasing age in both
groups of mice.   No difference in glucosyltransferase activity was observed between KK and control mice up to 20 days of
age.   However, the enzyme activity was found significantly
elevated in KK mice between 25 and 55 days of life, when
compared to that of age-matched control mice.   Later on, no
difference was found until 180 days of age between control
and KK mice.

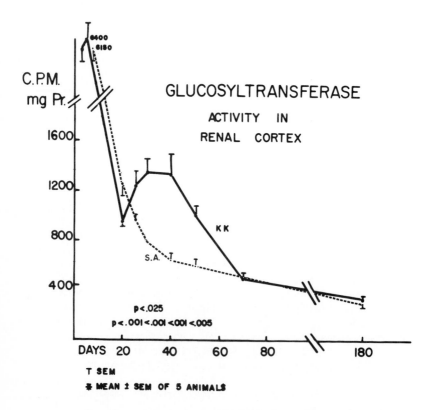

*Fig. 3.   Glucosyltransferase activity in the
renal cortex of control Swiss albino (SA) and KK
mice at different days of age.*

It is understandable from these biochemical studies that enzymes involved in the biosynthesis of the basement membrane glycoprotein in KK mice are elevated before there is an increase in mesangial matrix and normalized in the presence of glomerulosclerosis. Although the cause(s) for the normalization of protein synthesis, prolyl hydroxylase and glucosyltransferase activities in the presence of glomerulosclerosis is not clear, several possible explanations have been suggested (Oppermann et al., 1976, Reddi et al., 1977a).

Several questions arise as to the transient increase in glucosyltransferase activity in KK mice: Is the increase in enzyme activity specific to this species and is it genetically determined? Is the enzyme activity related to blood sugar levels? Is the increase in enzyme activity preceded or associated with an increase in the deposition of basement membrane-like material in the mesangium?

In order to investigate, whether the transient increase in glucosyltransferase activity is species specific, the enzyme activity was determined in KK mice and three other control mice, namely Swiss albino from our own colony, Swiss Webster and C57BL/6J, at 40 and 70 days of age. These studies have clearly shown that the enzyme activity was significantly increased in KK mice at 40 days of age, when compared to that of other control mice. However, at 70 days of age the enzyme activity was similar in all four groups of mice (Camerini-Davalos et al., 1977). This suggests that the observed increase in glucosyltransferase activity in the KK mouse is specific to this species.

In addition, the glucosyltransferase activity was measured in KK, Swiss albino and their F1 hybrid mice at 40 days of age, and their oral glucose tolerance tests were compared. The results are shown in Figure 4. As evident, the enzyme activity was significantly higher in KK than in Swiss albino mice. However, the glucosyltransferase activity in the F1 hybrid mice was found to be intermediate of both parental strains. Despite this difference, no conspicuous change in oral glucose tolerance tests (expressed as glucose area in the Figure) was observed among the three different groups of mice (Reddi et al., 1976a). These findings indicate that the glucosyltransferase is genetically determined and is dependent upon the diabetic gene-dosage of the animal. Furthermore, changes in glucosyltransferase activity can be observed without affecting the tolerance to oral glucose.

Recent experimental evidence indicated that the increase in enzyme activity is not necessarily associated with an increase in the deposition of the concerned product. Studies of experimental fibrosis (TAkeuchi and Prockop, 1969; Risteli and Kivirikko, 1974), pulmonary fibrosis in rats (Halme et al.

1970) and epinephrine-throxine induced arteriosclerosis in
rabbits (Fuller and Langner, 1970) have demonstrated an increase in prolyl hydroxylase activity before there was a
significant increase in tissue collagen content. This
suggests that the increase in enzyme activity precedes detectable structural abnormalities in these disease conditions.
Our data of increased prolyl hydroxylase and glucosyltransferase activities before the increased deposition of basement
membrane-like material in KK mice are, therefore, compatible
with these findings. Conversely, St Clair et al. (1975)
reported a reduction in prolyl hydroxylase activity in
plaque with increased hydroxyproline content from pigeon
aortas with naturally occurring atherosclerosis. This is in
good agreement with our decreased or normalized prolyl
hydroxylase and glucosyltransferase activities in kidneys
with diabetic glomerulosclerosis.

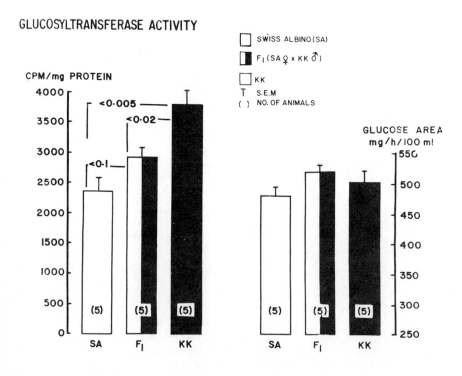

Fig. 4. Glucosyltransferase activity in the
renal cortex and glucose area in 40-day-old control
Swiss albino (SA), KK and their F_1 hybrid mice.

The question arises: How does genetics influence glucosyltransferase and associated biochemical parameters in KK mice? To answer this question, we performed quantitative morphometric analysis of glomeruli from 40 and 70-day-old KK and Swiss albino mice. These studies have shown that the KK mice have significantly increased number of mesangial cells at 40 days and an equal number of cells at 70 days of age, when compared to those from age-matched control mice (Camerini-Davalos et al., 1977). Since the mesangial cells probably lay down basement membrane-like material in the mesangium (Osterby, 1975), it can be suggested that the increased glucosylgransferase activity is associated with an increase in the number of mesangial cells. The normalization of enzyme activity may be related to the similar number of mesangial cells at 70 days of age. Although this is an indirect evidence, further experimental work is required to show the relationship between mesangial cell density and glucosyltransferase activity. These morphometric studies suggest that hyperplasia of mesangial cells at an early age of the animal may be an expression of inherited microangiopathy. However, the influence of other hormonal factors, if any, cannot be excluded.

In summary, our data suggest that the KK mouse is an ideal animal model for the studies of genetic diabetic microangiopathy, because this mouse develops glomerulosclerosis by two months of age. Biochemical changes in the kidney include elevated protein synthesis followed by an increase in prolyl hydroxylase and glucosyltransferase activities, which occur before the development of glomerulosclerosis and may be related to mesangial cell hyperplasia. The early increase in enzyme activity may be a sensitive indicator of glomerulosclerosis in those predisposed to diabetes. Further studies in this area are, therefore, indicated to elucidate the mechanisms responsible for the inheritance of diabetic microangiopathy in KK mice in particular and in genetic diabetes in general.

ACKNOWLEDGEMENTS

This study was supported in part by the General Research Support Grant RR-05398 from the General Research Support Branch, Division of Research Resources, National Institutes of Health, National Institutes of Health Training Grant AM 05617-05, Hope for Diabetics Foundation, Inc., and Diabetes Research Fund.

REFERENCES

Camerini-Davalos, R.A., Oppermann, W., Mittl, R. and Ehren-
    reich, T., *Diabetologia* 6:324, 1970.
Camerini-Davalos, R.A., Velasco, C.A., Reddi, A.S., Wehner, H.
    and Oppermann, W., *Diabetic microangiopathy in KK mice.
    I. Elevated kidney glucosyltransferase activity in pre-
    diabetes. (Submitted for publication)*
Duhault, J., Lebon, F., Boulanger, M. and du Boistesselin, R.,
    Bibl. Anat. 11:453, 1973.
Ehrenreich, T., Susuki, Y., Churg, J., Oppermann, W. and
    Camerini-Davalos, R.A., *In:* "Vascular and Neurological
    Changes in Early Diabetes", (R.A. Camerini-Davalos and
    H.S. Cole, eds.) Academic Press, New York, P. 271, 1973.
Fuller, G. and Langner, R.O., *Science* 168:987, 1970.
Grant, M.E., Harwood, R. and Williams, I.F., *J. Physiol.*
    257:56,, 1976.
Halme, J., Uitto, J., Kahanpaa, K., Karhunen, P. and Lindy, S.,
    J. Lab. Clin. Med. 75:535, 1970.
Kefalides, N.A. and Winzler, R.J., *Biochemistry* 5:702, 1966.
Kefalides, N.A., *Biochemistry* 7:3103, 1968.
Kefalides, N.A., *Int. Rev. Conn. Tiss. Res.* 6:63, 1973.
Kern, P., Laurent, M. and Regnault, F., *Rev. Europ. Etudes
    Clin. et Biol. XVII*, 882, 1972.
Oppermann, W., Treser, G., Ehrenreich, T., Lange, K., Levine,
    R. and Camerini-Davalos, R.A., *Clin. Res.* 16:347, 1968.
Oppermann, W., Ehrenreich, T., Patel, D., Espinoza, T. and
    Camerini-Davalos, R.A., *In:* "Vascular and Neurological
    Changes in Early Diabetes", (R.A. Camerini-Davalos and
    H.S. Cole, eds.) Academic Press, New York, Suppl. 2,
    p. 281, 1973.
Oppermann, W., Reddi, A.S., Velasco, C.A. and Camerini-Davalos,
    R.A., *Ann. N.Y. Acad. Sci.* 275:348, 1976.
Osterby, R., *Acta Med. Scand.*, Suppl. 574:1, 1975.
Reddi, A.S., Counts, D.F., Velasco, C.A., Oppermann, W. and
    Camerini-Davalos, R.A., *Clin. Res., XXIII*, 593A, 1975.
Reddi, A.S., Counts, D.F., Velasco, C.A., Oppermann, W. and
    Camerini-Davalos, *Endocr. Soc., 58th Ann. Meet.*, p. 230,
    1976.
Reddi, A.S., Oppermann, W., Reddy, M.P., Velasco, C.A. and
    Camerini-Davalos, R.A., *Experientia* 32:1237, 1976a.
Reddi, A.S., Counts, D.F., Oppermann, W. and Camerini-Davalos,
    R.A., *Prolyl hydroxylase activity in the kidney of normal
    and genetically diabetic KK mice (Submitted for publica-
    tion)*, 1977.
Reddi, A.S., Oppermann, W., Velasco, C.A. and Camerini-
    Davalos, R.A., *Exp. Mol. Pathol., (in press) 1977a.*

Risteli, J. and Kivirikko, K.I., *Biochem. J.* 114:115, 1974.

Spiro, R.G., *J. Biol. Chem.* 242:1915, 1967a.

Spiro, R.G., *J. Biol. Chem.* 242:1923, 1967b.

Spiro, R.G. and Spiro, M.J., *J. Biol. Chem.* 246:4899, 1971a.

Spiro, M.J. and Spiro, R.G., *J. Biol. Chem.* 246:4910, 1971b.

Spiro, R.G. and Spiro, M.J., *J. Biol. Chem.* 246:4919, 1971c.

Spiro, R.G. and Spiro, M.J., *Diabetes* 20:641, 1971d.

St. Clair, R.W., Toma, J.J., Jr., and Lofland, H.B., *Athero-sclerosis* 21:155, 1975.

Takeuchi, T. and Prockop, D.J., *Gastroenterology* 56:744, 1969.

Velasco, C.A., Oppermann, W., Marine, N. and Camerini-Davalos, R.A., *Horm. Metab. Res.* 6:427, 1974.

Wehner, H., Hohn, D., Faix-Schade, U., Huber, H. and Walzer, P., *Lab. Invest.* 27:331, 1972.

GLOMERULAR BASEMENT MEMBRANE

SYNTHESIS IN DIABETES

Margo P. Cohen

Wayne State University School of Medicine
Detroit, Michigan

ABSTRACT: The earliest microscopic change detectable
in the diabetic renal glomerulus consists of an increase
in the width of peripheral glomerular basement membrane
(GBM). This lesion, which progresses to the diffuse or
modular involvement of diabetic nephropathy represents
the ulstrastructural expression of alterations in GBM
biosynthesis in the diabetic state. Using glomeruli
prepared from rat renal cortex and incubated in the
presence of radioactive proline or lysine, the ability
of isolated glomeruli to synthesize and deposit basement
membrane in vitro has been established. GBM synthesis,
determined by the appearance of labeled hydroxylysine
in membranes obtained from sonicated glomeruli, is increased in preparations from streptozotocin-diabetic rats.
These changes are reversed in glomeruli obtained from
rats treated with insulin immediately following induction
of diabetes, but not in preparations from animals in which
insulin treatment was delayed. Glomerular lysyl hydroxylase activity, measured with chick embryo ^{14}C-protocolla-
gen substrate, is increased in diabetes, while glomerular
prolyl hydroxylase activity is unchanged. Administration
of insulin to streptozotocin-diabetic animals completely
restores glomerular lysyl hydroxylase to control levels.
These findings suggest that excess GBM in diabetes results

from the net effects of hyperglycemia and insulin
deficiency, and that metabolic control is an important
factor in allaying the progression of diabetic nephro-
pathy.

The earliest detectable microscopic change in the
diabetic renal glomerulus consists of an increase in the
width of the peripheral glomerular basement membrane and the
basement membrane-like material of the mesangial regions (1).
This lesion, which may progress to diffuse or nodular involvement, represents the ultimate ultrastructural expression
of as yet undefined biochemical alterations that promote the
accumulation of basement membrane in the diabetic state.
Prominent among the several lines of evidence suggesting
that the glomerular lesion results at least in part from
hyperglycemia and/or insulin deficiency is the demonstration
that peripheral basement membrane is normal at the onset of
juvenile diabetes and that measurable thickening becomes
electron microscopically manifest after 3-5 years of clinical
disease.   The extent of thickening also shows a positive
correlation with the duration of diabetes (2).   Furthermore,
glomerular changes have developed in kidneys transplanted
from non-diabetic donors to patients with diabetes (3).   These
studies argue against the concept of a separate genetic factor
in the pathogenesis of diabetic nephropathy, and suggest that
biochemical abnormalities resulting from the net effects of
hyperglycemia and insulin deficiency, which should precede
detectable microscopic changes, might reasonably be sought in
suitable animal models.

BASEMENT MEMBRANE SYNTHESIS *IN VITRO*

Glomeruli were isolated from rat renal cortex immediately
following sacrifice.   Details of the sieving process, and
collection and analysis of the glomeruli have been described
(4,5).   After timed incubation in modified Krebs' buffer and
in the presence of radioactive proline or lysine, samples are
centrifuged to separate the media from the glomeruli and the
basement membranes are collected by short, low-speed centrifugation after sonic disruption of the glomeruli in cold
buffered saline.   Washed sonicated basement membrane is
gently extracted with 0.1M acetic acid for separation of
acetic acid soluble and insoluble proteins.   This procedure
removes adherent interstitial collagen and separates newly
synthesized basement membrane, which is partially acetic

acid soluble (4,6) from mature basement membrane, which is
acetic acid insoluble (7). Labeled hydroxylysine and hydroxyproline isomers are determined by described methods on samples
subjected to acid or alkaline hydrolysis (6,8,9).

Amino acid analysis of acetic acid insoluble glomerular
basement membrane confirmed the adequacy of this preparation.
The ratio of 3-hydroxyproline to total hydroxyproline in these
samples was 1:5, and that of hydroxyproline to proline plus
hydroxyproline was 47%. After 2 hours of incubation,
3-hydroxy[14C]proline comprised 12% of total basement [14C]-
hydroxyproline. About 90% of the total [14C]hydroxylysine
in glomerular basement membrane obtained after 2 hours of incubation with [14C]lysine was glycosylated, with about 80%
of this in the disaccharide form.

When glomeruli from streptozotocin-diabetic rats were
incubated with [14C]lysine, a striking increase in basement
membrane synthesis was observed compared to control incubations (Figures 1 and 2). Diabetic rats were sacrificed 4-6
weeks after induction of diabetes, with a mean fasting blood
sugar over 400 mg%. Parallel increases in the incorporation
of labeled lysine and hydroxylysine into diabetic basement
membrane were observed at all time periods studied (10).

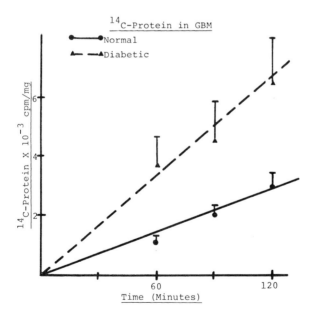

Fig. 1. Glomeruli were incubated for the times
indicated and basement membranes prepared after sonic
disruption. Results represent mean ± SEM of six
observations.

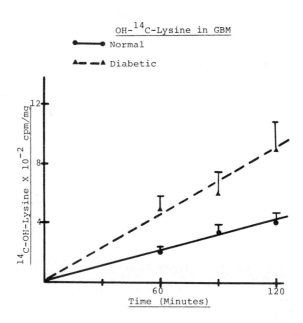

$OH-^{14}C$-Lysine in GBM
●——● Normal
▲– –▲ Diabetic

*Fig. 2. Values represent the mean ± SEM of
experiments described in Figure 1.*

Basement membrane synthesis was then determined in
glomeruli isolated from four groups of experimental animals
for which typical data is given in Table I. Diabetic animals
were sacrificed 5 to 6 weeks after streptozotocin injection.
Lente insulin was begun about 2 weeks after induction of
diabetes in one group of insulin-treated animals, and the day
following streptozotocin injection in the other group. The
degree of hyperglycemia was significantly diminished, although
not to euglycemic levels, by daily insulin administration.
Analysis of sonicated basement membrane obtained after incubation of glomeruli from these experimental groups with $[^{14}C]$-
lysine is seen in Figure 3. Basement membrane synthesis was
normalized in glomeruli from diabetic rats that had been
treated with insulin immediately following induction of
diabetes, but not in those in which insulin treatment was
delayed (11).

TABLE I

Body Weight and Blood Glucose Values
of Representative Animal Groups

	Body Weight (gm)	Blood Glucose (mg%)
Age-matched controls	280 ± 10 (9)	120 ± 10
Diabetic	140 ± 10 (9)*	450 ± 30*
Diabetic, immediate insulin therapy	290 ± 20 (8)#	230 ± 20#
Diabetic, delayed insulin therapy	270 ± 20 (8)#	270 ± 20#

Values represent mean ± SEM. Number of animals given
in parentheses.
* P < .01 comparing normal to diabetic values.
# P < .01 comparing diabetic to insulin-treated values.

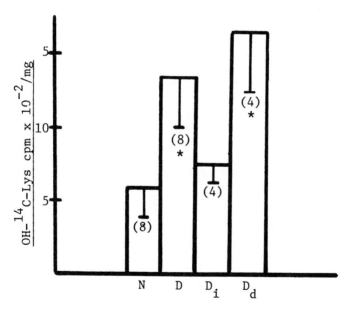

Fig. 3.   Basement membrane synthesis in glomeruli from
normal, diabetic and insulin treated rats. Glomeruli were incubated for 2 hours with [^{14}C]lysine and basement membranes
prepared as described in the text. Values represent the mean
± SEM of the number of observations given in parentheses.
* P <.05 compared to control; N = Normal; D = Streptozotocin
diabetic, untreated; D_i = Diabetic, immediate insulin treatment; and D_d = Diabetic, delayed insulin treatment.

GLOMERULAR PROTOCOLLAGEN HYDROXYLASE ACTIVITIES

In another series of experiments, corroboration of these
changes in basement membrane synthesis was sought by studying
the effect of experimental diabetes on glomerular protocollagen hydroxylase activities. Under-hydroxylated protocollagen
substrates were prepared after incubation of chick embryo
tibiae with either [^{14}C]lysine or [^{14}C]proline in the presence
of α,α'dipyridyl. The assay mixture contained α-ketoglutarate,
Tris-HCl buffer, ascorbate, ferrous sulfate, dithiothreitol
and bovine albumin in standard concentrations (12), and about
150 μg of ^{14}C-labeled hydroxylase substrate. Finally, 125-
250 μg/ml of glomerular 17,000 x g supernatant protein were
added as enzyme source. Optimum conditions for protocollagen
^{14}C-lysyl hydroxylation by rat glomerular supernatant fractions were established. Omission of co-factors or enzyme
resulted in markedly reduced levels of hydroxylation; activity
was proportional to time for up to 120 minutes of incubation
studied, and to enzyme protein concentration from 50-250 μg/ml.
Findings were similar with the radioactive prolyl-hydroxylase
substrate.

Lysyl hydroxylase activity was significantly increased
in the 17,000 x g supernatant prepared from glomeruli of
animals with streptozotocin diabetes (Figure 4). Mean blood
sugar in the experimental group was 300 mg%, with almost
undetectable immunoreactive insulin levels. In contrast,
lysyl hydroxylase in preparations from glomeruli obtained from
animals subjected to 95% pancreatectomy was not different from
that of control (Figure 5). The operated animals develop
overt hyperglycemia 6-10 weeks after pancreatectomy; in these
experiments, animals were sacrificed 4 weeks after operation.
Fasting blood sugars were thus not elevated in this experimental group at the time of sacrifice although insulin levels
were markedly diminished. These findings suggested that
chronic hyperglycemia, in addition to insulinopenia, is
requisite for changes in glomerular lysyl hydroxylase.

In other experiments, the effect of diabetes and insulin
therapy on protocollagen hydroxylase activities was determined (13). Glomerular supernatant protein was prepared from
control, diabetic 4 1/2 weeks post streptozotocin), and treated diabetic animals receiving daily subcutaneous injections
of insulin beginning the day after streptozotocin injection
and continued until the time of sacrifice. Blood sugar levels
diminished, although not to euglycemic levels, in animals
receiving insulin. The striking increase in glomerular lysyl
hydroxylase activity found in preparations from untreated
animals was completely normalized in samples prepared from
animals receiving insulin therapy (Figure 6).

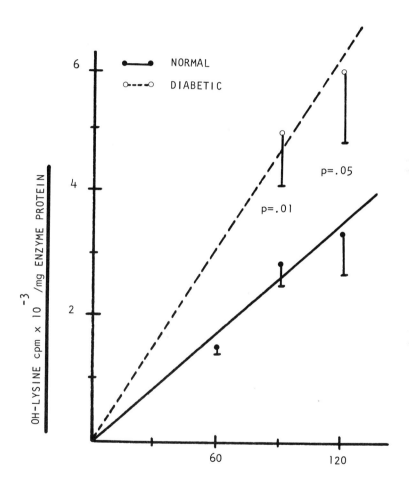

LYSYL HYDROXYLASE ACTIVITY IN 17,000 x g
SUPERNATANT FROM GLOMERULI OF NORMAL AND
STREPTOZOTOCIN-DIABETIC RATS

Fig. 4. *Values represent mean ± SEM of four*
experiments. Substrate concentration = 75 μg/ml.

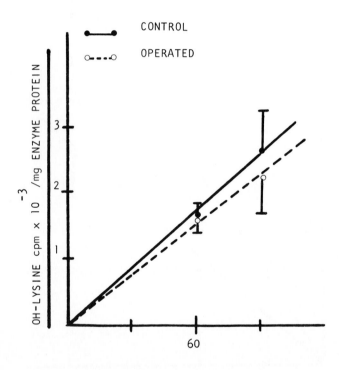

Fig. 5.   Values represent mean + SEM of four
observations; operated animals sacrificed four weeks
after 95% pancreatectomy with fasting euglycemia.
Substrate concentration = 150 µg/ml.

Lysyl Hydroxylase in Glomeruli from Normal,
Diabetic, and Insulin-Treated Rats

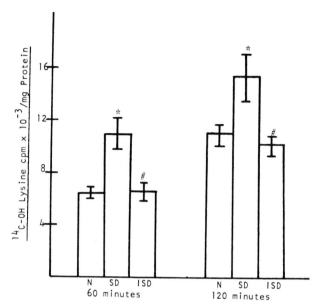

Fig. 6.   Results expressed as cpm ^{14}C-hydroxy-
lysine synthesized per mg. of enzyme protein.   Values
represent mean ± SEM of eight observations.
   N = Normal.
  SD = Streptozotocin-diabetic.
 ISD = Insulin-treated streptozotocin-diabetic.
 * P < .01, significantly different from normal.
 # P < .01, significantly different from strepto-
    zotocin-diabetic.

In contrast to these findings, glomerular prolyl hydroxylase activity in preparations from streptozotocin-diabetic
rats did not differ from that of control animals (Figure 7).
Daily insulin therapy to streptozotocin-injected rats also
did not significantly alter prolyl hydroxylase when compared
to activity in preparations from either normal or untreated
streptozotocin diabetic animals.

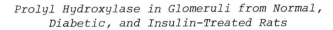

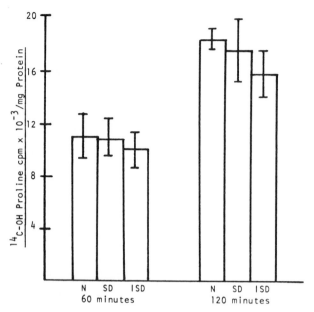

Fig. 7. Results expressed as cpm ^{14}C-hydroxy-proline synthesized per mg. of enzyme protein. Values represent mean ± SEM of four observations.
N = Normal.
SD = Streptozotocin-diabetic,
ISD = Insulin-treated streptozotocin-diabetic.

DISCUSSION

Elevated lysyl hydroxylase activity in diabetic glomeruli appears to reflect the increased basement membrane collagen synthesis, and its normalization with insulin therapy is a meaningful parameter with respect to correction of biosynthetic abnormalities which occur in the diabetic glomerulus. The lack of effect of streptozotocin-diabetes on glomerular prolyl hydroxylase, however, is puzzling. While it is possible that prolyl hydroxylase activity does not correlate with collagen biosynthesis in glomerular tissue, or that changes in hydroxy-proline-containing units of the glomerular basement membrane are not of particular importance in the diabetic state, the recent report by Grant et al. indicating that hydroxy[^{14}C]-proline synthesis is markedly increased in diabetic glomeruli makes this unlikely (14). Furthermore, no protein fraction with an unusually high hydroxylysine/hydroxyproline ratio has been isolated to date from glomerular basement membrane, and we have been unable to demonstrate lysine overhydroxylation in basement membrane isolated from incubated glomeruli. In a very recent report by Sato and Spiro, hydroxylysine/hydroxy-proline ratios range from 0.34 to 0.49 in less polar, colla-gen-like glomerular basement membrane subunits (15); it is thus possible that changes in a few critically occurring hydroxylysine residues are of importance.

Spiro has reported that renal cortical glucosyltransfer-ase, which effects the transfer of glucose to galactosyl-hydroxylysine units of basement membrane collagen, is in-creased in homogenates from alloxan-diabetic rats; this in-crease reverts towards normal with insulin therapy (16). Elevated glucosyltransferase probably occurs consequent to increased hydroxylysine synthesis, since lysyl hydroxylase and glucosyltransferase are sequentially-acting enzymes in the synthesis of basement membrane collagen. However, the in-creased lysyl hydroxylase observed in diabetic glomeruli may not represent the earliest change but may simply reflect an effect of diabetes on some more fundamental mechanism govern-ing basement membrane production. Nevertheless, the normali-zation of basement membrane synthesis, lysyl hydroxylase and glucosyltransferase all support the contention that metabolic control is am important factor in allaying the progression of diabetic nephropathy.

REFERENCES

1. Kimmelsteil, P., Kim, O.J., and Beres, J., *Amer. J. Clin. Path.* 38:270, 1962.
2. Osterby-Hansen, R., *Diabetologia* 1:97, 1965.
3. Mauer, S.M., Miller, K., Goetz, F.C., Barbosa, J., Simmons, R.L., Najarian, J. and Michael, A.F., *Diabetes* 25:709, 1976.
4. Cohen, M.P. and Vogt, C.A., *Biochim. Biophys. Acta* 393:78, 1975.
5. Westberg, N.G. and Michael, A.F., *Biochem.* 9:3837, 1970.
6. Grant, M.E., Kefalides, N.A. and Prockop, D.J., *J. Biol. Chem.* 247:3539, 1972.
7. Kefalides, N.A., *Arth. and Rheum.* 12:427, 1969.
8. Blumenkrantz, N. and Prockop, D.J., *Anal. Biochem.* 30: 377, 1969.
9. Askenazi, R. and Kefalides, N.A., *Anal. Biochem.* 47:67, 1972.
10. Cohen, M.P. and Khalifa, A., *Diabetes* 23:357, 1974.

11. Cohen, M.P. and Khalifa, A., *Diabetes* 25:349, 1976.
12. Khalifa, A. and Cohen, M.P., *Biochem. Biophys. Acta* 386: 332, 1975.
13. Cohen, M.P. and Khalifa, A., *Biochem. Biophys. Acta.* In press, 1976.
14. Grant, M.E., Harwood, O. and Williams, I.F., *J. Physiol.* 257:56, 1976.
15. Sato, T. and Spiro, R.G., *J. Biol. Chem.* 251:4062, 1976.
16. Spiro, R.G. and Spiro, M.J., *Diabetes* 20:641, 1971.

B. Immunologic Models of Disease

IMMUNOLOGIC AND NON-IMMUNOLOGIC ASPECTS

OF

EXPERIMENTAL GLOMERULONEPHRITIS

Seiichi Shibata

The Third Department of Internal Medicine
Faculty of Medicine
University of Tokyo
Hongo, Tokyo, Japan

SUMMARY: Efforts have been made to purify a nephrotoxic serum antigen in rats. The most purified substance was a glucose-containing glycopeptide which has no antigenic capacity to raise kidney-fixing antibodies in rabbits.

However, by a single injection of this substance into foot pads of Wistar strain rats, a spectrum of glomerular changes resembling human glomerulonephritis has been successfully induced.

Since this glycopeptide isolated from GBM interacted both in vivo and in vitro specifically with concanavalin A which can form precipitates with branched oligosaccharides containing α-D-glucopyranosyl units at the non-reducing terminus, non-immunologic pathogenesis of the present experimental glomerulonephritis is strongly suggested.

Taken together with the results of the chemical studies on the sugar portion of the nephritogenic glycopeptide (NEPHRITOGENOSIDE), this evidence also appears to contribute towards a better understanding of the chemical structure of the subendothelial surface of GBM.

535

I was initially asked to review the "immunologic models of diseases and basement membrane (BM)", but I think, now, this problem must be discussed from a wider aspect. I would like to designate this subject "disease inducing activity and BM", and describe it in the following order:

1. Immunologic aspects only.
2. Immunologic and non-immunologic aspects.
3. Chemical aspects.

All of the investigations on BM, including chemical studies, have, as is well known, started from Masugi nephritis (nephrotoxic serum nephritis). Masugi (1) introduced first a new aspect of "glomerulonephritis-inducing factor" to this nephrotoxic serum, in contrast with previous investigators.

Masugi nephritis is produced when heterologous antibodies to the renal tissue (i.e., the glomerular basement membrane (GBM)) are injected into an individual of the species which supplied the membrane antigen. This model is, generally considered as representative of classical immunology; however, the renal tissue antigen has the following unusual characteristics.

First, animals injected with anti-lung, anti-aorta, anti-liver or anti-muscle serum induce mainly glomerulonephritis, rather than the specific organ changes; that is, the nephrotoxic-serum antigen is distributed in many organs.

Second, with the aid of immuno-histochemical techniques, it was found that the nephrotoxic-serum antigen is distributed in many different structures, i.e., not only in the basement membrane, but also in the reticulum framework, the walls of capillaries and arterioles, sarcolemma, neurilemma and so on (2).

Main localization of the nephrotoxic-serum antigen in the renal tissue itself was first revealed to be in the glomerulus and specifically in the GBM, by Krakower et al. (3). This was followed by the isolation of GBM which has greatly stimulated many investigations, especially on the chemical composition of GBM.

Furthermore, as reported by Cole et al. (4), proteolytic enzymes, such as trypsin, made the renal tissue soluble. However, products of trypsin digestion of the renal tissue were poorly antigenic per se, and they were considered as only haptenic. Thus, even now in 1976, the insoluble GBM is used as the nephrotoxic serum antigen.

This problem, that "GBM cannot be solubilized without loss of nephrotoxic serum antigen" seems to leave a lasting effect, for example, on the immunopathological studies of this model.

As described above, the isolation of GBM has stimulated chemical studies of GBM, but it has been difficult to use chemical analyses as an indicator of biological activities. Therefore, chemical studies have been focused on the following three problems:

A. Chemical composition of whole GBM;
B. Comparative studies with other subjects, such as tendon collagen with similar chemical compositions; and,
C. Comparative studies with diseased GBM (diabetic, nephritic, nephrotic, etc.).

Materials obtained from GBM by the chemical procedures, such as proteolytic digestion (5) and 8 M urea extraction (6) did not have any nephrotoxic-serum antigen.

One additional but important problem is the evidence that antisera against epithelial basement membrane react with GBM, but have no ability to induce glomerulonephritis in rats by an intravenous injection (7). In other words, this evidence seems to suggest that although epithelial basement membranes such as lens capsule (8) and neoplastic epithelial basement membrane (9), are the principal components of BM they may not contain all of them.

Where did the nephrotoxic-serum antigen disappear?

Was the active principle destroyed or artificially discarded?

I. Isolation of Soluble Nephrotoxic Serum Antigen

In 1966, we successfully isolated nephrotoxic-serum antigen in a soluble state from insoluble GBM (10). As described above, products of trypsin digestion of the renal tissue have been considered not a complete antigen, but only as haptenic material (4). However, in our studies, the products of trypsin digestion of the renal tissue or GBM, have not been found to be a potent antigenic substance that induces nephrotoxic antiserum. That is, glomerulonephritis was induced only after an intravenous injection of antisera against trypsin-digest products of GBM.

Therefore, water soluble products of trypsin digests were further purified chemically to obtain the soluble glycoprotein (GP) by Geon block electrophoresis (11,12). Nephrotoxic-serum antigen activity was concentrated in only a few fractions, in which no nucleic acids were detected.

Figure 1 shows the distribution of the protein, hexose and nephrotoxic-serum antigen activity in each fraction of

Geon block electrophoresis. This figure strongly suggests a close relationship between sugar content and nephrotoxic-serum antigen activity.

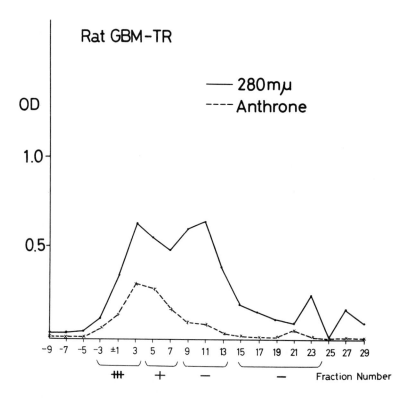

Fig. 1. *Distribution of the protein, hexose and nephrotoxic-serum antigen activity in each fraction of Geon block electrophoresis. The nephrotoxic-serum antigen activity was found in the fractions with high hexose content, especially in the ascending limb of the hexose peak.*

This active fraction was further purified by molecular sieving. It had the character of a glycoprotein which gave a single symmetric sedimentation boundary line. The yield was only 5.5 mg starting from 3 kg of the renal tissue.

It is noteworthy to point out that the monosaccharide composition of this active glycoprotein shows a predominance of glucose over galactose (glucose-dominant). This is contrary to Spiro's report (5). Also, the amino acid composition of this glycoprotein showed in some respects similarities to that of collagen, but in many respects was distinctly different.

Figure 2 seems to support our idea from another aspect. Column chromatography of the material extracted from GBM with 8 M urea resulted in the isolation of a component rich in carbohydrate. This fraction seems to coincide with "non-collagen fraction" named by Kefalides (6), and we found that only this fraction reacts specifically with the antiserum against soluble GP. Therefore, soluble nephrotoxic-serum antigen, GP, seems to be different from the GBM collagen.

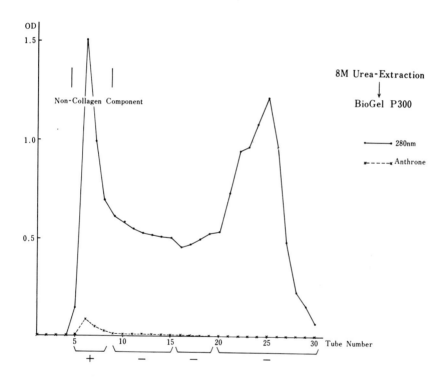

Fig. 2. Column chromatography of the material extracted from GBM with 8 M urea. The fraction rich in carbohydrate coincided with "non-collagen component" (6). Only these fractions reacted specifically with antisera against soluble GP by immunodiffusion.

The sediment of repeated trypsin-digestion was inactive, and Geon block electrophoretic fractions other than GP were all inactive. Thus, it seems logical to say that the water soluble GP represents the nephrotoxic-serum antigen in GBM.

In this model, severe proliferative glomerulonephritis was induced by a single intravenous injection of antisera against soluble GP. In immunofluorescent studies, a typical "linear" pattern was shown. Moreover, chronic glomerulonephritis, with contracted kidney, appeared three to four months after injection.

When glomerulonephritis induced by antisera against insoluble GBM was compared with that by antisera against soluble GP, various differences were noted (Table I). For example, the immunologic process of the latter cannot be explained by Kay's two-step theory (13) (combination of heterologous and autologous phases), but can be explained by a heterologous phase only.

TABLE I

Comparison of Glomerulonephritis by
Anti-insoluble GBM Serum and Anti-soluble GP Serum

	Anti-insoluble GBM	*Anti-soluble GP*
Latent period	7–8 days	–
Its immunologic explanation	1. Heterologous phase 2. Autologous phase	Heterologous phase only
Early morphologic changes:		
1. Exfoliation of endothelial cytoplasm	2.5 h	15 min
2. Polymorph infiltration	2.5 h	2 h
3. GBM itself	24–48 h	24–48 h

Furthermore, early morphologic changes of this model appeared as early at 15 minutes (immediately after a single injection of antisera) (14); they appeared before the marked infiltration of polymorphs (2 hours after injection). Therefore, the exfoliation of endothelial cytoplasm denuding the GBM is difficult to explain with the aid of leucocytic mediator (15). Moreover, morphologic, electron microscopic changes in the GBM itself appeared much later. It seems

difficult, therefore, to consider the Masugi model for the
nephritis mediated by anti-GBM antibody. Figure 3 is a
schematic comparison of the immunological explanation of
morphologic changes seen in the early stages of both groups.

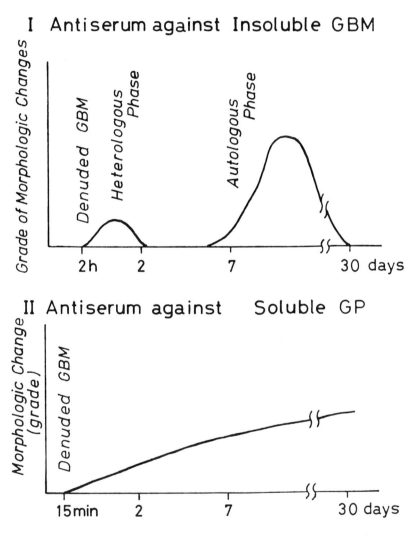

Fig. 3. Schematic demonstration of immunological events (glomerulonephritis) occurring in rats
injected with antiserum against insoluble GBM (I) and
in rats injected with antiserum against soluble GP (II).
Note the presence of autologous phase only in (I).

Labelled antisera against soluble GP reacted with not only GBM but also with BM of other organs, sarcolemma, neurilemma and others (16). This behavior is very similar to that of labelled antisera against insoluble kidney or lung.

In addition, as described above, the common, nephrotoxic-serum antigen exists in various organs such as kidney, lung, aorta, muscle and so on. This phenomenon can be explained chemically by the following results. Figure 4 shows Geon block electrophoretic patterns of the products of trypsin digests (kidney, lung, aorta, liver, heart, muscle, spleen and brain). A single hexose peak was obtained from all of them, except for brain, and only in the same ascending limb of the hexose peak we can find the common nephrotoxic-serum antigen (17). These results were approximately the same in products of collagenase digests from various organs. Further experiments by us revealed that almost all of the nephrotoxic-serum antigen activity was destroyed by the periodate oxication procedure (18).

Therefore, our idea that there is a close relationship between the sugar content and nephrotoxic-serum antigen activity seems to be supported.

II. Nephrotoxicity and Nephritogenic Activity

Now, let us examine the next problem. In 1971, we found unexpectedly evidence that soluble GP has another pro-perty in addition to that of nephrotoxic-serum antigen (19). Rats that were given a single injection in the hind footpads of homologous GP and Freund's incomplete adjuvant developed typical proliferative glomerulonephritis. We have referred to this activity (inducing proliferative glomerulonephritis by a single and direct injection) as "nephritogenic activity".

These experiments were attempted in the unfortunate period of "Zengakuren-struggle" - a time when we could not use our own laboratory except for housing animals for over a half a year.

This nephritogenic activity was found only in GP (the ascending part of the hexose peak in Geon block electrophor-esis corresponding to the distribution of the nephrotoxic-serum antigen). When renal tissue itself was used instead of GP, nephritogenic activity was, of course, not found.

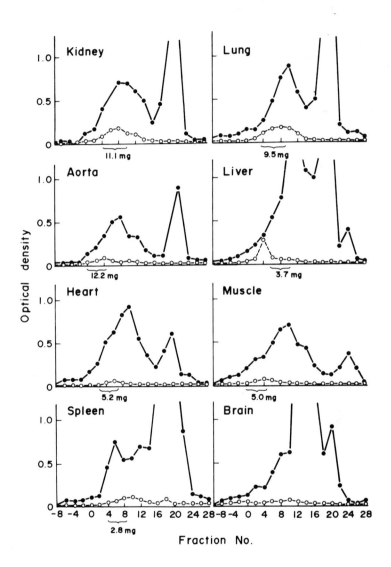

Fig. 4. Geon block electrophoretic patterns of trypsin digests of kidney, lung, aorta, liver, heart, muscle, spleen and brain. The nephrotoxic-serum antigen was exclusively found in the ascending limb of the hexose peak for each organ except brain by both immunodiffusion and bioassay. The same results were obtained for collagenase-digests of these organs.

III.   A New Experimental Model of Glomerulonephritis

After exhaustive (3 days) proteolytic digestion with
pronase and further purification of the above glycoprotein
a renal glycopeptide was isolated (20).   Antisera against the
glycopeptide no longer contained kidney-fixing antibody.   In
contrast, severe glomerulonephritis (clinically and morphologically) was induced by a single injection in the hind
footpads of this glycopeptide (GP-P).

As compared with GP, in GP-P far higher sugar content
was observed in Geon block electrophoretic preparation
Figure 5, lower).   The active slow-moving part of this
hexose peak was successfully separated as the first peak of
Sephadex G200 column (Figure 5, upper).   Monosaccharide

1 Sephadex G200

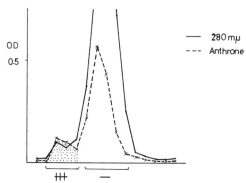

2 Zone electrophoresis

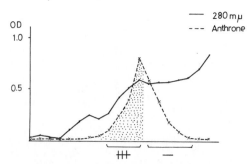

Fig. 5.   Preparation of renal glycopeptide from NTR(3)+
Pr(72) or NTR(72)+Pr(72).   Upper: the active slow-moving part
of the hexose peak was successfully separated as the first
peak of Sephadex G200 column.   Monosaccharide composition of
this active peak was glucose-dominant.   Lower: Geon block
electrophoretic pattern of GP-P.   Shaded area: Active fraction
glucose dominant.   Clear area: Inactive fraction: Galactose and
mannose dominant.

analysis of this active peak showed predominantly glucose
(glucose-dominant fraction).   In contrast, the monosaccharide
composition of the inactive peak consisted of galactose and
mannose.   Figure 6 is the well known scheme of Spiro (5) on
the isolation procedure of two kinds of glycopeptides from
insoluble GBM, the disaccharide unit in which glucose and
galactose are found in equivalent amounts and the heteropolysaccharide which contains no glucose.   In examining this
Figure, one can easily see that the void volume fraction of
Sephadex G50-column was discarded in his procedures.

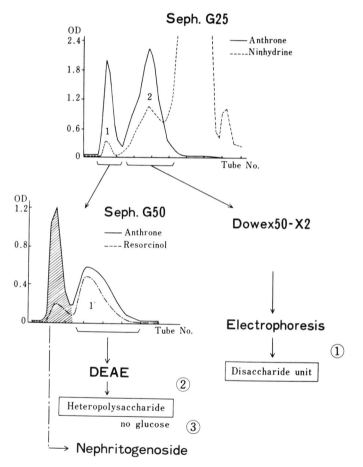

Fig. 6.   Isolation of nephritogenic glycopeptide (NEPHRI-
TOGENOSIDE) from GBM.   Schematic demonstration of nephrito-
enic glycopeptide is contained in the void volume fractions of
Sephadex G50-column, which has been discarded.   This is the
isolation procedure of two kinds of glycopeptides reported by
Spiro (5).

As a matter of fact, our nephritogenic glycopeptide (which I designate as NEPHRITOGENOSIDE) eluted in this void volume fraction. Comparing Figure 6 with former Figure 5, NEPHRITOGENOSIDE is the third glycopeptide constituent of GBM and the only one with biologic activity.

Very interesting observations were made from the morphologic examination (including electron microscopy) of kidneys with severe glomerulonephritis which was induced by a single injection in the hind footpads of the glycopeptide (Figure 7).

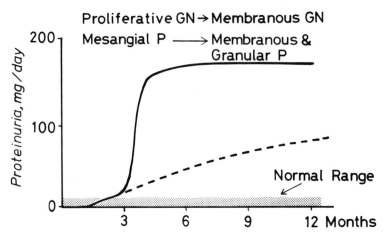

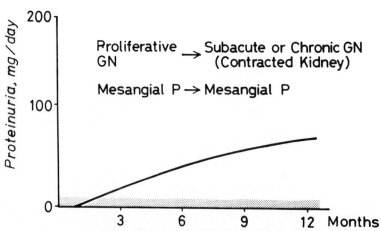

Fig. 7. Effect of TCA-treatment. Comparing Group II (given GP-P with TCA-treatment) with Group I (given GP-P without TCA-treatment), Group II may be the basal type of this model. In Group I, the basic pattern was shown by the dotted line.

A single injection of homologous renal glycopeptide produced proliferation and swelling of mesangial cells prior to the appearance of typical membranous glomerulonephritis. The morphologic changes of membranous glomerulonephritis occurred very slowly (4 to 5 months) after injection, and at the same time proteinuria abruptly increased. Regular and granular deposition of γ-globulin and β1c took place along the GBM, beside the mesangial deposits of protein.

In contrast, as shown in the lower part of this figure, when the same glycopeptide was treated with TCA or DEAE resin and then injected in animals, morphologic changes of membranous glomerulonephritis were not found and immunofluorescent staining of the kidney was of mesangial pattern, throughout. Seven to 8 months after injection, typical morphologic changes of subacute or chronic glomerulonephritis (contracted kidney), strikingly similar to those of human adult progressive glomerulonephritis, appeared (21).

Comparing the lower group with the upper group, the lower group may be the basal type of this laboratory model (in the upper group, shown by the dotted line). Consequently, I would like to designate the basal type of this laboratory model with a mesangial pattern of immunofluorescence as the third experimental model, next to that of Masugi nephritis and serum sickness nephritis.

Figure 8 is a schematic demonstration of the effect of TCA or DEAE resin treatment, in the production of typical chronic glomerulonephritis with contracted kidney, and in the complete prevention of the development of membranous glomerulonephritis. As shown in the center of this figure, glycopeptide without TCA treatment is composed of both glycopeptide and free amino acids or peptide. In contrast, the glycopeptide treated with TCA is completely devoid of these free amino acids or peptides. The dots surrounding the glycopeptide denote free amino acids or peptides which can induce immune complex reaction. These deposits are superimposed on the original morphologic changes of proliferative glomerulonephritis and a typical membranous glomerulonephritis is induced.

Furthermore, our new model is different from the models of Heymann's nephrosis (22) and Steblay's nephritis (23) (Table II). The main differences are in the solubility of the immunogen and in the immunofluorescent pattern. The former two models are induced by insoluble material (renal tissue or GBM), but our model is induced by soluble fraction of GBM. The immunofluorescent pattern in Heymann's model is granular, while in Steblay's it is linear, but in our model (basal type) it is mesangial.

Glycoprotein (GP)

Glycopeptide (GP-P)

without TCA-treatment

Glycopeptide (GP-P)

with TCA-treatment

∴ **free amino acids and peptides**

Fig. 8. Schematic demonstration of the effect of TCA-treatment. As shown in'the center of this figure, glycopeptide without TCA treatment is composed of both glycopeptide and free amino acids or peptides. In contrast, glycopeptide with TCA treatment discards completely these free amino acids or peptides. Amino acids or peptides surrounding glycopeptide can induce immune complex reaction. The deposits thus obtained superimpose the original morphologic changes of proliferative glomerulonephritis.

TABLE II

Glomerulonephritis Induced by Renal Tissue

Reported By	Injected Material	Solubility	Injec- tion	Immuno- Fluorescence Pattern
Heymann (1959)	Homologous Renal Tissue - Immune Complex Mechanism	Insoluble	Repeated	Granular
Stebley (1962)	Heterologous and Homologous GBM - Autoimmune Process	Insoluble	Repeated	Linear
Shibata	Homologous GBM (GP*) - Protein- Carbohydrate	Soluble Interaction	Single	Mesangial

* Glycoprotein fraction.

Heymann's model is now generally thought to be induced
by an immune-complex mechanism, and Steblay's by an autoimmune process. In that case, how does one explain the pathogenesis of our new model (basal type) with an immunofluorescent mesangial pattern?

The main troublesome point is the evidence that the
glycopeptide (GP-P) showed more intensive "nephritogenic"
activity in spite of losing its nephrotoxic-serum antigen
activity. In other words, even by heterologous immunization
kidney-fixing antibody could not be produced. Consequently,
it is difficult to consider the production of kidney-fixing
antibody by homologous immunization.

IV. Interaction of Nephritogenoside with Con A

Using chemical studies, we have recently come upon a new
clue for solving the pathogenesis of our model. It is well
known that Concanavalin A (Con A), a lectin isolated from
jack beans, binds specifically to α-D-glucopyranosyl, α-D-
mannopyranosyl and β-D-fructofuranosyl groups and forms
precipitates with branched polysaccharides and glycoproteins
containing these groups at the non-reducing terminus (24).
We also know that the monosaccharide composition of the
nephritogenic glycoprotein and of the glycopeptide is made up
of glucose, galactose and a trace of mannose (12).

Thus, active glycopeptide was applied to a column of Con
A affinity chromatography. As shown in the upper part of
Figure 9, the high and unretarded peak was shown to contain
no nephritogenic activity. However, in the material retained
by Con A and subsequently eluted with α-D-methylmannoside,
we found almost all of the principal nephritogenic activity.
As shown in the lower part of this figure, the active glycopeptide was sharply separated and purified from contaminated
α-D-methymannoside by gel filtration in a Bio-Gel P200
column.

These results strongly suggest that the active principle
of nephritogenic peptide is a glycopeptide with a branched
oligosaccharide having an α-D-glucopyranosyl or an α-D-manno-
pyranosyl unit at the non-reducing terminus.

A simple procedure, TCA-treatment, was successfully
introduced for removal of the mannose component without any
decrease of biological activity. The material thus obtained
was then applied to a column of Bio-Gel P200. Two hexose
peaks (A and B) were obtained, and only in the peak A (fast
peak) biological activity was demonstrated (Figure not shown).

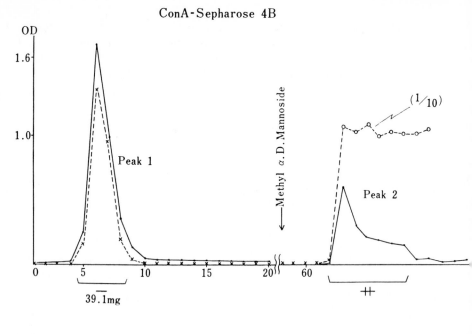

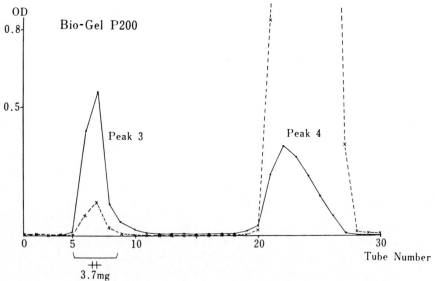

Fig. 9. Con A affinity chromatography of nephritogenic
glycopeptide. As shown in the upper graph, unretarded peak
was revealed to contain no nephritogenic activity, but the
material retained by Con A and subsequently eluted with α-D-
methyl-mannoside contained almost all of the principal nephro-
genic activity. As shown in the lower part of this figure,
the active glycopeptide was, through Bio-Gel P200 column,
sharply separated from contaminated α-D-methyl-mannoside.

Figure 10 shows the representative profile of peak A in the Con A column; on the left the non-retarded and on the right, the retarded fractions are shown. A high glycoprotein peak with biologic activity was detected only in the retarded fraction. The yield was about 10 mg, starting with 1 kg of renal tissue.

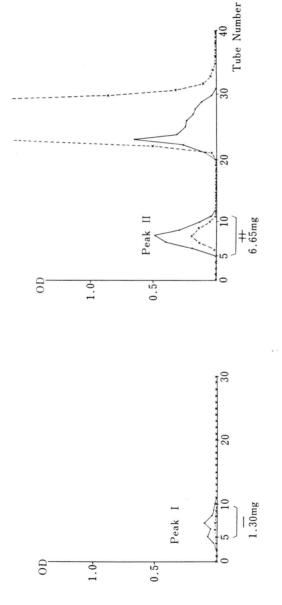

Fig. 10. Representative profile of peak A (purified active fraction) in the Con A column. (a) Con A - Sepharose 4B, non-retarded fraction. (b) Bio-Gel P200, retarded fraction. A high glycoprotein peak with biological activity was detected only in the retarded fractions.

On the contrary, when peak B was applied to Con A column, we could not find a significant glycoprotein peak in the retarded fraction (Figure not shown). Thus, it can be concluded that the nephritogenic glycopeptide contains a branched oligosaccharide having an α-D-glucopyranosyl unit at the non-reducing terminus.

The results noted above are summarized in Figure 11.

I. Studies on the Non-reducing Terminus of GBM (in vitro)

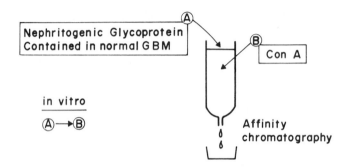

2. Studies on the Localization Form of GP in GBM (in vivo)

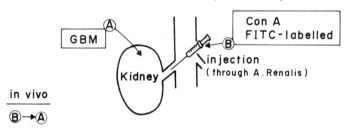

injected material	FITC in Glomerular Tufts
Labelled B only	⧣
Labelled B +α·D·Methyl-mannoside	—

Fig. 11. Schematic demonstration of Fig. 10. Top diagram, 1., shows glycoprotein (A) being applied to Con A (B). (A) is the substance which is contained in normal GBM. Therefore, if the surface of normal GBM is composed of renal glycoprotein, as shown in lower diagram, 2., the injected (B) (labelled Con A) must be, reversely found to bind with (A) (normal GBM). The injected Con A interacted specifically with the capillary wall and mesangium, as well as peritubular capillaries.

Here, the glycoprotein (A) is applied to Con A (B). (A) is
applied to Con A (B). (A) is the substance which is contained
in normal GBM. Therefore, if the surface of normal GBM is
composed of renal glycoprotein, as shown in the lower diagram,
the injected (B) (labelled Con A) must reversibly bind to (A)
(normal GBM).

As shown in Figure 12, the injected Con A interacted
specifically *in vivo* with the capillary wall and the mesangium
as well as the peritubular capillaries.

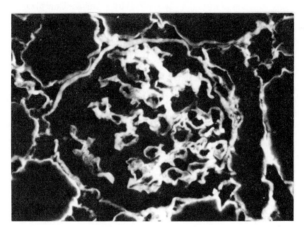

*Fig. 12. Immunofluorescent pattern of kidney
of a rat that received labelled Con A. Note specific
uptake of fluorescein in the capillary wall, mesangium and peritubular capillaries.*

Considering these results, as shown in the center of
Figure 13, the nephritogenic glycopeptides exist, at least
in the endothelial aspects of GBM. This glycopeptide has an
additional property: that is, it contains Con A receptor.
Since it is established that Con A can interact with the
glycopeptide only in the form of dimer or tetramer. It may
be suggested that a glycopeptide-Con A-complex will be formed
by the use of one of the two binding sites of Con A and then,
using the remaining binding site of Con A, this complex can
further interact with another Con A receptor. As a result,
the glycoprotein-Con A-complex can be expected to be deposited
on the GBM.

It is known that the deposited Con A cannot be visualized
by the usual electron microscopic techniques (25) and granular
deposits of these complexes cannot be detected by immunofluorescent or electron microscopic examination. Although
presently, Con A has been isolated only from jack beans, a

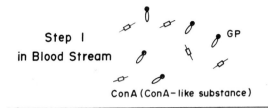

Step 1
in Blood Stream

GP

ConA (ConA-like substance)

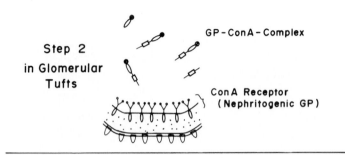

Step 2
in Glomerular
Tufts

GP-ConA-Complex

ConA Receptor
(Nephritogenic GP)

Step 3
in Glomerular
Tufts

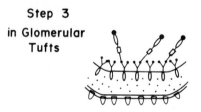

Deposition of
GP-Con A-Complex
in GBM

Fig. 13. Schematic illustration of deposition of glycopeptide-Con A-complex in GBM. Step 2: nephritogenic glycopeptides exist, as the proper component, in the endothelial
aspects of GBM, and the glycopeptide itself has the property
of Con A receptor. It is well established that Con A can
interact with the glycopeptide only in the form of dimer or
tetramer. Thus, glycopeptide-Con A-complex may be formed in
circulation by the use of one of the two binding sites of Con
A and then, using the remaining binding site of Con A, this
complex can further interact with another Con A receptor, see
Step 3. As a result, glycoprotein-Con A-complex can be
deposited in GBM.

mammalian lectin was isolated and purified from heaptic protein (26). Thus, our hypothesis should in the near future,
be confirmed in the human being. We have, recently, shown
the existence of a nephritogenic glycopeptide in $Zn-\alpha_2$-globu-
lin, and demonstrated that α_2-globulin reacts specifically
with Con A.

Even in the well established animal model of experimental glomerulonephritis, such as Masugi nephritis (1) and serum sickness type nephritis (27,28), the explanation of pathogenesis is, at present, limited only to the mode of deposition of antibody on GBM. Thus, our new model seems to be on a par with the two established models with respect to pathogenesis.

V.   Chemical Structure of the Sugar Portion

As a last problem, I would like to discuss briefly the chemical structure of the sugar portion of the nephritogenic glycopeptide (NEPHRITOGENOSIDE), because its biologic activity was shown, in a previous paper, to relate more closely with the oligosaccharide portion (29).

When the glycopeptide was applied to a Bio-Gel P100 column, the active portion was eluted, near the void volume, as Fraction I. Highly sansylated (DNS) leucine and poorly dancylated histidine were detected as N-terminals on a polyamide layer; thus Fraction I may be considered as a monocomponent.

Total sugar content was 24.3%. The monosaccharide contained glucose 23.0%, galactose 1.3% and N-acetyl-glucosamine 3.8%. Gas chromatography (Figure 14) and mass spectra of the methylated sugars yielded several products. Figure 14 shows the separation of peaks I through VIII. Peak I corresponds to 2-3-4-6-tetra methyl glucose; peak II to 2-3-4-6-tetra methyl galactose; peak III to 3-4-6-tri-methyl glucose; peak IV to the mixture of 2-3-6- and 2-3-4-tri-methyl glucose; peak VI to 3-6-di-methyl glucose; peak VII to 2-4-di-methyl glucose and peak VIII to 3-6-methyl glucosamine. As the shoulder peak, 2-3-4-6-tetra methyl galactose may be a contaminant, but this remains unresolved as yet. Peak I corresponding to 2-3-4-6-tetra methyl glucose, is the largest component, and peak VI , 3-6-di or peak VII, 2-4-di-methyl glucose are found in small amounts indicating the presence of only a few branched structures in the oligo-saccharide chain. The substitution of carbon 1 of (I) and the fact that tetramethyl makes up about one-half of the total amount of the free substitution suggests that the saccharide chain is, on the average, composed of three glucose residues. Amino acid analysis revealed a high glycine content followed by glutamic acid, aspartic acid, serine and threonine in that order. Since half-cystine and tyrosine were not detected, S-S linkages are not conceivable. Both hydroxylysine and hydroxyproline are detectable, but in a very low amount. The peptide-oligosaccharide linkage is not O-glycosidic, and studies in the future may show it to be N-glycosidic.

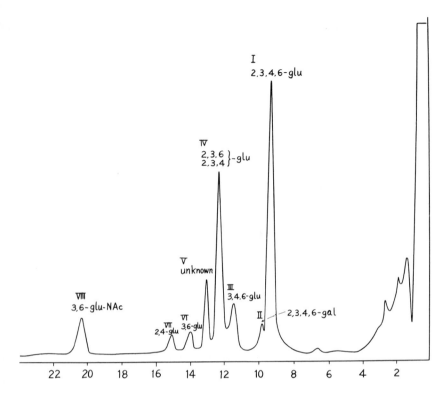

Fig. 14. Gas chromatography of NEPHRITOGENOSIDE (methy-lated sample). Mass spectra yielded several products. Peak I corresponds to 2-3-4-6-tetra methyl glucose; peak II to 2-3-4-6-tetra methyl galactose; peak III to 3-4-6-tri-methyl glu-cose; peak IV the mixture of 2-3-6- and 2-3-4-tri-methyl glu-cose; peak V is an unknown peak; peak VI to 3-6-di-methyl glucose; peak VII to 2-4-di-methyl glucose and VIII to 3-6-methyl glucosamine. As the shoulder peak, 2-3-4-6-tetra-methyl galactose may be a contaminated peak, but this remains unresolved. Peak I, 2-3-4-6-tetramethyl glucose is the largest component and peak VI, 3-6-di or peak VII, 2-4-di methyl glucose are found in small amounts, indicating the presence of only a few branched structures in the oligosaccha-ride chain.

From these results, the nephritogenic glycopeptide
(NEPHRITOGENOSIDE) is considered to be on (or in) the endothelial side of normal GBM, as shown in Figure 15. The non-
reducing terminus (α-D-glucoside) of each glycopeptide may
serve, naturally, as Con A receptor.

Endothelial Aspect of GBM (Our Scheme)

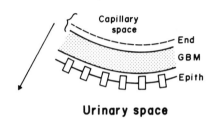

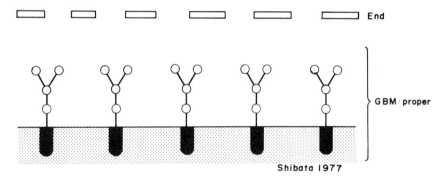

Shibata 1977

*Fig. 15. Schematic demonstration of the endo-
thelial aspects of GBM based on chemical studies of
NEPHRITOGENOSIDE which has both nephritogenicity and
Con A receptor activity.*

Immunopathologic studies of GBM, in general, have been
based on anatomic (pathological anatomy) and immunology, i.e.,
reactions with little or no chemical characterization of the
antigens used. Chemical characterizations of GBM have been
based on chemistry done without looking into its biological
activity. The most burning problem in the study of GBM is,
I think, the establishment of methodologies that will inte-
grate the immunologic, pathologic, chemical and clinical
studies.

REFERENCES

1. Masugi, M., *Beitr. Path. Anat.* 91:82, 1933.
2. Cruickshank, B. and Hill, A.G.S., *J. Path. Bact.* 66:283, 1953.
3. Krakower, C.A. and Greenspon, S.A., *Arch. Path.* 51:629, 1951.
4. Cole, L.R., Cromartie, W.J. and Watson, D.W., *Proc. Soc. Exptl. Biol. Med.* 77:498, 1951.
5. Spiro, R.G., *J. Biol. Chem.* 242:1923, 1967.
6. Kefalides, N.A., *Biochem. Biophys. Res. Commun.* 22:26, 1966.
7. Krakower, C.A. and Greenspon, S.A., *Arch. Path.* 66:364, 1958.
8. Dische, Z., *In:* "Small Blood Vessel Involvement in Diabetes", (M.D. Siperstein, A.R. Colwell and K. Meyer, Eds.), Amer. Inst. Biol. Sci., Washington, D.C., p. 201, 1964.
9. Pierce, G.B., Midgley, A.R., Jr. and Ram, T.S., *J. Exptl. Med.* 117:339, 1963.
10. Shibata, S., Nagasawa, T., Takuma, T., Naruse, T. and Miyakawa, Y., *Jap. J. Exptl. Med.* 36:127, 1966.
11. Shibata, S., Naurse, T., Nagasawa, T., Takuma, T. and Miyakawa, Y., *J. Immunol.* 99:454, 1967.
12. Shibata, S., Miyakawa, Y., Naruse, T., Nagasawa, T. and Takuma, T., *J. Immunol.* 102:593, 1969.
13. Kay, C.F., *J. Exptl. Med.* 72:559, 1940.
14. Shibata, S., *Proceedings of the 8th Congress of the International Association of Allergology,* Excerpta Medica, Amsterdam, p. 278, 1973.
15. Cochrane, C.G., Unanue, E.R. and Dixon, F.J., *J. Exptl. Med.* 122:91, 1965.
16. Nagasawa, T. and Shibata, S., *J. Immunol.* 103:736, 1969.
17. Shibata, S. and Nagasawa, T., *Immunology* 26:217, 1974.
18. Shibata, S., Naruse, T., Miyakawa, Y. and Nagasawa, T., *J. Immunol.* 104:215, 1970.
19. Shibata, S., Nagasawa, T., Miyakawa, Y. and Naruse, T., *J. Immunol.* 106:1284, 1971.
20. Shibata, S., Sakaguchi, H., Nagasawa, T. and Naruse, T., *Lab. Invest.* 27:457, 1972.
21. Shibata, S., Sakaguchi, H. and Nagasawa, T., *Nephron* 16: 241, 1976.
22. Heymann, W., Gilkey, C. and Salehar, M., *Proc. Soc. Exptl. Biol. Med.* 73:385, 1950.
23. Steblay, R.W., *J. Exptl. Med.* 116:253, 1962.
24. Goldstein, I.J., Hollerman, C.E. and Merrick, J.M., *Biochem. Biophys. Acta.* 97:68, 1965.

25. Bretton, R. and Bariety, J., *Ultrastructure Research*
     48:396, 1974.
26. Stockert, R.J., Morell, A.G. and Scheinberg, H.I.,
     Science 186:365, 1974.
27. Germuth, F.G., *J. Exptl. Med.* 97:257, 1953.
28. Dixon, F.J., Vasguez, J.J., Weigle, W.O. and Cochrane,
     C.G., *Arch. Path.* 65:18, 1958.
29. Saito, H. and Shibata, S., *Eighth International Symposium
     on Carbohydrate Chemistry, p. 104, 1976.*

Biology and Chemistry of Basement Membranes

GLOMERULAR PERMEABILITY IN MUNICH-WISTAR RATS

WITH AUTOLOGOUS IMMUNE COMPLEX NEPHRITIS

A TRACER STUDY

Eveline E. Schneeberger

Children's Hospital Medical Center
and
Harvard Medical School
Boston, Massachusetts

ABSTRACT: Glomerular barrier function is critically dependent on normal blood flow conditions (Kidney Int'l. 9:36, 1976). Thus, it is important that glomeruli be fixed in situ in the living animal to avoid artefactual diffusion of tracers or endogenous proteins. Under optimal conditions of fixation, endogenous proteins do not penetrate beyond the lamina rara interna (lri) of normal rats. Altered permeability of the glomerular capillary wall was examined in Munich-Wistar strain rats with AIC nephritis. The degree of proteinuria and immune complex deposition was not as extensive in these rats as in Lewis strain rats. Localization of endogenous albumin, or intravenously (IV) injected catalase or ferritin was used to assess glomerular permeability. Superficial glomeruli were fixed in situ by applying glutaraldehyde to the renal surface of the anesthetized rat. Endogenous albumin was localized by incubation with peroxidase labeled anti-rat albumin Fab fragments followed by a cytochemical reaction for peroxidase. The distribution of IV injected catalase was visualized by cytochemical means, while ferritin was identified directly in the electron microscope. In segments of the glomerular basement membrane (GBM)

containing immune complex deposits (ICD), albumin,
catalase and ferritin were present throughout the
thickness of the GBM, and in ICDs themselves. All
three proteins were present in the urinary space.
Rarely focal foot process detachment from an under-
lying ICD was present. The underlying space contained
protein. In segments of GBM free of ICDs, tracers
were largely confined to the lri with variable
penetration into the lamina densa. It appears that
in AIC nephritis, the GBM adjacent to ICDs is more
permeable to protein, and that the intervening GBM
may also show a slight increase in permeability.

INTRODUCTION

The glomerular capillary wall is composed of three
anatomical layers: (a) a fenestrated endothelium lining
glomerular capillaries, (b) the glomerular basement membrane
(GBM) consisting of the lamina rara interna, the lamina
densa and the lamina rara externa and, (c) the epithelial
cell layer having complex interdigitating foot processes.
The space between foot processes is spanned by a diaphragm
containing a regular array of rectangular pores (1,2).
While there has been some question as to which of these
layers primarily restricts the passage of plasma proteins
into the urinary space during normal ultrafiltration, it has
clearly been shown that glomerular barrier function depends
on the maintenance of normal blood flow conditions (3,4).
Thus when superficial glomeruli of anesthetized Munich-Wistar
rats are fixed *in situ*, and endogenous albumin localized by
means of anti-rat albumin Fab fragments conjugated to horseradish peroxidase (HRP), reaction product is largely confined
to the glomerular capillary lumen and endothelial fenestrae
with only a small amount being present in the lamina rara
interna. When the renal blood supply is temporarily interrupted, albumin diffuses across the glomerular capillary wall and
into the urinary space. However, this effect is reversible:
as soon as glomerular blood flow is reestablished, albumin
once again becomes confined to the capillary lumen.
These findings brought into question some of the earlier
tracer studies in which renal tissue had been fixed by
immersion, and in which diffusion of protein occurred between
the time of removal of the tissue from the animal and commencement of fixation. It thus became important to reexamine some
of the renal disease models as produced in Munich-Wistar rats
(5,6), so that *in vivo* fixation of subcapsular glomeruli could

be carried out. In the present study altered glomerular permeability in Munich-Wistar rats with autologous immune complex
(AIC) nephritis was examined by using three different proteins
as tracers: a) the rat's own circulating endogenous albumin
(69,000 daltons, PI 4.7, a_e 36A), b) intravenously injected
catalase (240,000 daltons, PI 5.7, a_e 52A) and c) intravenously injected ferritin (600,000 daltons, PI 4.5, a_e 61A).

MATERIALS AND METHODS

AIC nephritis was produced in 24 female Munich-Wistar
rats as previously described (7). By eight weeks all animals
had developed proteinuria up to 300 mg/24 hrs. as determined
by the Biuret method (8). The degree of proteinuria was not
as heavy as that usually observed in Lewis strain rats with
AIC nephritis. Endogenous circulating albumin was localized
by immunocytochemical means using rabbit anti-rat albumin Fab
fragments conjugated to HRP (3). Intravenously injected
catalase was visualized by cytochemical means (9,10) and
intravenously injected ferritin was visualized directly.
Rats were anesthetized with sodium pentobarbital, the
kidneys exposed and 2% phosphate buffered glutaraldehyde was
continuously dripped on the renal surface for 30 mins. followed by an additional 30 mins. fixation *in vitro* (3).
Following the immunocytochemical or cytochemical incubations,
the tissue was post-fixed in 2% aqueous OsO_4, dehydrated in
graded ethanol and embedded in Epon. Kidneys from ferritin
injected animals were post-fixed in reduced OsO_4 (11).
Thin sections were examined in a Philips 300 electron microscope.

RESULTS AND DISCUSSION

The following stringent criteria were applied in assess-
ing whether a glomerulus was well fixed. 1. The glomerulus
had to be within at least two to three tubules of the renal
capsule. 2. The lumens of glomerular capillaries had to be
fixed in a widely open position with no evidence of collapse.
3. The reaction product within the capillary lumens had to be
diffuse, finely granular and show no evidence of clumping.
In the case of ferritin the particles had to be uniformly
distributed in the capillary lumen. 4. The tubules surrounding the glomerulus had to be fixed with widely open tubular
lumens.

The glomeruli of Munich-Wistar rats with AIC nephritis
showed subepithelial deposits of immune complexes with spread
expanses of epithelial cytoplasm overlying them. In those
areas of the GBM in which the immune complexes were located,
endogenous albumin, seen as black reaction product, was
present in the full thickness of the GBM as well as in a high
concentration within the immune complex deposits themselves
(Figure 1). Reaction product was also present in epithelial
cell vacuoles as well as on the surface of these cells.
Rarely a portion of an epithelial podocyte was detached, and
in these regions albumin was present throughout the thickness
of the GBM, in immune complex deposits and in the space
beneath the detached foot process. In those areas of the GBM
where immune complex deposits were absent, albumin penetrated
a short distance into the lamina densa but not beyond, and
the lamina rara externa was free of reaction product.

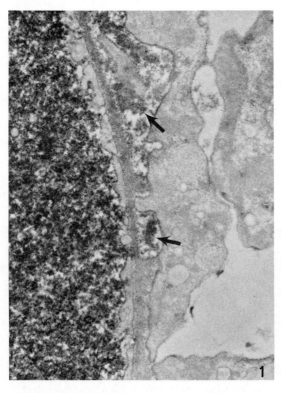

*Fig. 1. Portion
of a superficial glo-
erulus from a Munich-
Wistar rat with AIC
nephritis, fixed in
situ in the anesthetiz-
ed animal. Endogenous
albumin is visualized
as black reaction pro-
duct following incuba-
tion of the tissue with
anti-albumin Fab-HRP
conjugate. Reaction
product is present
within the GBM, the im-
mune complex deposits
(arrows) and on the
surface of epithelial
cells. Mag. X28,000.*

Similar observations were made with intravenously injected catalase. In the normal control Munich-Wistar rat,
catalase remained confined to the capillary lumen and was not
seen beyond endothelial fenestrae. In contrast, in the
Munich-Wistar rat with AIC nephritis reaction product was
present throughout the full thickness of the GBM, in the
immune complex deposits as well as in the urinary space
(Figure 2). As with albumin, catalase appeared to be present
in a higher concentration within the immune complex deposits
than in the surrounding GBM. Reaction product was present
in the lamina densa in those areas of the GBM which were free
of immune complex deposits.

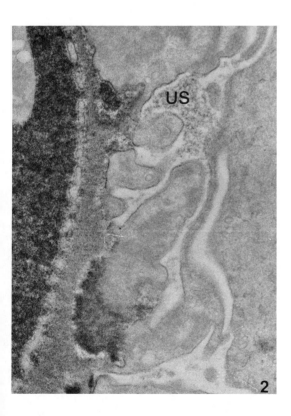

*Fig. 2. Portion
of a superficial glo-
merulus from a Munich-
Wistar rat intravenous-
ly injected with cata-
lase 45 mins. before
commencement of fixa-
tion in situ. Reaction
product is present
throughout the GBM and
in high concentration
within immune complex
deposits. Reaction
product is also seen in
the urinary space (US).
Mag. X25,000.*

In a well-fixed glomerulus of the Munich-Wistar rat with AIC nephritis, ferritin particles were uniformly dispersed in the capillary lumen and showed no clumping (Figure 3). Post-fixation in reduced OsO_4 allowed the more ready identification of ferritin particles and membrane structures, and under these conditions immune complex deposits appeared more electron lucent than the surrounding GBM. Large numbers of ferritin particles were present in some immune complex deposits, while in others there were very few. However, many ferritin particles were present in the GBM surrounding the immune complex deposits, as well as immediately adjacent to the overlying epithelial cell membrane (Figure 4). In those portions of the GBM which were free of immune complex deposits, ferritin particles did not extend beyond the inner third of the lamina densa.

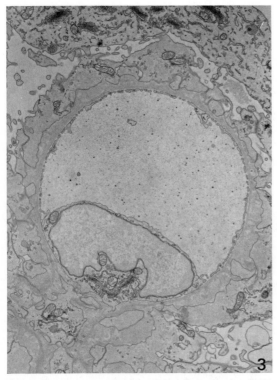

Fig. 3. Glomerular capillary loop from a Munich-Wistar rat intravenously injected with ferritin 10 mins. before initiation of fixation in situ. The capillary loop is fixed in a widely open position and the circulating serum proteins and ferritin are uniformly dispersed throughout the capillary lumen. Mag. X7,800.

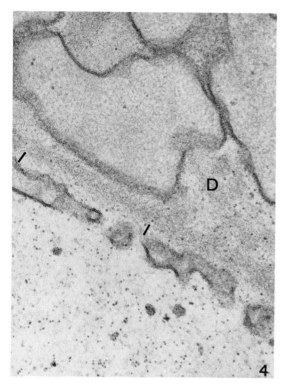

Fig. 4. Portion of a glomerular capillary wall from the same animal as shown in Fig. 3. Large numbers of ferritin particles are present in the GBM adjacent to and within the immune complex deposit (D). In that portion of the GBM (between small black lines) which is free of immune complex deposits, ferritin particles do not penetrate beyond the lamina densa. Mag. X79,200.

These findings indicate that in the area of immune complex deposits, the surrounding GBM permits the passage of protein molecules which are normally excluded from the GBM of control Munich-Wistar rats fixed under similar conditions. The mechanism whereby the GBM in these regions is rendered more permeable is not understood. It is possible that the passage and accumulation of antigen-antibody complexes in some way alters the spacing of the macromolecular components of the GBM. Alternatively their presence may result in a decrease in the intrinsic negative charge of the GBM. Such negatively charged sites within the normal GBM have been demonstrated by both indirect (12) and direct means (13), but their state in AIC nephritis remains to be investigated.

REFERENCES

1. Rodewald, R. and Karnovsky, M.J., *J. Cell Biol*. 63:423, 1974.
2. Schneeberger, E.E., Levey, R.H., McCluskey, R.T. and Karnovsky, M.J., *Kidney Int'l*. 8:48, 1975.
3. Ryan, G.B. and Karnovsky, M.J., *Kidney Int'l*. 9:36, 1976.
4. Ryan, G.B., Hein, S.J. and Karnovsky, M.J., *Lab. Invest*. 34:415, 1976.
5. Ryan, G.B. and Karnovsky, M.J., *Kidney Int'l*. 8:219, 1975.
6. Kuhn, K., Ryan, G.B., Hein, S.J., Galaske, R.G. and Karnovsky, M.J., *Lab. Invest. (in press)*.
7. Schneeberger, E.E., Leber, P.D., Karnovsky, M.J. and McCluskey, R.T., *J. Exp. Med*. 139:1283, 1974.
8. Henry, R.J., Cannon, D.C. and Winkelman, J.W., *IN:* "Clinical Chemistry, Principles and Techniques", Harper and Row, Publishers, New York, p. 411, 1974.
9. Venkatachalam, M.A. and Fahimi, H.D., *J. Cell Biol*. 42:480, 1969.
10. Herzog, V. and Fahimi, H.D., *IN:* "Electronmicroscopy and Cytochemistry", (Wisse, E., Daems, W.T., Molenaar, I. and Van Duijn, P., Editors), North Holland Publishing Company, Amsterdam, p. 111, 1973.
11. Karnovsky, M.J., *Eleventh Annual Meeting of the American Society for Cell Biology,* Abstract No. 284, New Orleans, La., 1971.
12. Rennke, H.G., Cotran, R.S. and Venkatachalam, M.A., *J. Cell Biol*., 67:638, 1975.
13. Caulfield, J.P. and Farquhar, M.G., *Proc. Nat. Acad. Sci.*, 73:1646, 1976.

Supported by U.S. Public Health Service Grant AM 16392.

COMPLEMENT-DEPENDENT CYTOTOXICITY INDUCED BY

ANTIBODIES TO GLOMERULAR BASEMENT MEMBRANE

COLLAGENOUS GLYCOPROTEINS*

Ph. Mahieu, Ch. Dechenne,

Ph. Graindorge and J. Foidart-Willems

University of Liège
Liège, Belgium

ABSTRACT: Anti-human GBM antibodies were obtained by immunization of rabbits with GBM collagenous glyco-proteins mixed with complete Freund's adjuvant. Their specificity was demonstrated by immunofluorescent microscopy and their titer determined by radioimmuno-assay. Human glomerular cells were grown to confluence in 199 Medium containing 20% fetal calf serum. Electron microscopy studies demonstrated that the growing cells were mostly epithelial in origin. The cytotoxic activity of 25 µl antiserum against glomerular cells (about 50,000) was appreciated in the presence of 25 µl of 1:5 dilution of guinea pig complement (C). After incubation for 45 min at 37°C, 0.2% trypan blue was added and viability was assessed.

In the presence of control rabbit sera and C, or C alone, at least 90% of the cells remained viable. In the presence of C and anti-GBM antibodies, only 10 to 15% of the cells remained viable. No cytotoxic effect

This work was supported by a grant from the Belgian Scientific Medical Research Fund.

was observed when positive sera were tested without C, with heat-inactivated C, or when using fibroblasts as target cells. Preincubation of anti-GBM antiserum with GBM glycoproteins or pretreatment of the cells with collagenase abolished the cytotoxic effect. On the contrary, preincubation of the antiserum with type I collagen or pretreatment of the cells with trypsin did not significantly reduce the cytolysis. Indirect immunofluorescence performed with anti-GBM antibodies showed a staining of glomerular cells. The data suggest that antibodies to GBM glycoproteins elicit C-mediated cytotoxicity and that some basement membrane collagenous antigens are present at the surface of the glomerular epithelial cells in culture.

The glomerular basement membrane (GBM) represents a particular form of connective tissue. It is characterized by its homogeneous amorphous appearance and by the absence of periodic structure by electron microscopy (1). Biochemical studies have shown that the GBM is mostly composed of collagenous polypeptides (type IV collagen) associated with non-collagenous polypeptides (2). The GBM collagen contains more hydroxyproline, hydroxylysine and glycosylated hydroxylysine than the interstitial collagens (3).

The biosynthesis of the collagenous component of GBM has been studied using either whole glomeruli (4,5) or cultivated glomerular cells (6,7). It has been demonstrated that the initially synthesized and secreted collagen by the glomerular cells consists of polypeptide chains exhibiting a molecular weight of about 140,000 (4,7). Recently, several studies have shown that the collagens (type I and M) synthesized *in vitro* by fibroblasts can not only be deposited as fibrils in the extracellular matrix, but also can be bound to the fibroblast membrane (8,9). Furthermore, antibodies to the type M collagen elicit complement (C)-mediated cytotoxicity against fibroblasts in culture (9). The present report is concerned with some immunological data indicating that GBM collagenous polypeptides are associated with the plasma membranes of glomerular cells in culture and that antibodies to these polypeptides can induce C-mediated cytotoxicity.

MATERIALS AND METHODS

Antigens

    GBM was isolated, according to the method of Krakower and
Greenspon (10), from normal human kidneys obtained at autopsies. The collagenous component was isolated, after limited
digestion of GBM with pepsin, according to the method of
Kefalides (2). Human type I procollagen and α-chains of the
type I collagen were a gift from Prof. Ch. M. Lapière (Liège);
Clq was purified by the method of Yonemasu and Stroud (11).

Antibodies

    Rabbit anti-human GBM collagen antibodies were raised by
subcutaneous immunization of rabbits with 5 mg of the purified
antigen mixed with complete Freund's adjuvant. Booster injections containing a similar amount of antigen were given at
3 week intervals. The development of antibodies was followed
by indirect immunofluorescence performed on normal human
kidney slices, according to the method of Unanue and Dixon
(12). Their titer was determined by radioimmunoassay (13),
after iodination of the collagenous polypeptides by the lactoperoxidase method (14). Rabbit IgG was prepared by ion
exchange chromatography on DEAE-cellulose (15).

Complement

    Sera from several adult guinea pigs were screened for
cytotoxicity to human glomerular cells in culture. Those
which were not cytotoxic at a dilution of 1 to 5 but still
exhibited a good complement activity at the same dilution were
pooled and stored at -70°C.

Target Cells

    Human kidneys, devoid of histologic lesions, were obtain-
ed at surgery from "cadaveric donors". Glomerular cells were
isolated using enzymatic digestion, sieving and differential
centrifugations (7); they were cultivated in 25 cm^2 Falcon
plastic bottles containing 10 ml of 199 medium supplemented
with heat-inactivated calf serum (20%), penicillin (200 U/ml)
and streptomycin sulfate (50 μg/ml), as previously described
(7). Electron microscopy studies have shown that the growing

cells were mostly epithelial in origin (16). These cells were
used as "target" cells between the first and fourth passage.
Cultivated human skin fibroblasts were a gift from Dr. R. J.
Winand (Liège).

Cytotoxicity Assay

For cytotoxicity experiments, cells were treated by a
0.25% trypsin solution for 30 min. at 37°C, washed and suspended in 199 Medium; 50,000 to 100,000 cells (contained in
25 μl of 199 Medium) were incubated with 25 μl of antiserum
at 37°C for 15 min. At that time, 25 μl of a 1:5 dilution of
guinea pig C was added. After incubation for 45 min. at 37°C,
0.2% trypan blue was added and viability assessed. The standard eosin exclusion test was also used, but because of the
toxicity of this dye, the counts were made only within the
first five minutes following staining. Cell counts were performed with a hemocytometer.

In order to study the specificity of the C-mediated
cytotoxicity, the cells were pretreated with purified bacterial collagenase (Worthington, types CLSP-A and CLSP-B) instead
of trypsin. They were then incubated for 30 min. at 37°C
with 1 ml of Hanks' balanced salt solution containing 1 mg of
collagenase, before the addition of antiserum and C. In order
to ascertain that some serine proteases contaminating the
commercial collagenase preparations were not responsible for
a "nonspecific" cell membrane proteolysis (17), the collagenase treatment of the cells was also carried out in the presence of 25 mM EDTA, an inhibitor of collagenase but not of
serine proteases.

Immunofluorescence Technique

Ten mg of normal or of anti-GBM collagen IgG, contained
in 5 ml of 0.05 M bicarbonate buffer, pH 9.6, were dialyzed
overnight at 4°C against 100 ml of the same buffer containing
20 mg of fluorescein isothiocyanate (FITC) (Nordic Pharmaceuticals and Diagnostics, Tilburg, The Netherlands). Free
FITC was separated from FITC-labeled IgG by overnight
dialysis at 4°C against 1 liter of 0.01 M phosphate buffer,
pH 7.4, followed by Sephadex G-25 chromatography using the
same buffer. Staining was achieved using 100 μl of trypsinized cells (about 100,000 cells) incubated with 100 μl of
FITC-labeled IgG. The cells were washed three times in 199
Medium containing 0.1% NaN_2, dried on cover slips and mounted
on slides in 90% buffer glycerol. They were observed in a
Leitz ortholux fluorescence microscope with epilumination.

RESULTS

Specificity and Titer of Antibodies to GBM Collagen

The specificity of anti-GBM collagen antisera used for
these studies was determined by immunofluorescent microscopy.
FITC-labeled anti-GBM collagen IgG stained the GBM, the TBM
and the Bowman's capsules linearly. Antisera were then
titrated by radioimmunoassay (13). Anti-GBM collagen antibody
titers averaged 1/10,000. Inhibition experiments were performed using a solution of anti-GBM collagen antibodies binding specifically 20% of labeled GBM collagenous polypeptides.
As shown in Table I, 100 ng of unlabeled GBM collagen strongly
inhibited the binding of 10 ng of radioactive antigens with
anti-GBM collagen antibodies, while 100 ng of Clq, procollagen
and α-chains of the type I collagen reduced the specific precipitation by only 10%. Finally, antibodies to type IV collagen bound C as assessed by standard C fixation techniques.

TABLE I

Specificity of Antibodies to GBM Collagen

Unlabeled Antigen (100 ng)	*Inhibition of $[^{125}I]$ GBM Collagen Precipitated* (%)*
GBM collagen	80
Clq	10
Type I procollagen	10
Type I collagen	10

*
*Specific activity of labeled GBM collagen was 2.3 x 10^6
cpm/μg. Ten ng (about 20,000 cpm) were used in each tube for
radioimmunoassay (13).*

Cytotoxic Activity of Antibodies to GBM Collagen

Since certain antibodies to the fibroblast membrane
collagens have been shown to elicit C-mediated cytotoxicity
(8,9), experiments were carried out in order to ascertain
whether the anti-GBM collagen antibodies were cytotoxic. In
the presence of C, anti-GBM collagen antibodies lysed about

90% of the cultivated glomerular cells at a 1:4 dilution.
Complement alone and control rabbit sera exhibited no cytotoxic effect. Similarly, no major cytotoxic effect was
observed when positive sera were tested without C, with heat-
inactivated C, or when using human fibroblasts as target
cells.

In order to control the specificity of the cytotoxicity,
inhibition experiments were performed with various collagen
antigens. Anti-GBM collagen antiserum was then incubated with
100 ng of unlabeled GBM collagen, C1q, procollagen and α1
chains of the type I collagen for 30 min. at 37°C, before the
addition of the trypsinized cultivated glomerular cells. As
shown in Table II, only the GBM collagen blocked the cytotoxic
effect of the anti-GBM collagen antiserum. This result is
consistent with the presence of GBM collagenous polypeptides
on the cultivated glomerular cell membrane.

TABLE II

*Inhibition of Cytotoxicity by Collagenous Antigens**

Unlabeled Antigen (100 ng)	Viable Cells (%)
GBM collagen	75
C1q	10
Type I procollagen	10
Type I collagen	15

*
*Antigens were mixed with anti-GBM collagen antibody for
30 min. at 37°C, before the incubation of antibody with the
cultivated cells, as described in the Materials and Methods.*

Further, treatment of the glomerular cells with collagenase prior to the addition of antiserum and C abolished the
cytotoxic effect (Table III). On the contrary, treatment of
the cells with EDTA-inactivated collagenase did not modify
the cytolysis, which remained quite similar to that observed
after trypsin treatment of the cells. Finally, direct immunofluorescence performed with FITC-labelled anti-GBM collagen
IgG showed a marked fluorescence of most cells in culture.

These results indicate that the GBM collagen present on the membrane of glomerular cells in culture elicits the cytotoxic response to the anti-GBM collagen antiserum.

TABLE III

*Inhibition of Cytotoxicity by Enzymatic Treatment of the Cultivated Glomerular Cells**

Antiserum	Dilution	Enzyme	Viable Cells (Percent)
Anti-GBM	1/4	Trypsin	10
-	-	Trypsin	90
+	1/4	Collagenase	75
-	-	Collagenase	90
+	1/4	Collagenase + EDTA	10
-	-	Collagenase + EDTA	90

The experimental conditions were described in the Materials and Methods.

CONCLUSIONS

1. Antibodies to GBM collagenous polypeptides can induce complement-mediated cytotoxicity against glomerular cells in culture.

2. The cytotoxic response is elicited by the presence of GBM collagenous polypeptides at the surface of the cultivated glomerular cells.

REFERENCES

1. Vernier, R.L., *IN:* "Small Blood Vessel Involvement in Diabetes", (M.D. Siperstein, A.R. Colwell, and K. Meyer, Eds.), American Institute of Biological Sciences, Washington, D.C., 1964.

2. Kefalides, N.A., *Biochem. Biophys. Res. Commun.*, 45:226, 1971.
3. Kefalides, N.A., *Connec. Tissue Res.* 6:63, 1973.
4. Grant, M.E., Harwood, R. and Williams, I.F., *Eur. J. Biochem.* 54:531, 1975.
5. Cohen, M.P. and Vogt, C.A., *Biochem. Biophys. Acta* 393: 78, 1975.
6. Killen, P.D., Quadracci, L.J. and Striker, G.E., *Fed. Proc.* 33:617, 1974.
7. Foidart-Willems, J., Dechenne, C. and Mahieu, P., *Diabète et Métabolisme* 1:227, 1975.
8. Lustig, L., *Proc. Soc. Exp. Biol. Med.* 133:207, 1970.
9. Lichtenstein, J.R., Bauer, E.A., Hoyt, R. and Wedner, H.J., *J. Exp. Med.* 144:145, 1976.
10. Krakower, C.A. and Greenspon, S.A., *Arch. Pathol.* 51:629, 1951.
11. Yonemasu, K. and Stroud, R.M., *J. Immunol.* 106:304, 1971.
12. Unanue, E.R. and Dixon, F.J., *Adv. Immunol.* 6:1, 1967.
13. Mahieu, P., Lambert, P.H. and Miescher, P.A., *J. Clin. Invest.* 54:128, 1974.
14. Marchalonis, J.J., *Biochem. J.* 113:229, 1969.
15. Sober, H.A. and Peterson, E.A., *Fed. Proc.* 17:1116, 1958.
16. Dechenne, C., Foidart-Willems, J., and Mahieu, P., *J. Submicr. Cytol.* 8:101, 1976.
17. Peterkofsky, B. and Digelmann, R., *Biochemistry* 10:988, 1971.

THE ROLE OF TUBULAR BASEMENT MEMBRANE

ANTIGENS IN RENAL DISEASES*

Boris Albini, Jan Brentjens, Elena Ossi

and

Giuseppe A. Andres

State University of New York at Buffalo
and
Buffalo General Hospital

ABSTRACT: Guinea pigs and rats immunized with homologous
or heterologous TBM develop an autoimmune tubulo-interstitial (TI) nephritis characterized by linear deposits
of IgG and C_3 in TBM, interstitial infiltration of lymphocytes, monocytes, macrophages, giant cells and destruction
of TBM. In man the condition in which TBM Ab is most
frequently found is in association with anti-glomerular
basement membrane (GBM) glomerulonephritis. In order to
evaluate the pathogenetic role of TBM Ab in man renal
tissues from 7 patients with anti-GBM/anti-TBM rapidly
progressive nephritis were studied, and the results were
compared with those obtained in 7 patients with rapidly
progressive anti-GBM nephritis and in 11 patients with
"crescentic" glomerulonephritis not associated with Ab
to renal basement membranes. All 7 patients with anti-
GBM/anti-TBM nephritis had TI lesions; five patients,

*This study has been supported by a USPHS Grant A1-10334
and Contract 91271 of the New York State Health Research
Council.

with higher levels of TBM Ab, had discontinuities in TBM, "increscents" of proximal convoluted tubules, proliferation and degeneration of tubular cells and accumulation of nuclear remnants in tubular lumina; three patients had giant cells in the interstitium and "adenomatous" transformation of the intra-medullary veins. Patients with anti-GBM or other varieties of "crescentic" glomerulonephritides had variable degree of interstitial mononuclear cell infiltration only. These findings show that in man TBM Ab may be associated with distinctive TI changes comparable to those observed in experimentally-induced anti-TBM nephritis.

In the last five years it has become evident that the immunopathogenetic mechanisms involved in glomerular injury may also cause tubulo-interstitial (TI) diseases (1,2). Indeed, antibodies reacting with tubular basement membrane (TBM) as well as immune complexes, may induce TI lesions in experimental animals and in man. In the present article we will review the role of TBM antigens and their antibodies in the development of TI renal disease.

There is no precise information concerning the origin of TBM. Since antibody to epithelial basement membranes (such as basement membrane of the lens capsule or basement membrane synthesized by yolk sac carcinoma) react with TBM, it is possible that the tubular epithelial cells may have a primary role in the synthesis and the maintenance of the TBM (3,4). In contrast to the glomerular basement membrane (GBM), the fine structure of the TBM is characterized solely by a "lamina densa" without contiguous "laminae rarae". Several methods of isolation of TBM have been proposed (see elsewhere in this volume). The best available preparations are 95-98% pure and contain a collagen moiety and glycoproteins. The amino acid composition and the oligo- and disaccharide moieties have been characterized.

The biochemistry of the TBM is discussed in another part of this volume. Several antigenic determinants have been tentatively described in the TBM. Some of them are common to all the epithelial basement membranes (3,4); others are shared by the TBM, the lung and the GBM (5); lastly, a few other antigenic determinants seem to be specific for the TBM (6).

EXPERIMENTAL ANTI-TBM DISEASE

It is generally recognized that heterologous anti-GBM antibodies induced experimentally by injection of kidney homogenates or more purified GBM preparations, are reactive with TBM as well as with basement membranes of other organs. It is difficult or impossible, however, to induce TI nephritis comparable to Masugi (or nephrotoxic) glomerulonephritis by using heterologous anti-TBM sera. Even the injections of high-titered heterologous antisera to rat TBM - which bind to TBM both *in vitro* and *in vivo*, fix complement along the TBM and generate a weak autologous immune response - do not induce detectable functional or histological changes (7).

Experimental anti-TBM nephritis has been induced in guinea pigs and rats. In the first model, described by Steblay and Rudovsky (8) guinea pigs are immunized with crude rabbit TBM preparation in Freund's adjuvant. After a few weeks, the guinea pigs develop antibodies which bind in a linear fashion to the TBM of proximal convoluted tubules (PCT)(Figure 1). Linear deposits of C3 are present in the same areas (9). These immunopathologic findings are associated with interstitial accumulation of mononuclear and giant cells (Figure 2), thinning and destruction of TBM, and tubular cell degeneration. In later stages of the disease, tubular atrophy and interstitial sclerosis are also present. Clinically, there is mild to severe glucosuria with normoglycemia, proteinuria and uremia. This disease has been passively transferred to normal guinea pigs by using serum from nephritic guinea pigs (10). For a successful transfer, C3 is necessary (11). After injection of anti-TBM antibodies into normal guinea pigs the antibodies first bind to the GBM and, to a smaller extent, to the TBM. With time, however, the intensity of GBM binding decreases whereas TBM binding increases (12). Since active immunization induces anti-TBM disease in C4 deficient guinea pigs (9), it seems that the alternate, rather than the classical, complement pathway is activated. The interstitial infiltrates consist mainly of lymphocytes and monocytes and only a few polymorphonuclear leukocytes. Experiments of irradiation and reconstitution (13) and studies performed with the rosette technique on tissue sections (14) indicate that virtually all the infiltrating lymphocytes belong to the B cell line. The cells which appear more directly responsible for tubular damage are the macrophages and the giant cells (Figure 3). The giant cells probably result from fusion of the macrophages and most of them have the same surface markers. Inhibition of macrophages by injection of silica significantly decreases the

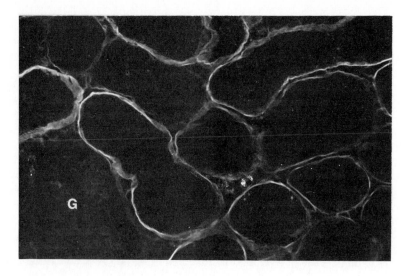

Fig. 1. Linear deposition of IgG along renal tubular basement membranes of a guinea pig with anti-TBM nephritis. The basement membrane of a glomerulus (G) does not stain for IgG. Direct immunofluorescence. X250.

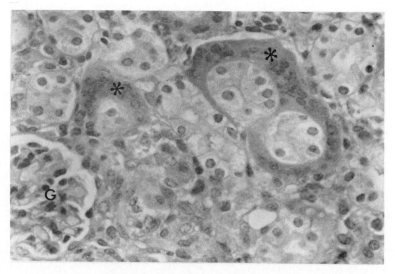

Fig. 2. Giant cells (asterisks) surround renal tubules in a guinea pig with anti-TBM nephritis. The glomerulus (G) appears normal. H and E stain. X250.

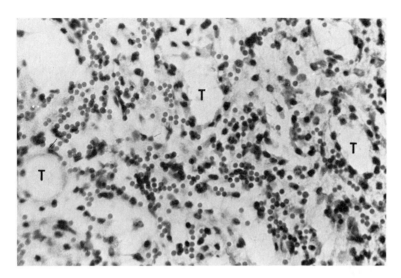

Fig. 3. Rosette formation of antibody (rabbit
IgG) coated sheep red blood cells with cells infil-
trated in the renal interstitium of a guinea pig with
anti-TBM nephritis. This characterizes these infil-
trated cells as macrophages. T: tubular lumen.
H and E stain. X250.

severity of the disease in spite of IgG and C3 deposition
along the TBM. The life span of nephritic guinea pigs
treated with silica is increased as compared to that of untreated nephritic controls (7). Guinea pig strain 13 is more
prone to develop anti-TBM nephritis than strain 2 (15).
A similar anti-TBM nephritis is induced in guinea pigs by
immunization with bovine TBM (16). This type of immunization
generates antibodies reacting to some extent with TBM. Binding of the antibodies to extra-renal basement membrane has
not been reported. In contrast, guinea pigs immunized with
rabbit TBM produce antibodies which bind to the TBM and to
the alveolar basement membranes (17). Accumulation of polymorphonuclear leukocytes and thickening of alveolar septa
is occasionally observed in the lung of these animals. Preliminary experiments show that anti-TBM antibody can cross
the placenta and bind to the TBM of the offspring (18). The
antibodies to TBM are auto-antibodies since they react with
the guinea pigs' own TBM. The disease, however, is not self-
perpetuating when the guinea pigs are immunized with low
doses of antigen (18). The frequency and severity of the TI
changes vary according to the amount of antigen injected.
Adjuvants are not necessary for a successful immunization (19).

In a second model, Brown Norway (BN) and BN x Lewis rats are immunized with Sprague-Sawley rat kidney homogenate incorporated into Freund's adjuvant. In addition, the rats are also injected with pertussis vaccine (20). Early in the course of the disease the rats develop antibody to TBM, which binds in a linear pattern, to TBM of PCT. C3 is also fixed to the same structures. Two to three weeks after the initial immunization, polymorphonuclear leukocytes, lymphocytes, monocytes and macrophages infiltrate the interstitium. Based on electron microscopy, the giant cells, which appear six to eight weeks later, result from fusion of macrophages. The TBM of PCT is disrupted and phagocytized by macrophages, and severe degenerative changes of tubular cells are evident. In a late stage, the disease is characterized by extensive tubular destruction, tubular atrophy and interstitial fibrosis. Furthermore, in this late stage, the rats develop autologous immune complex glomerulonephritis. Studies of the circulating anti-TBM antibodies show that the immunizing TBM antigens are present in BN, Lewis x BN F1 hybrids, Sprague-Dawley, August, Fisher, ACl, Wistar and Buffalo rats but not in Lewis rats (2). The antigen is also present in Hartley guinea pigs and in New Zealand white rabbits (2). The anti-TBM nephritis is passively transferable to normal BN rats by intravenous injection of relatively large amounts of circulating anti-TBM antibodies (20). A variant of anti-TBM nephritis is induced in BN or BN x Lewis rats by immunization with bovine TBM preparation (21). This model is similar to the one described previously but usually lacks the late development of Heymann's glomerulonephritis. After prolonged immunization, however, a few rats may show granular deposits of IgG in glomerular capillary walls. The use of pertussis vaccine enhances the severity of the disease. Linear deposits of IgG and C3 are localized along the TBM of proximal and, to a lesser extent, of distal convoluted tubules and Bowman's capsule (21). The interstitial infiltrates are formed by neutrophils, mononulclear cells and a few giant cells. This pathology leads to interstitial fibrosis and tubular atrophy. Sensitized cells appear to play a relative minor role in inducing these lesions (22). A marked difference exists between the pathology of female and male rats in both BN and BN x Lewis strains since males develop a milder disease with focal peritubular and periglomerular plasma cell infiltration (7). Other strains of rats immunized with bovine TBM develop antibodies to TBM. Nevertheless, *in vivo* binding of these antibodies to TBM is minimal in ACl, Buffalo, Wistar-Furth and DA rats and absent in Lewis and Max rats. The latter rat strains do not develop anti-TBM disease (21).

Another experimental condition characterized by formation of anti-TBM antibodies is renal transplantation. Rats with renal allografts transplanted across a weak histocompatibility barrier (Lewis x Fisher 344 Fl hybrid kidneys transplanted into Lewis rats) frequently develop antibody to TBM, linear deposits of IgG along TBM, interstitial cellular infiltration and tubular atrophy (2). Since linear deposits of IgG are found also in the TBM of the native kidneys left *in situ* it is proposed that the IgG contains auto-antibodies. This conclusion, however, is not supported by other studies performed in a similar model (Lewis rats receiving Lewis x BN Fl hybrid kidneys) (23). In this condition, linear deposits of IgG and C3 are found along the TBM of the grafted but not of the native kidneys. When kidneys from the same hybrids are transplanted into BN rats there is no development of anti-TBM antibody (23). The findings are interpreted as suggesting that foreign TBM antigens are introduced into the recipient through transplantation. The interpretation that anti-TBM antibodies - and not only allograft rejection - may have a role in the development of TI lesions is supported by the observation that rats in which rejection is suppressed, develop anti-TBM antibodies and a concomitant TI nephritis (24).

Rabbits injected with mercury chloride over prolonged periods of time, develop antibodies which react with TBM, GBM, vascular basement membrane, and peri- and endomysium. Later these animals develop immune complex glomerulonephritis (25).

The experimental models of anti-TBM nephritis have established certain immunopathologic features which are characteristic of this disease. These are: 1) linear deposits of IgG and C3 along TBM; 2) interstitial mononuclear cell infiltration, with occasional presence of polymorphonuclear leukocytes and giant cells, disruption of the TBM, tubular cell degeneration, tubular atrophy and interstitial fibrosis; 3) proteinuria, glucosuria with normoglycemia, aminoaciduria and progressive uremia. The antibodies to TBM are either auto-antibodies or antibodies directed against "foreign" TBM determinants.

HUMAN ANTI-TBM DISEASE

It is now well established that, although rarely, anti-TBM disease occurs in man. It seems that antibody to TBM may develop in at least five conditions: 1) in association with anti-GBM glomerulonephritis; 2) after an immune complex glomerulonephritis; 3) after treatment of patients with dimethoxyphenyl penicilloyl or diphenylhydantoin compounds; 4) after renal transplantation; and, 5) without apparent association with any of the conditions mentioned above.

The association of anti-TBM antibody with anti-GBM antibody is the most frequent, since 60-70% of patients with anti-GBM glomerulonephritis have antibody to TBM (Figure 4) (26,27). The antibodies present in the serum or eluted from the kidney react with the renal basement membranes (anti-renal basement membrane disease) and with the alveolar basement membranes as well (Goodpasture's disease) (26,28). Infiltration of mononuclear cells, with occasional giant cells, TBM "gaps", "splitting" and thickening of TBM are frequent features of this TI nephritis. In a few patients, proliferation of epithelial cells of PCT, tubular casts of nuclear remnants and proliferation of adventitial cells from the juxtamedullary veins have been observed (29).

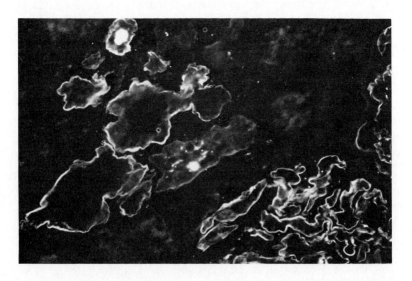

Fig. 4. Linear deposition of IgG glomerular as well as tubular basement membranes in a patient with anti-TBM nephritis. Direct immunofluorescence. X 250.

Scattered reports have described patients with anti-TBM antibodies and TI disease following an immune complex glomerulonephritis (30,31,32). Some of these patients had linear deposits of IgG and C3 in TBM, anti-TBM antibodies in the circulation and Fanconi's syndrome (31,32). Anti-TBM antibodies were also found in a patient with milk hypersensitivity and immune complex glomerulonephritis (33). In all these conditions it was proposed that the TBM may somehow have been altered by a primary immune complex injury, which in turn would trigger an autoimmune response.

A few patients treated with methicillin developed a TI nephritis and anti-TBM antibodies were detected in the circulation and bound to TBM. It is suggested that the dimethylphenyl penicilloyl group, which is largely secreted by the PCT, may bind to components of TBM. Then, the penicilloyl-TBM conjugate stimulates the formation of antibodies responsible for the disease (27,34,35). Similarly, IgG and diphenylhydantoin but not complement, have been found along TBM of a patient treated with diphenylhydantoin (36).

Another condition in which anti-TBM antibodies are relatively frequent is renal transplantation (about 15-20% of long surviving renal allografts) (27,37,38,39). The antibodies may react only with the TBM of the graft or may bind to the TBM of native kidneys as well. In renal transplantation it is particularly difficult to assess the pathogenetic role of anti-TBM antibodies since allograft rejection can account for interstitial cellular infiltration and renal functional impairment.

Finally, anti-TBM antibodies and severe TI nephritis may develop without apparent association with other diseases or causative agents. In a 6 year old boy with renal glucosuria, azotemia and aminoaciduria a renal biopsy revealed chronic interstitial nephritis with severe tubular damage. Linear deposits of IgG and C3 in TBM and circulating anti-TBM antibodies were found. Anti-GBM antibodies were not detectable (40).

In summary, these scattered observations show that, although lesions mediated by anti-TBM antibodies are quite rare, they can indeed occur in man. Certain histological and functional changes are reminiscent of pathology observed in guinea pigs or in rats with experimental anti-TBM disease. Studies of large series of patients are needed in order to evaluate the role of anti-TBM antibodies in human pathology.

REFERENCES

1. McCluskey, R.T. and Klassen, J., *N. Eng. J. Med.* 288:564-570, 1973.
2. Andres, G.A. and McCluskey, R.T., *Kidney Int'l.* 7:271-289, 1975.
3. Pierce, G.B., Jr. and Nakane, P.K., *Lab. Invest.* 17:499-514, 1967.
4. Pierce, G.B., Jr., *In:* "Chemistry and Molecular Biology of the Intercellular Matrix", (Balazs, E.A.), Academic Press, London, New York, Vol. 1, pp. 471-506, 1970.
5. Wilson, C.B. and Dixon, F.J., *In:* "The Kidney", Brenner, B.M. and Rector, F.C., Jr., Eds.), W. B. Saunders Company, Philadelphia, London, Toronto, Vol. II, Chapter 22, 1976.
6. Graindorge, P. and Mahieu, P., *Kidney Int'l.* 10:186, 1977.
7. Ossi, E., Albini, B. and Andres, G.A., *Unpublished observations.*
8. Steblay, R.W. and Rudofsky, U.H., *J. Immunol.* 107:589-594, 1971.
9. Rudofsky, U.H., McMaster, P.R.B., Ma, W.S., Steblay, R.W. and Pollara, B., *J. Immunol.* 112:1387-1393, 1974.
10. Steblay, R.W. and Rudofsky, U.H., *Science* 180:966-968, 1973.
11. Rudofsky, U.H., Steblay, R.W. and Pollara, B., *Clin. Immunol. Immunopathol.* 3:396-407, 1975.
12. Van Zwieten, M.J., Bahn, A.K., McCluskey, R.T. and Collins, A.B., *Am. J. Path.* 83:531-542, 1976.
13. Rudofsky, U.H. and Pollara, B., *Clin. Immunol. Immunopathol.* 4:425-439, 1975.
14. Szymanski, C., Albini, B. and Andres, G.A., *Manuscript in preparation.*
15. Hyman, L.R., Colvin, R.B. and Steinberg, A.D., *J. Immunol.* 116:327-335, 1976.
16. Lehman, D.H., Marquardt, H., Wilson, C.B. and Dixon, F.J., *J. Immunol.* 112:241-248, 1974.
17. Milgrom, M., Albini, B., Brentjens, J., O'Connell, D. and Andres, G.A., *Manuscript in preparation.*
18. Albini, B., Ossi, E. and Andres, G.A., *Unpublished observations.*
19. Rudofsky, U.H., *Clin. Exp. Immunol.* 25:455-461, 1976.
20. Sugisaki, T., Klassen, J., Milgrom, F., Andres, G.A. and McCluskey, R.T., *Lab. Invest.* 28:658-671, 1973.
21. Lehman, D.H., Wilson, C.B. and Dixon, F.J., *Kidney Int'l.* 5:187-195, 1974.
22. Lehman, D.H. and Wilson, C.B., *Int. Archs. Allergy Appl. Immun.* 51:168-174, 1976.

23. Lehman, D.H., Lee, S., Wilson, C.B. and Dixon, F.J., *Transplantation* 17:429-431, 1974.
24. Abbas, A.K., Corson, J.M., Carpenter, C.B., Strom, T.B., Merrill, J.P. and Dammin, G.J., *Am. J. Path.* 79:255-268, 1975.
25. Roman-Franco, A.A., Turiello, M., Albini, B., Ossi, E. and Andres, G.A., *Ninth Ann. Meeting Am. Soc. Nephrol.*, Washington, D.C., p. 63, 1976.
26. Koffler, D., Sandson, J., Carr, R. and Kunkel, H.G., *Am. J. Pathol.* 54:293-306, 1969.
27. Lehman, D.H., Wilson, C.B. and Dixon, F.J., *Am. J. Med.* 58:765-786, 1975.
28. McPhaul, J.J., Jr., and Dixon, F.J., *J. Clin. Invest.* 49:308-317, 1970.
29. Andres, G.A., *Unpublished observations.*
30. Morel-Maroger, L., Kourilsky, O., Mignon, F. and Richet, G., *Clin. Immunol. Immunopathol.* 2:185-194, 1974.
31. Tung, K.S.K. and Black W.C., *Lab. Invest.* 32:696-700, 1975.
32. Levy, M., Gagnadoux, M.F. and Habib, R., *Third Int. Symposium Pediatr. Nephrol.*, Washington, D.C., p. 13, 1974.
33. Harner, M.H., Nolte, M., Wilson, C.B., Talwalker, Y.B., Musgrave, J.E., Brooks, R.E. and Campbell, R.A., *Third Int. Symposium Pediatr. Nephrol.*, Washington, D.C., p. 8, 1974.
34. Baldwin, D.S., Levine, B.B., McCluskey, R.T. and Gallo, G.R., *N. Eng. J. Med.* 279:1245-1252, 1968.
35. Border, W.A., Lehman, D.H., Egan, J.D., Sass, H.J., Globe, J.E. and Wilson, C.B., *N. Eng. J. Med.* 291:381-384, 1974.
36. Hyman, L.R., Ballow, M. and Knieser, M.R., *Eighth Ann. Meeting Am. Soc. Nephrol.*, Washington, D.C., p. 54, 1975.
37. Klassen, J., Kano, K., Milgrom, F., Menno, A.B., Anthone, S., Anthone, R., Sepulveda, M., Elwood, C.M. and Andres, G.A., *Int. Arch. Allergy Appl. Immun.* 45:675-689, 1973.
38. Wilson, C.B., Lehman, D.H., McCoy, R.C., Gunnells, J.C., Jr. and Stickel, D.L., *Transplantation* 18:447-452, 1974.
39. Berger, J., Noël, H. and Yianeva, H., *Sixth Int. Congress Nephrol.*, Florence, Italy, p. 134, 1974.
40. Bergstein, J. and Litman, N., *N. Eng. J. Med.* 292:875-895, 1975.

GROUP A STREPTOCOCCAL MEMBRANES:

ISOLATION AND IMMUNOCHEMICAL STUDIES

Ivo van de Rijn and John B. Zabriskie

The Rockefeller University
New York, New York

ABSTRACT: *Components of membranes from Group A strepto-*
cocci have been shown to be cross-reactive with various
tissue antigens: i.e., sarcolemmal sheath of myo-
cardium, caudate nucleus, and glomerular basement mem-
branes. Group A streptococci were harvested in the
logarithmic phase of growth, and the cell wall was
removed under osmotically stabilized conditions using
purified Group C streptococcal phage-associated lysin.
After removal of the solubilized wall polymers, the
protoplasts were lyzed and treated with DNase and RNase
followed by extensive washings in phosphate buffered
saline. The purified protoplast membrane contained
74% protein, 2.5% carbohydrate, the remainder of which
was lipid. On sodium dodecyl sulfate slab gel acryla-
mide electrophoresis, the protein fraction of the proto-
plast membrane showed 35 major and 10 minor polypeptides.
A fraction containing four polypeptides has been iso-
lated which is cross-reactive with human sarcolemmal
sheath from myocardium and removes heart-reactive anti-
body from the sera of acute rheumatic fever patients.
This cross-reactive antigen is not involved with the
caudate nucleus or glomerular basement membrane cross-
reacting systems.

INTRODUCTION

Components of the Group A streptococcus have been demonstrated to cross-react with mammalian tissue components including skin (1), kidney (2), brain (3), heart valve glyco-proteins (4), and myocardium (5). There have also been reports of cross-reactions with the HLA system (6). However, Tauber et al. (7) have proposed that these observed cross-reactions with the HLA system were due to the activation of the alternate pathway of complement by components of the Group A streptococcal membrane. Antigens from three of the above cross-reactive systems have been localized in the protoplast membrane.

Perhaps the best characterized of these reactions has been the antigen which cross-reacts with mammalian heart tissue. Kaplan (8) demonstrated an antigen associated with the M protein of Group A streptococci thereby localizing this antigen in the cell wall. Zabriskie and Freimer (9), on the other hand, found a similar antigen in the membranes of all Group A streptococci studied. Recently, van de Rijn and Zabriskie (10) have purified this cross-reactive antigen from the membranes of group A streptococci 120-fold and have demonstrated this fraction to be a protein comprised of four polypeptides ranging from 22-32,000 daltons on sodium dodecyl sulfate slab gel gradient electrophoresis (SDS-PAGGE).

A second cross-reactive system associated with the membrane has recently been described. Husby et al. (3) demonstrated that patients with rheumatic chorea contained in their sera an antibody which bound to the caudate and subthalamic nuclei of the brain. These antibodies could be absorbed by the isolated neurons as well as the Group A streptococcal membrane. Finally, the third cross-reactive system which has been localized in the protoplast membrane is the one associated with the glomerular basement membrane. This cross-reaction is fully described in a paper by Zabriskie et al. also found in this volume.

The association of these cross-reactive antigens to certain disease states has spurred interest in the general chemical and serological identification of the various components of the streptococcal membrane. The present report discusses the methods used to grow the Group A streptococcus and isolate the protoplast membrane. In addition, some of the chemical and immunochemical properties of the membrane will be discussed. It is hoped that a complete understanding of the chemical and immunological nature of these cross-reactions will lead to a better understanding of the pathogenesis of these disease states.

MATERIALS AND METHODS

Bacteria and Media

All Group A streptococcal strains were obtained from
Dr. R. C. Lancefield of our laboratory. Preparation of the
growth medium has been described previously (3). Twenty
liters of dialyzed Todd-Hewitt broth was inoculated with a
1% inoculum of an overnight logarithmically growing culture.
The pH of the growing culture was allowed to drop to 7 and
titrated at this pH with 5 N sodium hydroxide. One the
culture stopped taking sodium hydroxide, it was sedimented
in a Sharples high speed centrifuge. The cells were then
stored at -70°C until further use.

Preparation of Cell Membranes

An aliquot of the bacteria were thawed and washed three
times in saline and finally once with distilled water. The
bacteria were then broken using a Vibrogen cell mill (Rho
Scientific, Comack, N.Y.) using a previously described
method (11). Bacteria (25 grams wet weight) were re-suspended
in distilled water, and breakage is permitted to occur in the
Vibrinogen cell mill until no gram positive organisms are
monitored. The cell walls are then removed by low speed
centrifugation, and the membranes are sedimented in the ultra-
centrifuge. The membranes are next washed in ribonuclease-
deoxyribonuclease buffer (RNase-DNase) and then treated for
two hours at 37°C with RNase and DNase. After centrifugation
of the membranes at 125,000 x g for an additional two hours,
the treatment is repeated. The membranes are re-sedimented
and washed three times in .01 M phosphate buffer saline pH 7.6
(PBS) and two times in distilled water. The membranes are
finally re-suspended in distilled water, and an aliquot is
lyophilized. The remaining membrane is brought up to the concentration of 20 mg per ml with .01 M phosphate buffered
saline pH 7.6, 0.02% sodium azide for further studies.

Protoplast membranes were prepared using Group C streptococcal phage-associated lysin. The enzyme was initially purified through the cellulose phosphate chromatography step
according to the procedure of Fischetti et al. (12). Lysin
purified to this step does not contain any Group C cell membrane or cell wall components. Group A streptococcal membranes were prepared using the purified Group C phage-associated lysin as described by Zabriskie and Freimer (9) with
slight modification as follows: after thawing, the organisms
were washed with saline and then resuspended in 0.06 M sodium

phosphate buffer (pH 6.1) containing 4% sodium chloride and
5×10^{-4} M dithiothreitol (DTT). Ten volumes of buffer were
used per gram wet weight of organism. To this was added DTT-
reactivated lysin at 10,000 units lysins/gram wet weight of
cells and five milligrams of DNase. The mixture was incubated
at 37°C for two hours and monitored for protoplast formation
by darkfield microscopy and gram stain. After what appeared
to be 100% conversion of intact streptococci to the protoplast state, the suspension was incubated for an additional
thirty minutes. The remainder of the treatment and the
procedures followed those of Zabriskie and Freimer (9) using
ten volumes buffer per gram wet weight of cells for all
volumes. The final protoplast membrane preparation was re-
suspenced in phosphate buffered saline at 20 mg/ml and
stored in the presence of 0.02% sodium azide.

Immunofluorescent Studies

Sera were assayed for the presence of heart-reactive
antibody or anti-caudate nucleus antibodies by the indirect
immunofluorescent staining technique of Zabriskie and
Freimer (9) with modifications by van de Rijn and Zabriskie
(13). Dessicated human heart sections or human brain
sections (4 μ) were first layered with the test serum and
then overlaid with fluoresceinated goat anti-human gamma
globulin. All slides were graded on a 0-4+ scale.

Absorption of Antibody

Five microliters of human test sera which gave a 4+
staining at a dilution of 1:10 was added to a microfuge tube
containing 45 microliters of a dilution of a sample containing
the absorbent. Contents were mixed and allowed to incubate
at 37°C for two hours and then overnight at 4°C. Just before
use, samples were centrifuged for five minutes in a microfuge
152 (Beckman), and the supernatant was used in the immunofluorescence assay. Unadsorbed sera which were incubated for
same periods of time were used as a 4+ stain control. Normal
sera were used as a negative control.

Polyacrylamide Slab Gel Electrophoresis

The protein profile of the Group A streptococcal membrane
was monitored by sodium dodecyl sulfate polyacrylamide gradient gel electrophoresis (SDS-PAGGE) by the method of Laemmli
(14). Using a discontinuous Tris-chloride buffer system,
electrophoresis was carried out on a 7.5-15% gradient slab
polyacrylamide gel at 25 milliamps for 18 hours.

Standard (bovine serum albumin, ovalbumin, carbonic anhydrase, and cytochrome C) were run on all gels for molecular weight determination. Samples and standards in 2% SDS, 5% mercaptoethanol, and 10% glycerol were boiled (2-5 minutes) in a water bath before loading onto the gels. After electrophoresis, the gels were placed in a solution of 7% acetic acid and 25% isopropanol for ten minutes to remove any excess SDS. The gels were incubated for two additional washings in the same solution. The gels were then stained using 0.2% Coomassie Blue and 7% acetic acid with 25% isopropanol for two hours at 45°C. For de-staining, the gels were placed in 7% acetic acid with 25% isopropanol.

Gels containing I^{125} labelled samples were run in a similar manner except that the gels were dried according to the procedure of Maizel (15) and exposed to Kodak X-ray film to determine their radiolabelled bands by radioautography.

RESULTS AND DISCUSSION

Mechanical versus Enzymatic Disruption of Group A
    Streptococci

In the past, two methods have been used to isolate Group A streptococcal membranes. Figure 1 depicts the procedures used in mechanical and enzymatic disruption of streptococcal cells. Mechanical disruption has the advantage that both the whole cell wall and cell membrane can be isolated from the preparation. If the prime purpose of the isolation is obtaining protoplast membranes, there are many disadvantages to this technique. Due to localized heating effects even in the presence of a carbon dioxide cooling system, many of the enzymes are denatured. Two other disadvantages are that the membranes are extremely fragmented, and there is a loss of up to 40% due to association of the membrane with the cell wall in the isolation procedure.

Protoplast membrane formation by phage-associated lysin leads to the isolation of essentially intact protoplast membrane. Since the purified phage-associated lysin contains no proteolytic activity, the membranes are isolated in the native condition which makes the study of its enzymes possible. Membranes isolated in either manner are essentially free of contaminating cell wall and cytoplasmic material (16). Finally, both preparations of membrane contain the cross-reactive antigens.

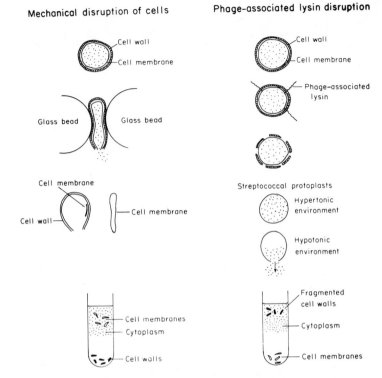

Fig. 1. Schematic diagram of mechanical and phage-associated lysin disruption of Group A streptococci.

Cross-reactions of the Protoplast Membrane
    with Human Tissue Components

When sera from patients with acute rheumatic fever are layered on heart sections and then counter-stained with fluoresceinated goat anti-human gamma globulin, an antibody is shown to be present in the sera which binds sarcolemmal sheaths (Figure 2a). If instead the sera is pre-incubated with membranes from Group A streptococci and the supernatant checked for the presence of the heart-reactive antibody after the membranes have been removed by centrifugation, the antibody is no longer present in the sera (Figure 2b). These results indicate that the Group A streptococcal membranes have an antigen cross-reactive with that of sarcolemmal sheath. These antibodies can also be removed by isolated sarcolemmal sheath preparations. We have demonstrated that these absorptions are not due to Fc receptors (13).

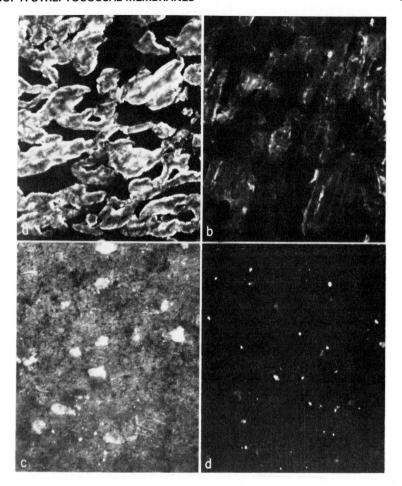

Fig. 2. Immunofluorescent staining of human myocardium (a,b) and caudate (c,d). Positive staining (a)
of sarcolemmal sheaths by antibodies from acute rheumatic
fever patients' sera and positive staining pattern (c)
of neurons by sera from rheumatic chorea patients. Both
these staining patterns are abolished if the sera is pre-
absorbed with Group A streptococcal membranes (b,d).
Remaining fluorescence (d) is due to nonspecific auto-
fluorescence produced by lipofuchsin granules in neurons
and nerve tissue.

Patients with rheumatic chorea also contain a second tissue antibody in their serum. When their sera is layered on
caudate and counter-stained with fluoresceinated goat anti-
human gamma globulin, an antibody is shown to be present which
binds caudate nuclei (Figure 2c). The antibody can be removed

by pre-absorption with Group A streptococcal membranes (Figure 2d) but not by human sarcolemmal sheaths, indicating that this is a different antibody than the heart-reactive antibody. Caudate neurons can also remove the antibody from the sera of patients with rheumatic chorea.

Chemical Studies on Streptococcal Membranes

Analysis of the streptococcal membrane demonstrates that it is composed mainly of protein and lipid with a low amount of carbohydrate also present (16). Table I depicts the chemical analysis of a Group A Type 6 streptococcal membrane.

TABLE I

Chemical Composition of Streptococcal
Protoplast Membrane S43/192*

	Percent
Protein	73.3
Lipid	24.0
Carbohydrate	2.2

*The chemical composition was taken as an average
of determinations on each of three prepared lots.

Each value is an average of three determinations on each of three prepared lots. The composition is in agreement with that demonstrated for Types 25 and 12 membranes described by Freimer (16). The carbohydrates of the membrane have been demonstrated to be mainly composed of glucose which is found in the form of glycolipid.

Purity of streptococcal membranes was monitored by using rhamnose and nucleic acid as constituents of contaminating cell wall and cytoplasm respectively. Good preparations of membranes contain less than 0.5% cell wall material. Normally only trace amounts of RNA have been found, indicating also that our preparations of membranes are essentially free of cytoplasm.

Analysis of the Protein Portion of the
    Streptococcal Membrane

    Since preliminary evidence has demonstrated that many
of the cross-reactive antigens are protein in nature, membranes were solubilized using sodium dodecyl sulfate and
electrophoresed in polyacrylamide gel. Figure 3 demonstrates
the protein pattern of membranes prepared from four Group A
streptococcal strains. Minor differences can be discerned

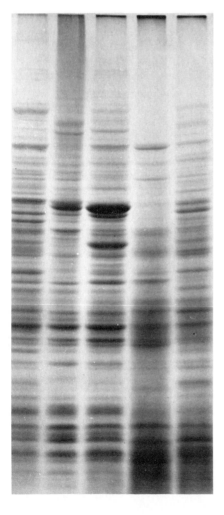

*Fig. 3. SDS-PAGGE of
five representative Group A
streptococcal membrane preparations. Membranes are from
(left to right) Type 18,
Type 5, Type 41, Type 14 and
Type 6 streptococci, respectively.*

in the protein patterns of the four Group A streptococcal membranes even though most of the polypeptides are in the range of 15-80,000 daltons. Each of the streptococcal membranes contained two very high molecular weight polypeptides which do not enter the gel in this system. In a 7-15% gradient gel system, the material at the top of the gel can be discerned as two high molecular weight polypeptides Each of the protoplast membrane preparations can be demonstrated to contain over 60 polypeptides using gradient slab gel electrophoresis.

We have isolated from these membrane preparations the antigens which cross-react with the myofibers of sarcolemmal sheaths. On 14% SDS polyacrylamide gel electrophoresis, our final product migrates as four polypeptides with molecular weights ranging from 22-32,000 daltons (Figure 4). This cross-reactive antigen comprises less than .6% of the dry weight of the membrane. At present, we do not know whether only a single one of these polypeptides contains the antigenic determinant or whether it is comprised of a combination of them.

Attempts to further separate each polypeptide have been unsuccessful to date. Through the use of specific immuno-absorbents, we hope to answer the above questions as to the specific nature of antigenic determinants of the streptococcal cross-reactive antigens of the heart as well as the heart antigen itself. Using the techniques devised for isolating the heart cross-reactive antigen, we are presently modifying them in order to obtain purified cross-reactive antigens to caudate nucleus as well as glomerular basement membrane.

ACKNOWLEDGMENTS

The authors are indebted to Ms. C. Eastby for invaluable and expert laboratory assistance.

This research was supported in part by Grant HL-03919 from the U.S. Public Health Service and in part by grants from the American Heart Association #70-1010 and from the New York Heart Association. I. van de Rijn is a recipient of a Senior Investigatorship from the New York Heart Association.

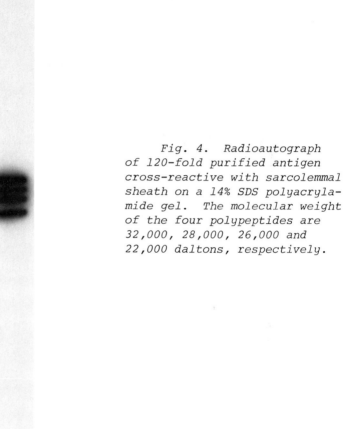

Fig. 4. Radioautograph of 120-fold purified antigen cross-reactive with sarcolemmal sheath on a 14% SDS polyacrylamide gel. The molecular weight of the four polypeptides are 32,000, 28,000, 26,000 and 22,000 daltons, respectively.

REFERENCES

1. Lyampert, I.M., Beletskaya, L.V., Borodiyuk, N.A.,
    Gnezditskaya, E.V., Rassokhina, I.I. and Danilova, T.A.,
    Immunol. 31:47, 1976.
2. Markowitz, A.S. and Lange, C.F., Jr., *J. Immunol.* 92:
    565, 1964.
3. Husby, G., van de Rijn, I., Zabriskie, J.B., Abdin, Z.H.
    and Williams, R.C., Jr., *J. Exp. Med.* 144:1094, 1976.
4. Goldstein, I., Rebeyrotta, P., Parlebas, J. and Halpern,
    B., *Nature* (London) 219:866, 1968.
5. Kaplan, M.H., *J. Immunol.* 90:595, 1963.
6. Rapaport, F.T. and Chase, R.M., Jr., *Science* (Wash.,
    D.C.) 145:407, 1964.
7. Tauber, J.W., Falk, J.A., Falk, R.E. and Zabriskie, J.B.,
    J. Exp. Med. 143:1341, 1976.
8. Kaplan, M.H., *In:* "Cross-reacting Antigens and Neo-
    antigens" (J.J. Trentin, Editor) Williams and Wilkins,
    Baltimore, p. 48, 1967.
9. Zabriskie, J.B. and Freimer, E.H., *J. Exp. Med.* 124:66,
    1966.
10. van de Rijn, I. and Zabriskie, J.B. Group A streptococ-
    cal antigens cross-reactive with myocardium. II. Puri-
    fication and characterization of the streptococcal
    antigen. Submitted for publication.
11. van de Rijn, I., Bleiweis, A.S. and Zabriskie, J.B.,
    J. Dent. Res. 55:C59, 1976.
12. Fischetti, V.A., Gotschlich, E.C. and Bernheimer, A.W.,
    J. Exp. Med. 133:1105, 1971.
13. van de Rijn, I. and Zabriskie, J.B., Group A streptococ-
    cal antigens cross-reactive with myocardium. I. Immuno-
    logical properties and purification of heart-reactive
    antibody. Submitted for publication.
14. Laemmli, U.K., *Nature* 227:680, 1970.
15. Maizel, J.V., Jr., *In:* "Methods in Virology" (K. Mara-
    morosh and H. Koprowski, Editors) Academic Press, N.Y.,
    9:179, 1971.
16. Freimer, E.H., *J. Exp. Med.* 117:377, 1962.

EPILOGUE

The First International Symposium on the Biology and Chemistry of Basement Membranes brought into focus a number of concepts on the nature of basement membranes through a series of excellent papers.

Evidence was presented to suggest that basement membranes perform a number of functions from filtration barriers to large solutes to barriers between cellular layers and connective tissue elements. It appears that the chemical nature, coupled with the ultrastructural organization of the protein components, determine the functional behavior of basement membranes in various tissues. There is good evidence to suggest that the net charge of basement membranes is a determining factor in the passage of solutes across the capillary wall. A number of papers stressed the role of basement membranes in embryonic development and differentiation.

The chemical and molecular composition of basement membranes was the subject of various papers. The existence in basement membranes of a collagenous component that exists in the form of procollagen was well documented. The proportion of this component varies among basement membranes, being highest in lens capsule, intermediate in glomerular basement membrane, and lowest in Reichert's membrane of the parietal yolk sac. One study conclusively showed that the collagenous component of basement membranes can be isolated as a triple-helical molecule composed of three identical α-chains each having a molecular weight of 95,000. The procedure involves pepsin digestion followed by reduction and alkylation and a second pepsin digestion at low temperature. Two collagenous components, the so-called A and B chains, have been isolated from whole tissues and organs. Although it was initially thought that they derive from the same collagen molecule and that they represent a component of basement membrane, subsequent studies failed to substantiate this assumption. Those collagenous components whose molecular weight is about 55,000 are thought to represent cleavage products of the larger molecular weight chains.

Several studies established the immunogenicity of basement membranes and its components. The antigenic determinants of the procollagen component appear to depend at least on the integrity of disulfide bonds and on the hydroxylation of proline. A small molecular weight component isolated from glomerular basement membrane and kidney cortex and tentatively given the name nephritogenocide has nephritogenic properties.

Studies on the biosynthesis of basement membranes established the fact that newly synthesized basement membrane procollagen is secreted in the extracellular space as a 480,000 molecular weight triple-helical molecule without undergoing a time-dependent reduction in its size.

Changes of basement membranes in disease have been attributed to immunologic mechanisms and to mechanisms whose nature still remains unknown. The mechanism responsible for basement membrane changes in diabetes remains unclear. Data were presented suggesting a deficiency in the catabolic removal of glomerular basement membrane in diabetes.

Work in the future will have to continue in an attempt to better understand the chemical nature of these structures, their immunochemical properties, and the mechanisms responsible for observed changes in disease.

INDEX

T

A
B
C 8
D 9
E 0
F 1
G 2
H 3
I 4
J 5

DATE DUE

JUL 2 7 2000		
DEC 2 9 2004		